Ordering the Human

RACE, INEQUALITY, AND HEALTH

R I H

RACE, INEQUALITY, AND HEALTH
Edited by Samuel Kelton Roberts Jr. and Michael Yudell

The Race, Inequality, and Health series explores how forms of racialization have created a wide range of phenomena, from producing inequities in health and healthcare to inspiring social movements around health. The goal of this series is to publish field-defining works across history, the social sciences, the biological sciences, and public health that deepen our understanding of how claims about race and race difference have affected health and society.

Sebastián Gil-Riaño, *The Remnants of Race Science: UNESCO and Economic Development in the Global South*

Rob DeSalle and Ian Tattersall, *Troublesome Science: The Misuse of Genetics and Genomics in Understanding Race*

Michael Yudell, *Race Unmasked: Biology and Race in the Twentieth Century*

ORDERING THE HUMAN

The Global Spread of Racial Science

EDITED BY
ERAM ALAM, DOROTHY ROBERTS,
AND NATALIE SHIBLEY

COLUMBIA UNIVERSITY PRESS *NEW YORK*

Columbia University Press
Publishers Since 1893
New York   Chichester, West Sussex
cup.columbia.edu
Copyright © 2024 Columbia University Press
All rights reserved

Library of Congress Cataloging-in-Publication Data
Names: Alam, Eram, editor. | Roberts, Dorothy E., 1956– editor. | Shibley, Natalie, editor.
Title: Ordering the human : the global spread of racial science / edited by
Eram Alam, Dorothy Roberts, and Natalie Shibley.
Description: New York : Columbia University Press, [2024] | Series: Race, inequality,
and health | Includes bibliographical references and index.
Identifiers: LCCN 2023043173 (print) | LCCN 2023043174 (ebook) | ISBN 9780231207324
(hardback) | ISBN 9780231207331 (trade paperback) | ISBN 9780231556927 (ebook)
Subjects: LCSH: Scientific racism—Case studies. | Racism in medicine—Case studies. |
Racism—Case studies. | Eugenics—Case studies. | Science—Social aspects—Case studies.
Classification: LCC Q172.5.S35 O73 2024 (print) | LCC Q172.5.S35 (ebook) |
DDC 500.89—dc23/eng/20231115
LC record available at https://lccn.loc.gov/2023043173
LC ebook record available at https://lccn.loc.gov/2023043174

Printed and bound by CPI Group (UK) Ltd, Croydon, CR0 4YY

Cover design: Chang Jae Lee
Cover image: © Science Museum Group

Note from the Series Editors

FIVE YEARS AGO, we set out to curate a series exploring the intersections of race and health. We sought out manuscripts across disciplinary and topical subjects to build a list that would elevate historical scholarship while building bridges between the past and present. In this way, we thought, we can understand the persistence of racism and inequality in all areas of health.

Ordering the Human: The Global Spread of Racial Science, this thoughtful and provocative collection of essays edited by Eram Alam, Dorothy Roberts, and Natalie Shibley, represents all that we have hoped for our series—and all packed into one volume! This book brings together a group of scholars whose collective expertise cuts across disciplines, from public health to anthropology, from history to medicine. Together, they draw connections between the historical roots of racism and a range of contemporary health issues. They globalize the interconnections of race, racism, and health, linking the Global South to a health literature that has long ignored non-Western matters. The book also makes a case for how concepts of science, the globe, and race have shaped our modern world.

We hope that this volume inspires other scholars to build on these disciplinary links, which are essential to characterizing the relationships between race, racism, inequality, and health.

Samuel K. Roberts, Columbia University
Michael Yudell, Arizona State University

Contents

CONTENTS

CONTENTS

Preface

DOROTHY ROBERTS

WHEN THE HUMAN Genome Project completed mapping the entire human genetic code, the biological concept of race seemed to have finally met its end. The scientists who led the project, as well as President Bill Clinton, all declared that human genes could not be classified by race. Yet instead of hammering the last nail into the coffin of a devastating ideology, the science that emerged from sequencing the human genome was shaped by a resurgence of interest in race-based genetic variation. My dismay at this development led me to write *Fatal Invention: How Science, Politics, and Big Business Re-create Race in the Twenty-First Century*, published in 2011, which investigated the emergence of a new biopolitics in the United States that relies on reinventing race in biological terms using cutting-edge genomic science and biotechnologies. One of my main objectives when I moved from Northwestern University to the University of Pennsylvania the following year was to launch an initiative that addressed the resilient science of racial difference and its harmful consequences for society.

With generous support from Penn's president and provost, I founded the Penn Program on Race, Science, and Society (PRSS), housed in the Center for Africana Studies. PRSS brings together faculty, fellows, students, and visitors across campus to share ideas for transformative and interdisciplinary approaches to the role of race in scientific research and biotechnological innovations, aiming both to promote social justice and to dispel the myth that race is a natural division of human beings. The symposium The Future of Race and Science: Regression or Revolution?, held at Penn on April 11, 2014, marked PRSS's inauguration. The event convened eight prominent visionary scholars from a wide range of disciplines in the United States to discuss their ideas about future approaches to race and science.

For PRSS's second symposium, I wanted to broaden the scope of our conversation about race, science, and society to a global scale. By bringing together an international group of biological and social scientists, historians, and legal scholars, PRSS aimed to uncover how race as a unit of analysis is defined, operationalized, and reconstituted through scientific and biomedical practices in various national and political contexts. As the legacies of racial classifications continue to influence and circumscribe lives, the symposium provided a vital starting point to analyze the global role of racial science and to strategize possible ways out of the naturalization of race.

I was fortunate to have appointed Eram Alam as PRSS's postdoctoral fellow for 2016–2018. Dr. Alam earned her PhD from Penn's Department of the History and Sociology of Science and brought a keen understanding of the movement of racial concepts in scientific knowledge production and the operation of race as a political technology in various times and places—an understanding reflected in her introduction to this book and the themes that structured the symposium. Dr. Alam's successor as PRSS postdoctoral fellow, Natalie Shibley, was a PhD student in Penn's Department of Africana Studies and Department of History, and her proximity and expertise enabled her to attend the symposium and participate in planning its outputs. Dr. Alam, Dr. Shibley, and I continued our generative collaboration as coeditors of this book.

The Ordering the Human: Global Science and Racial Reason Symposium was held on April 12, 2018, at Penn's Perry World House. The day following the symposium was devoted to a workshop attended exclusively by the participants to delve more deeply into the symposium's themes. We also brainstormed how to construct an anthology from the papers presented at the symposium. Most of the original participants remained on the book project, and their revised papers are included here. Regrettably, several of the non-U.S. scholars, from Belgium, Mexico, Singapore, and South Africa, were unable to contribute chapters. Fortunately, we were successful in our efforts to recruit additional papers from scholars writing from a range of disciplinary and global approaches.

Ordering the Human is the final product of six years of collective effort. The symposium and book project would not have been possible without the excellent assistance of Penn's Center for Africana Studies administrators and staff, including director Camille Z. Charles, former deputy director Gale Garrison, and associate director Teya Campbell. The PRSS 2021–2024 postdoctoral fellow, Dr. Hafeeza Anchrum, offered tremendous help with the final stages of the anthology's production. We are deeply grateful for their outstanding and consistent support for every aspect of this initiative.

Ordering the Human

Introduction

ERAM ALAM

What modalities of order have been recognized, posited, linked with space and time, in order to crease the positive basis of knowledge as we find it employed in grammar, philology, in natural history and biology, in the study of wealth and political economy.

—MICHEL FOUCAULT, *THE ORDER OF THINGS*

Let's face it. I am a marked woman, but not everybody knows my name. . . . I describe a locus of confounded identities, a meeting ground of invest-ments and privations in the national treasury of rhetorical wealth. My country needs me, and if I were not here, I would have to be invented.

—HORTENSE SPILLERS, "MAMA'S BABY, PAPA'S MAYBE:
AN AMERICAN GRAMMAR BOOK"

Biology is a metaphor for the destiny imposed on the other.

—ALBERT MEMMI, *RACISM*

THIS VOLUME WAS inspired by the difficulty in defining the following terms: *modern science, globality,* and *race*; each term necessarily enrolls the other two, revealing their co-constitutive entanglements. Modern science as a power/knowledge enterprise could only come into existence by creat-ing a unified, racialized object called the globe; the global as a whole entity was a necessary condition for modern science to make universal claims and categorize people into races; and race was the foundational classificatory schema upon which modern science organized the globe. In this dizzying circularity, one point must be made unequivocally clear: there was no pure science that was then contaminated by racial thinking. Modern science, race, and the globe emerged simultaneously during the Enlightenment as mutually reinforcing, justificatory alibis for violent outcomes and irrational desires. This is the starting point for the collection.

A recent, illustrative example of this trifecta's operations is the COVID-19 pandemic. During this viral emergency, the globe was quickly partitioned into Euro-America and the rest of the world, and people were classified into either viral perpetrators or victims based on racial categories. These discursive strategies once again merged science, the global, and race in a way that produced significant material effects that differed among groups of people. For example, by May 2020, the United States, India, and Brazil had the highest infection and death rates. And within this statistic, in all three countries, those who suffered most were poor, lower caste, Black, or Indigenous people. These statistics are not surprising and could have been predicted well in advance of the spread of SARS CoV-2; the question is: Why? In other words, how has scientific rationality constructed the globe in such a way that the disproportionate deaths of nonwhite bodies are the expected norm?

In different ways, the authors in this collection investigate aspects of the "why" question through their specific empirical case studies. Collectively, they highlight tensions in scientific universalism and fixity versus racial malleability and flexibility, attend to the mechanisms that consolidate racial ways of knowing, and trace the forces and flows that influence the global movements of racial concepts in scientific knowledge production. Contributions span the eighteenth century to our present pandemic moment. And authors are disciplinarily diverse, with expertise in genetics, forensics, public health, history, sociology, and anthropology, and geographically expansive, with projects based in South Africa, India, Brazil, Argentina, New Zealand, South Korea, Iran, Lebanon, and the United States. In compiling chapters with such range, we aim to excavate the different mechanisms by which global science and racial reason are recruited to propel projects of power and domination.

Although modern science, globality, and race are intertwined, I temporarily and artificially separate them to provide a brief genealogy of each concept. This exercise is undertaken with the knowledge that no singular, linear understanding of these concepts can be written that captures the contours and complexities of these ideas. Instead, my aim is to provide a foundational backdrop and establish a shared vocabulary moving forward.

Modern Science

Modern or *modernity* has become such a ubiquitous qualifier that it can obscure rather than amplify meaning. In this volume, by adding it before *science*, it is intended to invoke the Enlightenment in Europe—a period that marked a new understanding of the human vis-à-vis nature and assembled different tools for making knowledge claims. In using the terms *Enlightenment* and *science* to characterize this era, I take liberties that require mention. Scholars have long debated the periodization between modern and premodern, the role of the Scientific Revolution in catalyzing the Enlightenment, and the appropriateness of discussing the Enlightenment as a deliberate, cohesive intellectual project without contingency and chaos, for example. These interventions are a necessary corrective to the linear progress narratives that center European rationality and exceptionalism. However, when using these terms, I collapse this complexity and use *modern* as a stand-in for an attitudinal or epistemic shift that influenced European ways of being and knowledge production over many centuries, and I anachronistically deploy *science* in lieu of *natural philosophy*.

Before the Enlightenment, God was the ultimate referent, and scientific pursuits were oriented toward reflecting on God's grand design. The visible world was simply all that was accessible to man's limited faculties, and through careful, deliberate study, a human might gain a glimpse of the invisible, transcendental beyond. Put differently, Christian man was the passive recipient of godly nature, and science was a knowledge practice embedded within this worldview, constrained by parameters set by the church. The Enlightenment disrupted God's jurisdiction by transferring authority to man and making him responsible for ordering the world using his mental faculties. To be clear, this was not necessarily a secular pursuit that rejected God's provenance; it was a shift in the mechanism of knowledge acquisition. Reason displaced unexamined religious belief, and the visible world was a thing-in-itself—the only thing one could truly know and understand—and contained causal, deterministic laws, rules, and patterns waiting to be uncovered through empiricism. In this Enlightenment ethos, man existed because he thought himself to exist; his existence was self-referential. Man became the starting point, and his mental organization was projected onto nature. Although this clean, coherent version of modern science with an identifiable break from a less rational past is contested, of importance is the fact that Enlightenment Europeans from the eighteenth century onward were invested in constructing and narrativizing a shift using these terms.[1]

Developments in physics and mathematics were foundational to this new orientation and set the field for this epistemic shift. Examples include the following universal, reproducible statements:

> Motion is change of place. No motion or direction of motion takes precedence over another. Every place is equal to every other. No point in time has precedence over any other. Every force is defined as—is, that is, nothing but—its consequence as motion within the unity of time.... Every natural event must be viewed in such a way that it fits into the ground plan of nature. Only within this perspective of this ground-plan does a natural event become visible as such.[2]

In this lengthy quotation, we find a vital reconceptualization of space and time; they are uniform, stable, portable, and predictable. From this starting point, science was equipped with a grid or a table upon which to project universal statements about nature, construct hierarchies, develop developmental logics, and make claims of progress.

This new, modern science was capacious and voracious in scope and design, and key to its practice was a commitment to universal, objective, value-neutral ideals and observations. Given the impossibility of this mission, inevitably what transpired was an equation of western European bourgeois ideals as universal truths.[3] The practitioners of modern science set out to gather, organize, and systematize the globe—an object that came into existence only through this logic—and impose an order upon it. Examples of this "attitude of modernity" at work include Carl Linneaus's *Systema Naturae*, a vast taxonomic project whereby all known animals, plants, and minerals were named, classified, and placed into hierarchies. No less ambitious was the *Encyclopedie*, edited by Denis Diderot and Jean le Rond d'Alembert, a multivolume compendium of all the world's knowledge in a format that could circulate among a certain kind of man.

This consuming quest for a total knowledge also produced a new calculative logic. Inspired by the universal laws determined by the physicists, astronomers, and mathematicians, researchers such as Adolphe Quetelet sought to uncover a "social physics," a kind of mathematical understanding of social life. To this end, he developed concrete data practices across networks of practitioners and is most famously known for his calculations of the average man, a graphic rendering of "normal" human variation that statistically identified an ideal mean. Quetelet's statistical legacies include the precursor to the contemporary Body Mass Index, a measure that continues to be used widely. Many massive and rigorous data-collection efforts such as Quetelet's were oriented toward the nation-state and institution building—a reminder that modern science and statecraft were intimately linked from their inception.[4]

Armed with science and statistics, the modern European nation-state sought to fashion itself using reason as the founding and guiding principle. No longer solely under the rule of a divinely ordained sovereign, these post-revolutionary societies (English, American, and French) were faced with governing "We the People" and a "Declaration of the Rights of Man."[5] But who are the People, and how do we identify Man? Using statistics, the nation-state developed appropriate management strategies to contend with these categories. More specifically, the nation-state implemented statistical normalization or the establishment of the "norm," which was taken up in nearly every domain as an essential governance strategy. Different from the contemporary colloquial usage of the *norm* as average, this mathematical metric identified the norm as a range of optimal conditions, "a bandwidth of the acceptable that must not be exceeded."[6] Using distributions around the norm, this new calculative logic was regulatory and ultimately invested in the optimization of a certain kind of life. For example, probabilistic rationale and cost calculations influenced decision making about questions such as what level of illness was permissible in a population and how those who fall ill will affect the economy. In tandem, various apparatuses of the nation-state collected massive amounts of data from which to "qualify, measure, appraise, [and] hierarchize" and "distribute the living in the domain of value and utility."[7] This was a new mode of power based on the scientific management of life, or what Michel Foucault has called biopower, and its application, biopolitics, and was crucial for modern forms of governance.

Biopolitics was the coming together of two kinds of power. The first focused on "the body as a machine" and was an "anatomo-politics of the human body."[8] This meant ensuring that a body was functioning to its maximum capabilities, contributing socially and economically in the way that it should, and generally conforming to expectations. This is why, for example, bodies that are disabled and cannot enter the labor market and people with reproductive capacities who choose not to bear children challenge normative expectations. The other pole was more abstract and focused on the concept of a "population." This version of population was not simply a multiplicity of individuals—or not only a multiplicity of individuals—it was a novel political entity and the fundamental object for a nation-state's "economic-political actions."[9] It was a form of regulatory control that intervened on the species level and centered questions of health, life expectancy, longevity, births, and deaths, to name a few. Overall, biopolitics was concerned with the "administration of bodies and the calculated management of life" and converting social randomness into predictable risk.[10]

In this abbreviated genealogy of modern science, two dilemmas perpetually disturbed the seamless application of these new principles. If the human could be ordered and made predictable according to statistics and mathematical logics, then what of free will? Was the human simply machinic and no different than any other animal? And the second, "what constitutes the humanity of human beings?"[11] Was it a capacity for reason, and if so, did all humans across the globe have equal access to reason?

Globality

Modern science was predicated on a globe, and the globe was invented during the European "Age of Exploration" and the inauguration of the transatlantic slave trade. This conquering drive had a different scale and orientation from past imperial visions, which operated without a mathematical, all-encompassing spatial concept of the earth.[12] This new drive began as a cartographic mission and quickly activated an imaginary that overlaid a mythical grid on land and sea and ushered a dangerous, rapacious sensibility that was deterritorialized and highly expansive. With the addition of this calculative framework, all lands and their inhabitants—human and nonhuman—could be disciplined and categorized to construct a cohesive, unified entity.

In more concrete historical terms, this new global order was made possible by "the legendary and unforeseen discovery of a new world"[13] and the godly permissions granted by the Roman Catholic Church that gave

Portugal a monopoly on trade and plunder in West Africa and gave Spain the blessing to colonize the Americas to increase its wealth. In this papal decree, violence and subjugation were sanctioned to achieve economic gains, thereby intertwining missionary zeal and economic accumulation. All that was encountered during the Age of Exploration was assimilated into this new paradigm.

The globe had—and continues to have—an enigmatic character. It was simultaneously all encompassing yet also a view from nowhere. And to contend with this paradoxical unknowability, a Eurocentric framing and cosmology were imposed on this cartographic entity. The European globe began with "a common spatial order encompassing the whole earth."[14] This was an exercise in compression, making the vast diversity of human and nonhuman life conform to universal, linear principles of scientific reason and its corollary, progress. In other words, all objects encountered in this new global framework could now be compared and arranged along a hierarchical gradient—a gradient in which European Man was the apex and the exclusive entity with full access to reason. Only European Man could be transparent to itself; it was the only kind of human that could claim full access to interiority.

Simultaneous with the birth of the globe was the "dawn of the era of capitalist production" catalyzed by primitive accumulation. "The discovery of gold and silver in America, the extirpation, enslavement, and entombment in mines of the indigenous population of the [New World], the beginnings of the conquest and plunder of India, and the conversion of Africa into a preserve for the commercial hunting of blackskins" were the necessary preconditions for a new economic system that demanded movement and connectivity like never before.[15] The cartographic globe morphed into lucrative trade routes where objects of all kinds, including human bodies, were compressed into commodities with an exchange value. This exercise in spatial compression accelerated rapidly as European nations vied with one another to capture and conquer as much land as possible, turning this new globe into a battlefield.[16] And crucially, whereas the physical land was finite, the global imaginary was infinite—a belief that fueled capitalist, consumptive desires.

Although the global imaginary was an idealistic projection, it nonetheless had consequences in multiple domains. In the realm of the ideological, there was an instantiation of a new consciousness wherein the earth was turned into a whole. As mentioned previously, science was instrumental in ushering in this perspective. The wholeness of the globe meant that every-thing that existed was a piece of this whole and had a function in relation to everything else. The globe was an interdependent entity, and all who existed on the globe were compelled to accept this consciousness in the same way. This way of ordering the globe is what eventually allowed for ideas of development, civilizational standards, and modernization theories to become the metrics of progress. Embedded in these logics were two impor-tant premises: first, development had an end that was achieved and could be judged by the "superior values of a particular [Euro-American] collective," and, second, it was "a natural outcome of the operations of the laws of nature."[17] Using this rubric, development as a structuring discourse judged non-Europeans as unable to access the highest levels of developmental per-fection because the laws of nature prevented them full access.[18]

Land and sea were no longer impediments to the dissemination and pro-liferation of a European macronarrative that spread rapidly and whose effects continue to influence the present. In material terms, this globe was connected like never before and resulted in novel social relations—some vio-lent and genocidal and others premised on different types of exchange. Common to all, however, was a new relationship to the movement of people, ideas, things, and institutions—a reimagined circulatory system in which abstract universals about political systems, economics, history, science, and culture, for example, could be projected all over the world; "each spreads through aspirations to fulfill universal dreams and schemes."[19]

Despite the force of these global designs, their uptake and application across the world was never even or without friction. The friction framework sets aside the local/global distinction and instead reveals how "abstract claims about the globe can be studied as they operate in the world."[20] Put another way, universals can only ever be understood as "local knowledge" because they require the interpretive matrix of "historically specific cul-tural assumptions."[21] Precisely because of this characteristic, they are always already engaged in the "sticky materiality of practical encounters," even if unwillingly.[22] One can never see the globe in its wholeness from within the structure; there is no such thing as an abstracted view from nowhere. The orderly grid of abstracted, universal truths that constitutes this imaginary entity must always contend with the land and people that its designers wanted to dominate. And in this grounding, its operations emerge in both unpredictable and expected ways.

Race

One of the most dangerous manifestations of global ideologies or the global-as-ideology was the drive to define and catalogue the human and develop a science of Man—a practice that was and remains integral to modern science. This new index was a break from previous moments during which difference was experienced as spectacle, something featured in "fairs" and "tournaments." In the modern period, there was "a new way of connecting things" whereby a knowledge of the human species required the "continuous, ordered, and universal tabulation of all possible differences."[23] There emerged an obsessional drive to enumerate and measure all the characteristics of an entity and its possible variations in order to generate a schematic of value.

When this logic was oriented toward the human, Enlightenment scientists crafted the foundation for modern-day conceptions of race. "The expansion of the slave trade in the 1700s necessitated an expanding conceptual racial system of governance," and European naturalists transported the theological belief in a natural creation of races into science.[24] Often considered one of the first inventors of racial science, Johann Blumenbach developed craniometry as a technique to group and categorize human difference during the eighteenth century. Using skull size along with other visible traits, he concluded differences between humans could be organized into five "natural," "biological" categories: Caucasian, Ethiopian, American, Mongolian, and Malay, which were correlated to global regions. Within this schema, Blumenbach posited a hierarchy with Caucasians as the high point and all other races as degenerate forms. His philosophically minded contemporaries, while less concerned with the mechanistic explanation for racial difference, agreed with the conclusions. Nearly all emphasized fixity, or extreme difficulty of change, arguing that the races have "a predetermined ability or natural disposition" that was most obviously manifest in external traits but also had an inner dimension that affected the moral, mental, and political trajectories of racial groups.[25]

As time went on and scientific practice became more formalized and institutionalized, fervor around understanding the sciences of Man intensified. Scientists published and disseminated manuals with detailed instructions on how to gather, handle, measure, and ultimately systematize anthropological information.[26] This kind of orientation had the net effect of translating human difference into particularized, fixed racial categories defined against a Caucasian, Christian, male ideal with full access to the category of Man.[27] But if this was Man and all else was less than Man, were there any ethical commitments between Man and non-Man? If so, what were they? This remained an open question.

To reiterate, race and modern science had a simultaneous inception. Classificatory logics that were the foundation of scientific reason were confronted with disordered human variation, and racial ordering was deployed as powerful political technology to tame difference. Once physical descriptors of skin color, hair, and physiognomy were demarcated, upon this scaffolding a whole range of mental, psychic, social, and political variables could be assembled.[28] And armed with this race-based arsenal, violence, civilizing missions, and colonial domination proceeded with scientific justifications. Biological explanations are a useful alibi and have incredible staying power because they absolve those in power of responsibility. This insidious racial logic continued well into the twentieth century and motivated eugenic visions, sterilization laws, immigration policy, apartheid, medical experimentation, segregation, and child kidnapping, for example, across the globe.

Despite attempts to discount racial typologies as natural, irrefutable truths, the weight of these sedimented descriptors still yields significant force in the world, a troubling reality made apparent in responses to the Human Genome Project. The map of the human genome made unequivocally clear that "race" is not genetically encoded; there is no specific DNA sequence for race. Instead, humans are 99.9 percent genetically the same, and the genetic variation in the human species cannot be divided according to any of the definitions used to identify races in other species. "Biologically, there is one human race."[29] Then what was this thing, and how and why did it yield so much real, material power? "Race applied to human beings is a *political* division: it is a system of governing people that classifies them into a social hierarchy based on invented biological demarcations."[30] It facilitated the creation of a modern world in which Euro-American social, cultural, and economic success was considered a natural outcome and where all others inevitably fell short of this ideal. And even irrefutable DNA evidence was unable to dislodge this belief.

After the human genome was sequenced, racial logics remained firmly in place, only assuming more subtle and insidious manifestations. In fact, a new language of precision has arisen around racial science. Newly armed with genetic and genomic capabilities, scientists claim they can reconfigure race with even more exactitude than before. Paradoxically, the realization that race is not a genetic category has unleashed a desire to make sure that it remains so. Racial scientists in the genomic age employ two main strategies to circumvent the old, obviously racist typological approach: statistical probability and geographic ancestry.[31] In one example of statistical probability masking racial logics, scientists used computer programs to sort vast collections of genetic data from people across the globe into genetic clusters based on human-designed algorithms. Their findings showed that five of the main genetic clusters identified corresponded to major geographic regions—Africa, Eurasia, East Asia, Oceania, and America—which coincidentally aligned with the five major racial divisions. Upon further scrutiny, however, it became evident that inputs determine outputs; the way that scientists posed questions to unstructured DNA data and the way they interpreted the outcomes depended on and reified familiar racial categories. Later, they conceded that different methodological choices would reveal different kinds of groupings.[32] These same issues of data selection and organization bias exist in genetic-ancestry assessments. Scientists select and modify data to create ancestry groupings that essentially generate a one-to-one overlap with racial typologies, sometimes using continental ancestry and race interchangeably.[33] Scientists employing even the most cutting-edge technologies can inadvertently defer to Enlightenment racial ideas to produce and interpret the findings.

The inability to define and stabilize the science fiction of biological race has been one of its political strengths as a force in the world. The cultural theorist Stuart Hall posited that "race works like a language." In any language system, the meaning of a concept or a signifier does not have an essence and cannot be defined using positive terms (e.g., race is . . .) but rather gains its meaning "in the shifting relations of difference, which they establish with other concepts and ideas in a signifying field":

Their meaning, because it is relational, and not essential, can never be finally fixed, but is subject to the constant process of redefinition and appropriation: to the losing of old meanings, and appropriation and collection and contracting of new ones, to the endless process of being constantly resignified, made to mean something different in different cultures, in different historical formations at different moments of time.[34]

Race is a "floating signifier." To say that race is relational, akin to a language, does not mean to deny its very real consequences in the world; we are all made in and through language systems. It is to emphasize that racial logic was a scientific system invented to govern people and contend with the anxiety of difference. And once the system was created, it yielded an incredible power to choreograph human action.

*   *   *

Because we are inheritors of an Enlightenment worldview, these ideas—science, the globe, and race—inform every aspect of our present. Examining how they have come together to mean something different in different places, during different historical moments, and at different times constitutes the main objective of this edited volume. Although the authors of each chapter employ various methodologies, theoretical frames, and literatures, all share a commitment to mapping out the political operations of race and science against the backdrop of globality. In so doing, they show how these entangled ideological commitments produce material consequences that continue to unevenly affect lives and possibilities. Admittedly, in doing this kind of research, at times there might be a reification of the very categories we are trying to demystify. This is the pernicious nature of these concepts—there is no outside or abstracted place of purity. Instead, through these various case studies, the authors dissect the linguistic, bureaucratic, institutional, and embodied ways that race, science, and the globe structure knowledge production and practices in the world. By excavating these mechanics, we hope to make space for new imaginaries that engender novel political formations and ways of being.

The volume is divided into three thematic parts. The first, "Stability and Circulation," explores how racial logics are stabilized, exported, and circulated. Difference is disguised, covered over to produce commodities, markets, and objects of knowledge divorced from the conditions of making that are expected to travel without friction. Most often, this occurs by aggregating individuals into populations and then converting populations into composites. In these serial translations, erasures and new classifications are imposed to produce recognizable data for purposes of surveillance, criminalization, health policy, and immigration, for example, that can circulate all over the globe. In the production of these stable composites, homogenizing narratives, social and political assumptions, and cultural slippages are surreptitiously embedded.

Paul Wolff Mitchell explores this idea in its most material form in chapter 1, on eighteenth-century Dutch and German skull collecting. Naturalists at the time were obsessed with collecting and circulating these objects, arguing that differences in bone structures suggested innate racial typologies and, most importantly, differences in brain function. These racial differences were presumed to be most pronounced in terms of reasoning capability and access to rationality. Eric Reinhart's chapter 2 shows how a psychocentric definition of Enlightened Man—a white, male civilized being who can access the "I" statement, know itself, and be free—continues to operate as a standard against which "reason" and "sanity" are calibrated in contemporary U.S. medical and psychiatric practice.

In chapter 3, Sebastián Gil-Riaño and Julia Rodriguez raise a question often ignored when ordering the human: What is the place of the child in racialized visions of the future? Motivated by the drive toward national progress, Latin American scientists focused their attention on *puericultura* (child care) and mental hygiene. Combining economic development discourse with human development, they made the child an important object through which nations could attain civilizational progress. In chapter 4, Amade M'charek examines two recent forensic genetic technologies, familial searching and DNA phenotyping, to show how individual DNA is extrapolated to create suspected populations of interest. While much scholarship uses difference as its analytic angle, this chapter uses the idea of sameness to better understand the operation of racialization, forensics, and criminalization. And finally, in chapter 5 Denise Ferreira da Silva reflects on the COVID-19 pandemic to understand how supposedly abstract scientific tools such as models, maps, and curves obscure the underlying racial logics of their creation and organization. Because of their success in persuading publics of their value neutrality, models and graphs move with ease, normalizing the death and maiming of nonwhite persons around the world.

In part 2, "Purity and Mixture," scholars investigate how and why racial purity and mixture are contingent categories with deeply political meanings. In the United States and South Asia, for example, purity was enforced culturally and legally, resulting in the need to categorize and draw rigid boundaries around racialized bodies. In contrast, many Latin American and South American regions promote mixture as a positive attribute of the national polity. However, this discursive appreciation is often superficial, symbolic, and limited. Material and social inequities continue to manifest according to racial hierarchies, and individuals considered "mixed" are socially disadvantaged by their nonwhite status. Moreover, "mixture" is often interpreted as the combination of pure races, thereby failing to contest the concept of purity. Older methods of identifying pure races based on phenotypic characteristics have been replaced and recast in technoscientific terms in the current genomic age. Using DNA sequences and large databases of genetic samples, scientists, such as those affiliated with genetic ancestry testing, portray populations and individuals as percentages originating in fixed biogeographical populations. This new way of identifying and estimating mixture based on abstract ancestral proportions divorces people from their lived and relational circumstances, replacing these relationships with imagined ancestral communities. While sometimes innocuous, the underlying drive to know, locate, and differentiate ancestral roots can have violent and material consequences.

In chapter 6, Projit Bihari Mukharji argues against the postcolonial use of hybridity as a liberatory possibility. Instead, by examining the scientific studies of hybridity in India, or what he calls "biometric hybridity," the firm, unyielding commitment to identifying the pure components of the hybrid entity in question become clear. In South Korea, Jaehwan Hyun explains in chapter 7, political celebrations of multiculturalism and multiethnicity reveal a fixation with purity. Referring to this as "biocultural purity," South Korean governance regimes apply microbiological aspects of cultural differences among human groups as criteria to classify pure "Koreans" and non-Koreans. These classificatory strategies are used for citizenship status, criminal investigations, and medical and forensic genetic research, with adverse outcomes for those labeled as non-Korean. In chapter 8, João Luiz Bastos and Ricardo Ventura Santos wrestle with the paradox of the Brazilian nation, which simultaneously prides itself on its *mestizaje* (mixed) character while reifying purity in its daily operations and practices. Although set in vastly different political, social, and historical contexts, these case studies make clear the shortcomings of hybridity or mixture as a counterproject to racial purity.

Part 3, "Past and Promise," explores how scientists and the broader public in various nations navigate the use of racial categories to link the deep past with future imaginaries. Implicit in these investigations is the desire for an authentic self or collective rooted in the biological and unencumbered by the social. The social is seen as the site that produced nefarious racial hierarchies and devastation, whereas the biological is understood to be "beyond race" or a strategy to "circumvent racial thinking." In practice, however, race simply operates as a repressed object, temporarily sublimated yet bound to resurface. It is discursively recoded in nationalistic, ethnic, linguistic, religious, or genetic terms to serve political purposes. On an individual level, knowledge of a genetic past holds the potential for a knowable future, a new level of preparedness in the face of uncertainty and risk. In assembling various national cases, temporal investments in race and its relationship to understandings of individual and collective history and futures become apparent.

In chapter 9, Elise Burton reminds us that while race and genetics have occupied much scholarly space over the last two decades, the importance of the visuality of race must not be underestimated. Using cases of forensic facial reconstruction in the Middle East, this chapter shows how an ancient "face" is scientifically and aesthetically resurrected to fuel racialized national projects. In chapter 10, Isaac Warbrick explores the enduring colonial legacies of weight discourse on Māori bodies in New Zealand. Historically deemed deviant by Western standards, Māori bodies are turned into pathologized objects in need of repair. In response to these derogatory depictions, Māori activists make room for Indigenous cosmologies and practices recognizing that these are essential to their future. In chapter 11, Noah Tamarkin focuses on the concept of "indigenous DNA" in South Africa—a sought-after commodity imagined as a window into humanity's distant past. Despite sustained and concerted efforts to stabilize "indigenous DNA," this slippery construct always managed to mutate and morph to suit the arguments of various stakeholders. In chapter 12, Alyssa Botelho and David Jones explore the claim that South Asians have the greatest risk of coronary artery disease (CAD). They show how both South Asian and CAD are contingent, fluid entities inflected with various social, cultural, and political economic motivations at different historical periods. Nevertheless, "South Asian" susceptibility to CAD continues to capture the imagination and resources of clinicians and researchers invested in solving this potentially nonexistent "racial" disparity. And in chapter 13 Banu Subramaniam concludes the volume with a critique of racism-cum-nationalism and argues that these concepts always underwrite understandings of the biological. Comparing COVID-19 responses in the world's largest democracy, India, and the world's "strongest" democracy, the United States, reveals that in both nation-states ideas of "democracy" obscure the vast racialized infrastructures that shape pandemic statistics and inequities.

By placing these case studies alongside one another, this compilation highlights the ways racial logics, the concept of a globe, and modern science are deeply codependent; each is ancillary to the others. The resilience and mutability of race across time and space remind us that race is a "floating signifier," a technology or set of tactics and strategies, not a natural division of humanity. It is a political, plastic technology that perpetuates ideological commitments in diverse contexts—a point that must be made emphatically as we analyze the pandemic and its consequences. Once again, race as nature was deployed as a fixed, biological fact and an explanatory framework for higher disease burden and death. These chapters provide a necessary counterargument to this deadly line of thinking.

Notes

The epigraphs in this chapter are from Michel Foucault, *The Order of Things: The Archaeology of the Human Sciences* (New York: Pantheon, 1970), xxi; Hortense Spillers, "Mama's Baby, Papa's Maybe: An American Grammar Book," *diacritics* 17, no. 2 (1987): 65; Albert Memmi, *Racism*, trans. Steve Martinot (Minneapolis: University of Minnesota Press, 2000), 71.

1. Peter Dear, *Revolutionizing the Sciences: European Knowledge and its Ambitions, 1500–1700*, 2nd ed. (Princeton, NJ: Princeton University Press, 2009), 1–9.
2. Martin Heidegger, "The Age of the World Picture," in *Off the Beaten Path*, ed. and trans. Julian Young and Kenneth Haynes (Cambridge: Cambridge University Press, 2002), 60.
3. Steven Shapin, *The Scientific Revolution*, 2nd ed. (Chicago: University of Chicago Press, 2018).
4. M. A. Quetelet, *A Treatise on Man* (New York: Burt Franklin, 1842).
5. The Haitian Revolution was an important test of the limits of these "universal" Enlightenment ideals of Man and Freedom. For more on how and why, see C. L. R. James, *The Black Jacobins: Toussaint L'Ouverture and the San Domingo Revolution* (New

York: Vintage, 1989); and Michel-Rolph Trouillot, *Silencing the Past: Power and the Production of History* (New York: Beacon, 2015).

6. Michel Foucault, *Security, Territory, Population: Lectures at the Collège de France 1977–1978*, ed. Michel Senellart, trans. Graham Burchell, 1st ed. (New York: Picador, 2009), 21.

7. Michel Foucault, *History of Sexuality*, vol. 1: *An Introduction* (New York: Vintage, 1990), 144.

8. Foucault, *History of Sexuality*, 1:139.

9. Foucault, *Security, Territory, Population*, 64.

10. Foucault, *History of Sexuality*, 1:140.

11. Michel Foucault, "What is Enlightenment?," in *The Foucault Reader*, ed. Paul Rabinow (New York: Pantheon, 1984).

12. Carl Schmitt, *The Nomos of the Earth in the International Law of Jus Publicum Europaeum*, trans. G. L. Ulmen (New York: Telos, 2006), 53.

13. Schmitt, *Nomos of the Earth*, foreword.

14. Schmitt, *Nomos of the Earth*, 55.

15. Karl Marx, *Capital*, vol. 1: *A Critique of Political Economy* (New York: Penguin, 1992), 915.

16. Marx, *Capital*, 1:915.

17. Denise Ferreira da Silva, "Globality," *Critical Ethnic Studies* 1, no. 1 (Spring 2015): 35.

18. Da Silva, "Globality," 35.

19. Anna Lowenhaupt Tsing, *Friction: An Ethnography of Global Connection* (Princeton, NJ: Princeton University Press, 2005), 1.

20. Tsing, *Friction*, 6.

21. Tsing, *Friction*, 7.

22. Tsing, *Friction*, 1.

23. Michel Foucault, *The Order of Things: An Archaeology of the Human Sciences* (New York: Vintage, 1994), 131, 144.

24. Dorothy E. Roberts, "Race and the Enlightenment," in *Four Hundred Souls: A Community History of African America, 1619-2019*, ed. Ibram X. Kendi and Keisha N. Blain (New York: One World, 2021), 119–20; Terence Keel, *Divine Variations: How Christian Thought Became Racial Science* (Stanford, CA: Stanford University Press, 2018).

25. Denise Ferreira da Silva, *Toward a Global Idea of Race* (Minneapolis: University of Minnesota Press, 2007), 119.

26. Da Silva, *Toward a Global Idea of Race*, 122–23.

27. Da Silva, *Toward a Global Idea of Race*, 127.

28. The power of classification is also operative in sex and gender categories and was always co-constitutive with racial categories. As Londa Schiebinger powerfully shows, sexual science *and* racial science were deployed to conclude that the white, European male body was the epitome of development. Following this "scientific" conclusion, assumptions about behavior, social roles, and appearance all became entangled in the classificatory schemes. For more, see Londa Schiebinger, *Nature's Body: Gender in the Making of Modern Science* (Boston: Beacon, 1993), esp. chaps. 4–5.

29. Dorothy Roberts, *Fatal Invention: How Science, Politics, and Big Business Re-create Race in the Twenty-First Century* (New York: New Press, 20120), x.
30. Roberts, *Fatal Invention*, x.
31. Roberts, *Fatal Invention*, 58.
32. Roberts, *Fatal Invention*, 61–62.
33. Roberts, *Fatal Invention*, 67.
34. Stuart Hall, "Race, the Floating Signifier: What More Is There to Say About 'Race'?," in *Selected Writings on Race and Difference*, ed. Paul Gilroy and Ruth Wilson Gilmore (Durham, NC: Duke University Press, 2021), 362.

PART ONE
Stability and Circulation

ONE

Origins of Races, Organs of Intellect

Polygenism, Political Order, and the Enlightenment Construction of Cranial Race Science

PAUL WOLFF MITCHELL

APPROPRIATE TO THE fatal invention of race, the skull has become iconic of racial science. In the eighteenth century, anatomists in European metropoles began to systematically amass skulls of racialized, colonized, and enslaved people from across the globe for comparative description and measurement.[1] By the late nineteenth century, European and North American private collectors, anthropological societies, and anatomical museums had assembled many thousands of human skulls, divided and labeled according to racial taxonomies.[2] Into the twentieth century, reams of cranial measurements in medical and anthropological journals evidence the skull's enduring centrality in the construction of racial science.[3] An influential 1914 German-language anthropometric guide, Rudolf Martin's *Lehrbuch der Anthropologie*, included over four hundred pages on cranial description and measurement, over twice that devoted to the rest of the skeleton.[4] Around the same time, confidence in the skull's reliability as a marker of race began to erode, and skulls ceded the cutting edge of racial science to brains, blood, and genes as the primary objects through which anatomist-anthropologists constructed claims about racial difference.[5]

This chapter traces how skulls came to matter in racial science, attending to how anatomists increasingly fixed race in the body's depths in the late eighteenth century, placing the seat of racial difference in the bony foundation of the face and vault around the brain. The logic of cranial race science was decisively formed within tensions between two anatomists trained at the University of Göttingen in present-day Germany, Samuel Thomas Sömmerring and Johann Friedrich Blumenbach, in the 1780s. Their differences concerned the depths of racial difference in the body and over time, hanging on the question of whether all humans shared common ancestry (monogenism) or if racialized differences evidenced creation of races as separate species (polygenism). The bodies of Black Americans recently arrived in central Europe after the American Revolutionary War were an immediate prompt for this debate, which catalyzed Blumenbach's project of amassing a global human skull collection. The subsequent French and Haitian revolutions were crucial political contexts in which a metric of difference proposed by Sömmerring—"cranial capacity"—became a pervasive, persistent, and pernicious tool in discourses naturalizing hierarchical, racialized social order in the development of scientific racism into the nineteenth century and well beyond.

Race in the Body's Foundation: Enlightenment Monogenism and the Depths of Difference

Understanding race as a concept crafted toward the naturalization of social discriminations Europeans applied through "direct, political, social and cultural domination" in colonialism and enslavement, racialized bodies became crucial sites for the articulation of political projects in scientific parlance within Enlightenment anatomical discourses.[6] Through the long eighteenth century, racial classifications collated colonial knowledge, prejudice, and presumption about non-European peoples, interweaving descriptions of skin color, stature, hair texture, and facial features to construct racial categories.[7] These classifications hung heavily on associations of embodied differences with geography, behavior, and mind. The tenth, definitive edition of the *Systema Naturae*, in 1758, Linnaeus's typology of four color-coded races in four corners of the earth, with four typical humors and four forms of social organization, is an icon of the period's classificatory paradigm: black, phlegmatic Africans ruled by caprice; red, choleric Americans ruled by custom; yellow, melancholic Asians ruled by opinion; white, sanguine Europeans ruled by rite.[8] With professional authority over knowledge of the human body allowing them a privileged role in the construction of racial science, medically trained anatomists

[22]

dissected flesh and bone under pen and knife for foundations of racial classifications and traces of the fundamental nature and origins of racial difference.[9] Although absent in Linnaeus's mid-eighteenth-century taxonomy, cranial form became essential in racial classification decades later. The waxing primacy of skulls in racial science was first founded on the necessary assumption that bones were, beyond other tissues, deeply and durably marked by race.

Finding race in the skeleton became important in the face of doubts about the sturdiness of skin color to hold the weight of influential Enlightenment environmentalist explanations for bodily difference. Through the mid-eighteenth century, typified in Comte de Buffon's *Histoire naturelle*, the effects of climate, food, and mode of life on the body over lifetimes and generations were commonly presumed ultimate causes of racial differentiation, termed "degeneration," from Adam, whether taken as literal or figurative. Buffon, and later others like Johann Friedrich Blumenbach, argued that the first humans were created fair-complected and beautiful.[10] This monogenism, or theory of shared human ancestry, explained racial difference by positing bodies and ultimately "varieties" or "races" as inscribed by place and habit over time as humans dispersed from their site of origin. Although circumscribing divine action to creation and positing naturalistic explanations for racial differentiation, this dominant strain of Enlightenment monogenism bore Eurocentric prejudices and rhymes with Genesis, as light-skinned originals degenerated into darker-skinned others in the harsher climes outside of Eden.[11]

The rub was that skin pigmentation, central to eighteenth-century racial classification, was unruly. Debates on albinism, concern that colonial settlement could lastingly darken skin, and anxieties about the heritability of complexion complicated any stabilization of color-coded racial classifications.[12] Other features presumed racially distinct suggested instability, especially those without obvious efficient environmental causation. In these cases, naturalists proffered cultural customs as explanations: for example, Buffon wrote that African mothers flattened the noses and lips of their children, compared to European features he presumed natural.[13] Explanatory resort to variable cultural practices shaping the body was commonplace, even for cranial form: the hands of mothers and midwives; the consistency of diet; and hats, helmets, and headscarves served as skull form's primary causes.[14]

But prevailing presumptions of bony plasticity strained as anatomical fixation on skeletal minutiae met concerns about the stability of racial features. In the 1720s–1740s, anatomists including Alexander Monro in Edinburgh, William Cheselden in London, and Bernhard Siegfried Albinus in Leiden published meticulously illustrated texts describing and depicting bony anatomy in unprecedented detail.[15] These exhaustive osteological studies were the backdrop to reframing the significance of racialized bones, occasionally trafficked through colonial violence, warfare, or grave robbing and displayed in European curiosity cabinets and anatomical theaters over prior centuries.[16]

In 1764, Petrus Camper, a student of Albinus, delivered a lecture on the origins of racial differences at the University of Groningen in the Dutch Republic. Camper argued against Buffon's theory of Adam's white skin color. Contending that skin color could change significantly over time, Camper reasoned that Adam could have been created any shade. Arguing further against Buffon regarding the cultural fashioning of racialized facial features, Camper demonstrated in the Netherlands a fetus taken from a woman from Angola through networks of Dutch enslavers: "They are shaped like already that in the womb, the form depending only on the projection of the upper and lower jaws, and that is how the nose becomes flat and small by itself, and the lips thick."[17] Camper's student, Samuel Thomas Sömmerring, later wrote, rife with the chauvinism of a professional anatomist, that "fundamental differences" distinguishing Africans and Europeans were not only skin color, hair texture, or a nose supposedly "depressed during tender childhood" but that "differences sufficient for the physiologist must not be incidental, brought about by custom; rather, more convincingly, they must be irrefutably found in the body's foundation, in its most solid parts, in the skeleton itself."[18] Insulated in theory from climatic and cultural forces shaping outward appearance, skeletons became keys to rooting race deep in the body.

Racial Face to Racial Skulls: Physiognomy and Facial Lines

Anatomists sought signatures of bony racial difference through, inter alia, limb proportions, foot form, and pelvic architecture by the early nineteenth century.[19] But skulls were core to skeletal racialization, as millennia-old physiognomic discourses purporting the revelation of character in the face were a ready discursive bridge from body to mind, a connection fundamental to the racial project. Medieval and Renaissance physiognomic texts often commented on minority communities in Europe, such as Jews and "Moors," associating stereotyped facial features with social distinctions, moral worth, mental ability, and racialized aesthetics.[20] The notion that the face's signification of internal states could be systematically understood was revived in Johann Caspar Lavater's popular physiognomic writing in the 1770s.[21] Contemporary colonial historical and ethnographic accounts of bodies differing from those of European authors wove physiognomic commentary, overburdened with aesthetic judgments and moral connotations, into descriptions of facial form.[22]

Early racial-cranial comparisons, such as Johann Benjamin von Fischer's 1743 medical dissertation at Leiden University, read as anatomical commentary within the established genre of racialized physiognomy. Fischer described and published engravings of three skulls, each respectively representative of Europe, Asia, and Africa, in portrait-like, three-quarter turns. For Fischer, cranial form reflected both varied environments and the expression of interior qualities into bone, writing that "habits of the body and soul" and "visiting foreign lands" effected changes in the skull, formed by "nature and the soul."[23] Trained at Leiden alongside Fischer, Petrus Camper recast Fischer's cranial trinity in a widely reproduced diagram purporting to demonstrate innate racial differences through metrical physiognomy.

In the 1750s, Camper was appointed to the Amsterdam Guild of Surgeons, allowing him to collect "heads and other bones, of humans of all ages" in dissections in Amsterdam's anatomical theater.[24] By the 1760s, Camper had turned his attention to skulls as stable signatures of racial difference. By 1770, Camper sketched a series of skulls and associated faces in standardized profiles and frontal views: a monkey, an orangutan, an eleven-year-old boy from Angola, a Kalmyk man from Central Asia, and a man from an unspecified part of Europe, followed by Greek and Roman statuary.[25]

Camper's novelty was a measurement: the facial angle, which defined the jaw's projection relative to the forehead. According to Camper, this angle both demonstrated differences among continentally divided human races and differences between humans and apes, with the facial angle increasing from ape to African and Asian and from these to Europeans.[26] Although he did not explicitly infer mental states from the face, Camper arranged a racialized physiognomic aesthetic hierarchy. Following then-influential neoclassical aesthetics, Camper proposed that Greco-Roman statuary possessed ideal beauty, with facial angles up to 100 degrees. Camper asserted that Europeans, among living humans, most nearly approached this ideal.[27] To Camper, the greater the facial angle, the more beautiful the face: "What is a beautiful face? We answer: one such that its facial line . . . makes an angle of 100 degrees."[28]

Camper's theory of the facial angle, about which he lectured during life but was only posthumously published and widely circulated, was a new metrical formalization of old ideas within racialized aesthetic physiognomy. Its primary intervention was the popularization of the notion of innate racialized craniofacial difference. Camper's student, Samuel Thomas Sömmerring, would fatefully propose the ultimate site of racial difference in a yet deeper stratum of the body, in the brain.

Racial Skulls to Racial Brains: "Organs of the Intellect" and Cranial Capacities

Anatomists first sought race by color in the brain. By the 1750s, Johann Friedrich Meckel in Berlin and Claude-Nicolas Le Cat in Rouen dissected bodies of Black people to allege later disproven racial differences in the colors of bodily fluids and brain tissue.[29] Around the same time, an anatomist at the recently founded University of Göttingen, Albrecht von Haller, experimented on the sensitivity of nerves and their reception in the brain, prompting subsequent decades of contention about relations among the brain, nerves, and mind, particularly on the neuroanatomy distinguishing humans and animals and their respective mental functions.[30] By the 1780s, these ideas widened in circulation, such as in Johann Gottfried Herder's *Ideen zur Philosophie der Geschichte der Menschheit* (*Ideas for a Philosophy of Human History*), published in March 1784: "The larger and finer the collecting point of all sensations, the brain, is, the more intelligent and finer this type of organization becomes. . . . Who would not be pleased if a philosophical dissector took it upon themselves to give a comparative physiology of several animals, especially those close to humans, according to these forces."[31]

Sömmerring, who wrote a medical dissertation on cranial nerves at Göttingen in 1778, fancied himself just such a philosophical dissector. In

November 1784, five years after his appointment as professor of anatomy at the Collegium Carolinum in the city of Kassel in the Landgraviate of Hesse-Kassel, Sömmerring published *Über die körperliche Verschiedenheit des Mohren vom Europäer (On the Bodily Difference of the Moor from the European)*.[32] This thirty-page text was the first to systematically assert racial differences in brain size, to claim that differences in brain size related to differences in mental capacity, and to ascertain brain size through cranial measurements. Sömmerring's opening lines outlined his prejudices: "We Europeans" appear to "possess a privilege over Black people," which Sömmerring proposed to explain in bodies. Citing Camper's facial angle as inspiration, Sömmerring reported on dissection the bodies of "African Moors" to determine whether the "organs of the intellect" rather than "conceited pride" justified his presumptive white privilege, conjecturing that Black people were anatomically nearer to apes than were Europeans.[33]

Although he surveyed the entire body, Sömmerring's core argument was that the brain's mass compared to that of the nerves separated humans from animals: a small brain compared to the mass of the nerves was suited for "mere animal life," while a relatively greater mass of the brain held the "most excellent organic construction for mental ability." Larger sensory organs, and corresponding aspects of craniofacial architecture, like the size of the jaws, eye orbits, and nasal aperture, implied larger nerves. Sömmerring measured brains and also outlined measurements to demonstrate from the skull that the larger the bony "brain capsule," or cranial vault around the brain, the smaller the face. He alleged that his measurements of the skulls and brains of a handful of Black people compared to those of white Europeans demonstrated white cerebral superiority on average, musing that these observations might explain what he called "historical facts" of Black inferiority.[34] The dehumanizing racism was not new, but its manifestation in the practice of measuring cranial contours and claiming racial differences in brain size and mental power were.

At the end of his book, Sömmerring commented that other peoples may be more anatomically "animal-like" than "African Moors," but he limited his comparisons to those of Black people and white Europeans.[35] This limitation was at least in part a practical matter. Although European anatomists had occasionally observed and collected the remains of racialized people caught in colonial networks and the trade in enslaved people in prior years, the movement of a group of Black Americans to Hesse after the American Revolutionary War was a contingent but crucial context to the formation of Sömmerring's novel cerebral race science.

Revolutionary Racial Subjects: The Black Hessians

In the 1770s, the landgrave of Hesse-Kassel, Friedrich II, uncle of Britain's King George III, sent thousands of Hessian mercenaries to fight for the British in the American Revolutionary War.[36] By 1784, the mercenaries returned, with scores of Black Americans.[37] Some were among the Black Americans who had escaped enslavers following British proclamations promising freedom in exchange for military service; others were conscripted by Hessians as aides or servants. Following U.S. victory, these Black Americans fighting for the British were sent as prisoners of war to New York City. After negotiations formally ending the war in 1783, some of these Black soldiers and their families stayed in the United States; others settled elsewhere in the British Empire. Some journeyed to Hesse, a land without race-based slavery, as subjects of Friedrich II. The landgrave eagerly received them. Since at least the late sixteenth century, "chamber Moors" or "court Moors"—like opulent gardens, menageries, and curiosity cabinets—were tokens of wealth and prestige among central European nobility. While "Moor" was an expansive term for racialized others from Africa, Asia, or the Iberian Peninsula since the fifteenth century, by the eighteenth century, those termed "Moors" in European courts are traceable to the transatlantic trade in enslaved people. A few such "Moors" were already in the court of Friedrich II before 1783.[38]

In October 1783, Friedrich II ordered the newly arriving Black soldiers transferred into to his royal guard.[39] Those who would become known as the "Black Hessians" came to feudal Europe as court subjects and objects of racial spectacle: images from the 1780s depict Black drummers in Hessian corps dressed in ornate uniforms topped with feathered turbans.[40] Records of Black members of the landgrave's regiment, listed with Germanified names and birthplaces across the former American colonies, Caribbean, and Africa, are as diverse as the fragmentary traces of their fates or those of their kin, sporadically recorded in marriage and baptism records. Some died of illness, some disappeared, and some, or their children, married local Germans. By about 1840, explicit traces of Black presence in and around Kassel wane.

Records of "foreign racial" marriages between Germans and "Moors at the noble houses" were of concern to Nazi eugenicists over a century later.[41]

By late 1783, Sömmerring was perhaps the largest community of people of African descent in central Europe in the late eighteenth century. He wrote of seeing dozens of Black people in Kassel; by 1784, Sömmerring recorded dissection of at least four of them.[42] Sömmerring's dissections and descriptions of their bodies were the basis of *On the Bodily Difference*. Their bodies may have been furnished to Sömmerring under the landgrave's official medical orders deeming dissection for anatomical study and training the fate of those in Hesse-Kassel who died of suicide or were found dead; who died in orphanages, prisons, or almshouses; or who accepted medical care without payment, among others.[43] However, Sömmerring did not only rely on sanctioned channels for bodies: in a letter to Camper, Sömmerring wrote about digging up "incognito" the buried body of a Black woman, "as quickly as possible," presumably anticipating reputational consequences or resistance to his grave robbing, if he were discovered.[44]

Göttingen's Golgotha Takes Form

One of Sömmerring's colleagues and mentors at Göttingen, Johann Friedrich Blumenbach, published a withering review of *On the Bodily Difference*.[45] Blumenbach wrote his medical dissertation at Göttingen in 1775, *De generis humani varietate nativa* (*On the Natural Varieties of Humankind*). Published in 1776, Blumenbach wrote on the anatomical basis for Linnaeus's fourfold racial taxonomy and, following Buffon, argued for a monogenetic white origin for all humans.[46] Blumenbach's first edition of *De generis* was commentary on existing literature: Blumenbach summarized previously published reports on human natural history, including cranial anatomy, and relied on already published illustrations rather than on direct observation of racialized skulls or other body parts. Although he managed Göttingen's new Royal Academic Museum, Blumenbach possessed only a single human skull, from the local graveyard, when he finished his dissertation in 1775.[47]

Like Camper and Fischer, Blumenbach was attentive to faces rather than brains, idealized beauty rather than mental capacity, fitting this discourse into a monogenetic account of racial origins and degeneration. In the first edition of *De generis*, Blumenbach was hesitant about skull morphology's significance as the defining marker of racial difference, writing that published reports and illustrations of skulls allegedly of the same people varied too much and that skulls likelier reflected conditions of life rather than innate forms: "almost all the diversity of the form of the head in different nations is to be attributed to the mode of life and to art."[48] Reflecting the widening sphere of European colonialism in the South Pacific, Blumenbach updated the second edition of De generis in 1781 to include a fifth variety—of people from Southeast Asia and Oceania—to the first edition's Linnaean four-part classification.[49] Blumenbach's expansion of Linnaeus's classification was not spun from direct observations of cranial anatomy but from colonial travelogues.

The third and final edition of De generis, published in 1795, was dramatically different: this edition represented the five racial varieties Blumenbach described in the 1781 edition with exemplars from a collection of racialized skulls he had systematically amassed beginning in 1784. Between 1784 and 1795, Blumenbach collected about a third of the approximately 240 skulls he would amass from across the globe through the first decades of the nineteenth century, forming the largest anthropological collection of human skulls in the world at the time.[50]

Why did Blumenbach suddenly start to collect human skulls for comparative racial study in 1784, and why did skulls play a key role only in Blumenbach's third edition of De generis? Some have suggested that Blumenbach "discovered skulls as crucial evidence for the natural history of man only at the suggestion of Camper and Sömmerring."[51] Others have noted tensions, in both private correspondence and publication, in the previously friendly relationship between Blumenbach and Sömmerring, beginning around 1784, when each began to write about comparative racial cranial anatomy.[52] Close attention to the circulation and collection of racialized skulls to and between Göttingen and Kassel in the 1780s supports a more precise claim: Blumenbach's collection and study of skulls for racial science was in response to Sömmerring's claims in On the Bodily Difference.

Excavating the chronology of accessions into Blumenbach's skull collection is telling. By early 1784, Blumenbach had only collected three skulls: one disinterred from a Göttingen graveyard, another sent by a student from Switzerland, and one of a mummified Egyptian purchased from a merchant.[53] Blumenbach's racial classification through the second edition of De generis was neither illustrated with cranial anatomy nor informed by

Blumenbach's direct observation of skulls or of living people within the racial classifications he constructed.[54] Unlike Sömmerring in Kassel, Blumenbach did not have ready access to racialized bodies for dissection in Göttingen. In 1784, Blumenbach began to solicit colleagues for skulls for racial comparison.[55] Blumenbach's first of many recorded solicitations of non-European skulls from colleagues and associates was to Camper in September 1784, asking for the skull of a Khoikhoi person from the Dutch Cape Colony. In the same letter, Blumenbach mentioned that he had recently requested skulls from the Göttingen alumnus Georg Thomas von Asch, court physician in St. Petersburg, who would send about a quarter of the skulls Blumenbach would ultimately amass.[56]

The first skulls Blumenbach received in 1784 were from the United States, provided by another Göttingen-trained anatomist, Christian Friedrich Michaelis. Michaelis served as a Hessian military physician among Friedrich II's mercenaries, based in New York City but cultivating contacts with naturalists and physicians in Philadelphia. Michaelis—who published observations on the grave-robbed remains of a Black teenager in New York City during his time there—may have been prompted to collect skulls by Camper, who corresponded with Michaelis about acquiring mastodon fossils from the United States in the 1780s.[57] After the war's conclusion, Michaelis was appointed a professor at Kassel, alongside Sömmerring, in June 1784; he traveled between Kassel and Göttingen in subsequent months.[58] In this period, Blumenbach received two skulls from Michaelis: that of an Indigenous man allegedly executed for murder in Philadelphia decades prior, collected by one of Philadelphia's first anatomists, Abraham Chovet, and of a Black man named Hamden who died in New York City (per information in Blumenbach's notes, apparently relayed by Michaelis).[59]

Blumenbach did not only begin soliciting racialized skulls from colleagues in 1784; he began writing about them shortly thereafter. Blumenbach's January 1785 review of Sömmerring's *On the Bodily Difference* includes Blumenbach's first direct comparative observations of racialized skulls. Blumenbach compared the head of a mummified Egyptian and a white European with the two skulls recently "possessed through the goodness of Dr. Michaelis." Comparing sizes of braincases and facial features, Blumenbach observed that Hamden's skull was no more different from that of a white European than the others and that no human skulls were closer to the form of apes than others. Blumenbach characterized Sömmerring's comparisons between

Black people and apes and any inferences about racial difference drawn from them as specious; he affirmed unequivocally that all human variations "flow into one another" within a "single human family."[60]

Notably, Blumenbach began to collect skulls months before Sömmerring's publication of On the Bodily Difference in November 1784. But Sömmerring's publication was not a surprise: Blumenbach and Sömmerring exchanged letters regarding Sömmerring's ideas about neuroanatomy throughout the early 1780s, and Sömmerring had contact with Blumenbach in the middle months of 1784.[61] Although only partial correspondence between Sömmerring and Blumenbach survives, a few lines in a letter Sömmerring wrote to Camper in May 1784, just months after the arrival of Black Americans in Kassel, hint at reasons for Blumenbach's reaction to Sömmerring's text. Sömmerring wrote to Camper that he believed his dissections and theorizations of race in the brain had far-reaching consequences that would have negated the entire framework of Blumenbach's De generis: Sömmerring wrote that he was "more and more" convinced that "Africans and Europeans" were not "two simple varieties" but rather "basically different species . . . that there were two Adams."[62]

Anatomies of Other Adams

Sömmerring's increasing conviction of polygenism—that racial differences were not signs of varying customs and environments etched onto peoples over time but rather that they traced to separate creations, that human races were separate species—was dangerous. In the 1650s, the lawyer and scriptural exegete Isaac La Peyrère enraged Protestant, Jewish, and Catholic authorities across western Europe with his heretical theory of separate racial creations in Prae-Adamitae (Men Before Adam), published in Amsterdam.[63] His books were publicly burned, and he was arrested to recant before the pope. As Terence Keel has observed, "In the wake of La Peyrère's pre-Adamite scandal, orthodox visions of common human descent and recent human creation would continue to be reaffirmed as the true account of the origin of racial differences and the proper framework from which to view human history."[64]

Such controversy was especially unsavory for anatomists. Enlightenment racial anatomists relied on professional appointments achieved through proximity to political power, whether in the form of monarchs, nobles, or governing assemblies, clear in Buffon's position as director of the Jardin du Roi (the king's garden, including the royal natural history collection), or Sömmerring's and Blumenbach's titles as both professor and *Hofrath* (court counselor), or the many memberships in "royal societies" listed under author's names in publications. This proximity not only secured professorships but also access to bodies for dissection and study. Anatomists achieved this access either through regulations supplying bodies of the socially marginalized or those convicted of capital offenses to appointed medical faculties for dissection, or, in cases of collecting the racialized dead, through privileged connections to colonial networks, which could be plied for remains.[65]

Entanglement of political and religious institutions meant that questioning common human ancestry was, minimally, a professional liability for physician-anatomist racial scientists in northwestern European states where racial-anatomical discourse was most influentially crafted in the eighteenth century, including ancien régime France, the Holy Roman Empire, and the United Kingdom. Even in more tolerant states such as the Dutch Republic, reputations were on the line. Unsurprisingly, the few polygenists who published their claims after La Peyrère and before Sömmerring's *On the Bodily Difference*—including David Hume, Lord Kames, Voltaire, Edward Long, and John Atkins—were not professors or, except for Atkins, anatomically trained.[66] Perhaps the first university professor to openly profess polygenism in print was the historian and philosopher Christoph Meiners in Göttingen, in 1785. Although protected by Göttingen's remarkably liberal stance on academic freedom, Meiners's reputation suffered.[67] Among the few others to openly express polygenism in the 1780s was a close friend of Sömmerring, the naturalist Georg Forster, to whom Sömmerring dedicated the second edition of *On the Bodily Difference*.[68]

Polygenism was an extreme insistence on the profundity of racial difference, elevating race to the level of separate creations. This insistence was especially gratuitous, as monogenism was malleable enough to accommodate a wide spectrum of convictions about racial hierarchy without implying heresy. Monogenism's assumption of common ancestry and transformation over generations allowed for human equality in principle, as, for example, the Black abolitionist Olaudah Equiano argued in 1789.[69] However, in no sense was monogenism ipso facto a defense of political equality, a call for abolition, or a cry against the violence of European colonialism, despite its potential deployment toward these ends. As was clear in the primacy of whiteness in Buffon's and Blumenbach's theories of degeneration and in the equation of whiteness with aesthetic perfection in these and Camper's accounts, much of Enlightenment monogenism was built from familiar materials of Eurocentric racism and was readily conceptually concordant with the naturalization of whiteness as ideal and white supremacist racial order as obvious.[70] Nonetheless, polygenism was a radical rejection of even the possibility of human unity or equality, the limit case in reifying the reality and significance of race. Unsurprisingly, into the nineteenth century polygenism would provide the most direct rhetorical support to proslavery arguments in the United States.[71]

Professorial prudence dictated that polygenism was simply a bridge too far by the time Sömmerring entertained its crossing. Camper's response to Sömmerring in 1784 was blunt: "I have dissected five Black people," Camper wrote, "but I have found no trait that would indicate a difference of species between them and Europeans."[72] Sömmerring dropped the topic in correspondence, and he published his position as monogenism in *On the Bodily Difference* a few months later—despite obvious tensions between this declaration and the substance of the text and legitimate questions regarding whether this declaration was primarily sincere or strategic.

Sömmerring was at pains to underscore that he did not espouse polygenism in a revised edition of the text in 1785, including passages copied ad verbatim from Blumenbach's review.[73] Sömmerring kept his core argument but added a caveat-laden preface admitting that "misunderstanding" of his views might "somewhat excuse the tyrants under which [enslaved Black people] heave in both the Indies" but underscoring that Black people are "true humans, as good as we are."[74] Blumenbach published a review of the second edition of what he described as Sömmerring's "strange text," skipping much of the content except Sömmerring's affirmation of monogenism.[75] Shortly thereafter, Sömmerring complained to Blumenbach's brother-in-law about Blumenbach's "dissatisfaction" with *On the Bodily Difference*; elsewhere, Sömmerring described Blumenbach's criticisms as rooted in jealousy and professional ambition.[76]

For his part, Sömmerring could not shake suspicions of polygenism. He was appointed to a professorship at the Catholic stronghold of Mainz in 1784, before the publication of *On the Bodily Difference*. Following his inaugural lecture there, which included the dissection of the head of a Black man recently deceased in Kassel, a cleric accused Sömmerring of denigrating the biblical

Magi from the Gospel of Matthew, widely represented in religious artwork as dark skinned, to the level of apes.[77] A local newspaper spread the threat of an inquisition against the anatomist. Sömmerring turned to other topics in anatomy for publication in the coming years, although he continued compiling notes for a never-published third edition of *On the Bodily Difference*.[78]

Blumenbach, in the meantime, continued to amass skulls for racial comparisons, soliciting them from students, colleagues, and associates.[79] The union between the Electorate of Hanover (where Göttingen was located) and Great Britain and Ireland, the founding of the University of Göttingen by King George III's grandfather, and the university's standing as a respected site of medical education all provided Blumenbach with links to British colonial agents and physicians who could send Blumenbach skulls. In late 1789, months after Camper's death and with a collection of twenty-four skulls, Blumenbach lectured in Göttingen for the first time on the cranial forms of human "natural varieties" within his fivefold classification.[80] His prior hesitancy about the relevance of skulls in understanding the natural history of the human species was gone. He criticized Camper's facial angle for not differentiating among "skulls from the same people" and that the skulls Camper depicted were not representative of characteristic "national forms." Blumenbach openly questioned the use of any measurements to classify these differences, preferring the approach of minute anatomical description, illustrations of skulls in portrait-like turns, and aesthetic commentary on supposedly representative exemplars of cranial-racial difference. Certainty about "national forms," Blumenbach wrote, only came with "comparison of many skulls of the same people with another."[81] Impetus for collection of ever more skulls was a methodological shift after 1784, an attempt to empirically demonstrate the monogenism foundational to Blumenbach's conceptualization of racial difference since his dissertation in 1775. It was this turn to cranial anatomy, already evident in his commentary on Sömmerring and lectures in 1789, which characterized the third and final edition of *De generis* in 1795 and differentiated it from its prior editions.

Blumenbach continued collecting and describing skulls within the same framework throughout the remainder of his career. His defense of monogenism may have been motivated by political convictions in addition to his defense of the core assumption of monogenism structuring his dissertation and entire subsequent theorizations of human natural history in later editions of *De generis*. Notably, however, Blumenbach never criticized slavery in print. This reticence may have been influenced by Göttingen's connection to the British Empire, where slavery in British colonies was only abolished in 1833. By that time, Blumenbach was nearly retired, and scores of other anatomists across Europe and the United States had amassed comparative racial skull collections—although not necessarily motivated to demonstrate monogenism.

New Orders, Different Constitutions: Polygenism and Cranial Capacity Into the Nineteenth Century

Years after leaving Mainz, Sömmerring published a long review of a polygenist text by the Dutch naturalist Jacob Elisa Doornik, printed in Amsterdam in 1808, who had suggested that human races were separately created, that these had transformed from an animal-like to a human state at different rates, and that Europeans were furthest in this progression. Sömmerring wrote with evident sympathy for Doornik's "doubts about this more or less universally accepted unity of the species."[82] That Doornik would dare to write such a book, suggesting both polygenism and the transformation of species, and that Sömmerring would commend it were signs of much of what had changed since the mid-1780s.

The arrival of Black Americans in Kassel was not the only aftermath of the American Revolution pertinent to racial science's development. By the final decade of the eighteenth century, the stability of European political orders founded on divine monarchical right and the feudal system was in question. Liberal political movements found vindication in the United States' victory against the British, erupting into anticlerical and secularizing forces in the French Revolution beginning in 1789. These ultimately served to loosen the cultural and political binds of Christian orthodoxy across Europe.[83] European, and, later, U.S. naturalists increasingly departed from fitting the natural history of the human species into canonical monogenetic readings of the Mosaic creation account. And, crucially, presumptions of unassailable white political supremacy became vulnerable in the wake of the successful revolt of enslaved people in Haiti beginning in the 1790s. The Haitian Revolution's threat to racial order structuring the Atlantic world inspired further abolitionist movements but also deepened racial anxieties, whetting white appetites for anti-Black scientific racism.[84]

In this ferment, Sömmerring's arguments circulated well beyond their initial reception in the 1780s, propelling a mania to not only collect skulls demonstrating supposed racial physiognomies but to measure brains toward the bulwarking of white supremacist racial hierarchy. The British anatomist Charles White included the first and only printed English translation of Sömmerring's *On the Bodily Difference* in a 1799 publication of his polygenist 1795 address in Manchester, the budding center of the British cotton industry, *An Account of the Regular Gradation in Man*.[85] The British surgeon Richard Saumarez replicated Sömmerring's claims about racial brain size in 1798, filling skulls of white people and a single Black person with water and comparing their volumes, surmising white cerebral superiority from skull capacity.[86] In 1817, Georges Cuvier's report on the dissection of Sarah Baartmann referenced "Sömmerring's rule" in alleging "signs of inferiority" on her skull.[87] In the same decade, the anatomist Julien-Joseph Virey measured the cranial capacity of skulls of white and Black people in Paris. The U.S. physician Samuel Cartwright included Virey's measurements in an appendix to reprints of U.S. Supreme Court Chief Justice Roger B. Taney's arguments in *Dred Scott v. Sanford* (1857), which denied rights of U.S. citizenship to people of African descent.[88]

Sömmerring's argument about brain size and mental power found another propulsive current. In the late 1790s, the Viennese physician Franz Joseph Gall formulated a "doctrine of the skull," later known as "phrenology," expanding from associations between behavior and the brain paved by von Haller, Herder, and Sömmerring. However, contrary to Sömmerring's theorization of brain function, that the entire brain operated in unity, Gall posited precise connections among parts of the brain, their supposed outward manifestation on the surface of the skull, and specific mental functions or aspects of character.[89] Phrenologists tended to focus on the relative sizes of different parts of the brain rather than total brain size and on reading the skull's external surface rather than measuring its contents. The possibility of phrenological measurement on the heads of the living made for a more popularly accessible practice than the cranial race science constructed from skull collections that, in general, only medically trained anatomists in privileged positions could amass. Phrenology's popularization spawned a variety of interpretations and deployments, from predictable justifications of slavery and colonialism to endorsement and interest from abolitionists in the United States.[90] However, most fundamentally, phrenology circulated beyond academic circles the notion that the brain was the organ of the mind and that its measurement—through the skull—revealed profound differences in innate ability.

In the 1830s, William Hamilton in Edinburgh and Friedrich Tiedemann in Heidelberg countered the swelling tide of racist arguments articulated through brain size by filling ever greater numbers of skulls, racially sorted, with sand and millet seed, reporting no racial differences.[91] These works were overshadowed by the U.S. race scientist Samuel George Morton in Philadelphia. Morton amassed the first global racial skull collection larger than Blumenbach's—enlarged near the outset of Morton's project with his purchase of Doornik's skull collection.[92] Morton's comparative racial measures of hundreds of skulls by cranial "internal capacity" from 1839 through 1849 forwarded white supremacist polygenism founded on claims about racial differences in brain size.[93] These measures were widely circulated through the 1840s and 1850s in the United States, popularized by Josiah Nott and George Gliddon in the Types of Mankind, criticized by Frederick Douglass in The Claims of the Negro Ethnologically Considered, and served as fundament to the proclamation of Alexander H. Stephens, vice president of the Confederate States of America, that the enslavement of Black people was based upon the "physical, philosophical, and moral truth" of innate inequality: "This truth has been slow in the process of its development, like all other truths in the various departments of science."[94]

Darwinian evolutionary thought rendered debate between polygenism and monogenism effectively obsolete: in an evolutionary paradigm, all humans share common ancestry. However, the preferred polarity of change over time was reversed. For evolutionists, the pinnacle of racial hierarchy rested on maximal progression from the ape rather than minimal degeneration since creation. Evolutionists absorbed the measurement of cranial capacity for their own articulations of racial difference in this vein, beginning with Thomas Henry Huxley's comparisons of ape and racialized human skulls, recycling Morton's data, in the 1860s.[95] Throughout these years and into the first decades of the twentieth century, anatomist-anthropologists collected hundreds of thousands of human skulls for racial science, with the large majority of these still in museums and university collections today. While skulls have long since lost their primacy in scientific racism, "the organs of the intellect" have not.[96]

Coda

The centrality of bodily features in the construction of Enlightenment racial classifications lent anatomists perceived expertise in crafting discourses on race. The search for race "in the body's foundation" was not merely a matter of skin color's instability or anxieties about what environments might do to Europeans in settler-colonies but also an effect of increasingly minute attention to osteological detail in the early and mid-eighteenth century. Anatomists, as dissectors and collectors, could assemble privileged observations on racial difference, rationalizing and systematizing what had previously been the occasional collection of the remains of racialized others. Skull collection, either in elaborations of racialized physiognomy or through novel claims that mental power could be measured from brain size, allowed the shortest plausible rhetorical distance from body to behavior to political order. However, the emergence of the skull as racial science's principal object is only partially understood in tracing explicit logics of difference in published accounts of European racial anatomists.

Focus on differences and debate between Sömmerring and Blumenbach in the period of 1784–1785 draws attention to the consolidation of cranial race science in central Europe through the movement of bodies after the American Revolutionary War, both through Black Hessians arriving in Kassel and remains trafficked from Philadelphia and New York City to Göttingen at the start of Blumenbach's global racial skull collection. Connections to the British Empire made Kassel and Göttingen endpoints in the circulation of bodies, living and dead, through globalizing colonial-imperial networks. These bodies became grist for debates about polygenism, monogenism, and racial difference increasingly invested in the systematic collection and study of racialized human remains, amassed by any means. The remains of the Black Hessians Sömmerring dissected in Kassel, among other remains he collected through his career, were transferred to Frankfurt am Main by the time of Sömmerring's death. They were destroyed by American bombs in World War II.[97] Other remains of Black Hessians were sent to Camper by Sömmerring and to Blumenbach by Michaelis.[98]

The revolutions of the late eighteenth century opened space for European anatomists to safely shed prior adherence, sincere or otherwise, to commitments to human unity. They also charged racial science with heightened significance in naturalizing racialized political order as movements for abolition spread across the Atlantic world, situating Blackness and the brain as recurrent preoccupations of scientific racism in European and North American metropoles through the greater part of the nineteenth century. Among the clamor of proclamations of liberty and equality and demands for abolition, estimations of "cranial capacity" first articulated through Sömmerring's polygenist suspicions in the 1780s spread as a means to carve racialized inequality deep into the body and, with enduring effect, into the mind.

Notes

1. Miriam Claude Meijer, *Race and Aesthetics in the Anthropology of Petrus Camper (1722-1789)* (Amsterdam: Rodopi, 1999); Wolfgang Böker, "Blumenbach's Collection of Human Skulls," in *Johann Friedrich Blumenbach: Race and Natural History 1750-1850*, ed. N. Rupke and G. Lauer (Abingdon: Routledge, 2019), 80–95; Nicolaas Rupke and Gerhard Lauer, "Introduction: A Brief History of Blumenbach Representation," in *Johann Friedrich Blumenbach: Race and Natural History 1750-1850*, ed. N. Rupke and G. Lauer (Abingdon: Routledge, 2019), 3–15.

2. Andrew Zimmerman, *Anthropology and Antihumanism in Imperial Germany* (Chicago: University of Chicago Press, 2010); Ann Fabian, *The Skull Collectors: Race, Science, and America's Unburied Dead* (Chicago: University of Chicago Press, 2010); Ricardo Roque, *Headhunting and Colonialism: Anthropology and the Circulation of Human Skulls in the Portuguese Empire, 1870-1930* (New York: Palgrave Macmillan, 2010); Samuel J. Redman, *Bone Rooms: From Scientific Racism to Human Prehistory in Museums* (Cambridge, MA: Harvard University Press, 2016); Paul Turnbull, *Science, Museums, and Collecting the Indigenous Dead in Colonial Australia* (Cham: Palgrave Macmillan, 2017); Paul Wolff Mitchell, "The Fault in His Seeds: Lost Notes to the Case of Bias in Samuel George Morton's Cranial Race Science," *PLoS Biology* 16, no. 10 (2018): e2007008; Laurens de Rooy, "The Shelf Life of Skulls: Anthropology and 'Race' in the Vrolik Craniological Collection," *Journal of the History of Biology* (June 2023).

3. For example, T. Wingate Todd and Margaret Russell, "Cranial Capacity and Linear Dimensions, in White and Negro," *American Journal of Physical Anthropology* 6, no. 2 (1923): 97–194; Iris Clever, "The Lives and Afterlives of Skulls: The Development of Biometric Methods of Measuring Race (1880–1950)," PhD diss., University of California–Los Angeles (2020).

4. Rudolf Martin, *Lehrbuch der Anthropologie in systematischer Darstellung* (Jena: Gustav Fischer, 1914), 475–890.

5. Dorothy Roberts, *Fatal Invention: How Science, Politics, and Big Business Re-create Race in the Twenty-First Century* (New York: New Press, 2011); Jonathan Marks, "The Origins of Anthropological Genetics," *Current Anthropology* 53, no. S5 (2012): S161–S172; Claudio Pogliano, *Brain and Race: A History of Cerebral Anthropology*

(Leiden: Brill, 2020), 233–96; Amade M'Charek, "Tentacular Faces: Race and the Return of the Phenotype in Forensic Identification," *American Anthropologist* 122, no. 2 (2020): 369–80.

6. Aníbal Quijano, "Coloniality and Modernity/Rationality," *Cultural Studies* 21, nos. 2–3 (2007): 67; Sylvia Wynter, "Unsettling the Coloniality of Being/Power/Truth/ Freedom: Towards the Human, After Man, Its Overrepresentation—an Argument," *CR: The New Centennial Review* 3, no. 3 (2003): 257–337.

7. Thierry Hoquet, "Biologization of Race and Racialization of the Human: Bernier, Buffon, Linnaeus," in *The Invention of Race: Scientific and Popular Representations*, ed. N. Bancel, T. David, and D. Thomas (London: Routledge, 2014), 17–32.

8. Staffan Müller-Wille, "Linnaeus and the Four Corners of the World," in *The Cultural Politics of Blood, 1500–1900*, ed. K. A. Coles, R. Bauer, Z. Nunes, and C. L. Peterson (London: Palgrave Macmillan, 2014), 191–209.

9. Londa Schiebinger, "The Anatomy of Difference: Race and Sex in Eighteenth-Century Science," *Eighteenth-Century Studies* 23, no. 4 (1990): 387–405; Andrew S. Curran, *The Anatomy of Blackness: Science and Slavery in an Age of Enlightenment* (Baltimore, MD: Johns Hopkins University Press, 2011), 1–28, 74–116; Rana A. Hogarth, *Medicalizing Blackness: Making Racial Difference in the Atlantic World, 1780–1840* (Chapel Hill: University of North Carolina Press, 2017).

10. Hoquet, "Biologization of Race and Racialization of the Human," 17–32; John H. Zammito, *The Gestation of German Biology: Philosophy and Physiology from Stahl to Schelling* (Chicago: University of Chicago Press, 2018), 107–14; Terence Keel, *Divine Variations: How Christian Thought Became Racial Science* (Stanford, CA: Stanford University Press, 2018), 39–40.

11. Keel, *Divine Variations*, 23–53.

12. Schiebinger "Anatomy of Difference," 389–91; Meijer, *Race and Aesthetics*, 68–81; Hogarth, *Medicalizing Blackness*, 34, 50, 70–71; Curran, *The Anatomy of Blackness*, 74–116.

13. George LeClerc, Comte de Buffon, *Histoire naturelle, générale et particuliére: avec la description du Cabinet du Roi, Tome Troisième* (Paris, 1749), 459.

14. Alexander Monro, *The Anatomy of the Humane Bones* (Edinburgh, 1726), 61–63; Schiebinger, "Anatomy of Difference," 392–95; Meijer, *Race and Aesthetics*, 154–58.

15. Monro, *The Anatomy of the Humane Bones*; William Cheselden, *Osteographia: Or the Anatomy of the Bones* (London, 1733); Bernhard Siegfried Albinus, *Tabulae sceleti et musculorum corporis humani* (Leiden, 1749).

16. For example, Gerardus Blancken, *A Catalogue of All the Chiefest Rarities in the Publick Theater and Anatomie-hall of the University of Leyden* (Leiden, 1704), 9; M. Winslow, "Confirmation particuliere du crâne d'un sauvage de l'Amérique septentrionale," in *Histoire de l'Academie Royal des Sciences de 1722* (Paris, 1724): 322–24.

17. Petrus Camper, "Redevoering over den Oorsprong en de Kleur der Zwarten," *De Rhapsodist* 2 (1772): 387–88; Dienke Hondius, "Access to the Netherlands of Enslaved and Free Black Africans: Exploring Legal and Social Historical Practices in the Sixteenth–Nineteenth Centuries," *Slavery & Abolition* 32, no. 3 (2011): 377–95.

18. Samuel Thomas Sömmerring, *Über die körperliche Verschiedenheit des Mohren vom Europäer* (Mainz, 1784), 6–7.

19. For example, Charles White, *An Account of the Regular Gradation in Man, and in Different Animals and Vegetables; and from the Former to the Latter* (London, 1799); Gerardus Vrolik, *Beschouwing van het verschil der bekkens in onderscheidene volkstammen* (Amsterdam, 1826).

20. Meijer, *Race and Aesthetics*, 115–23; Martin Porter, *Windows of the Soul: Physiognomy in European Culture, 1470-1780* (Oxford: Oxford University Press, 2005); Irven M. Resnick, *Marks of Distinctions: Christian Perceptions of Jews in the High Middle Ages* (Washington, DC: Catholic University of America Press, 2012), 13–19; Pogliano, *Brain and Race*, 2, 9–61.

21. Johann Caspar Lavater, *Physiognomische Fragmente zur Beförderung der Menschenkenntnis und Menschenliebe* (Leipzig, 1775–1778); Richard Gray, *About Face: German Physiognomic Thought from Lavater to Auschwitz* (Detroit, MI: Wayne State University Press, 2005), 1–56.

22. Bronwen Douglas, "Science and the Art of Representing 'Savages': Reading 'Race' in Text and Image in South Seas Voyage Literature," *History and Anthropology* 11, nos. 2–3 (1999): 157–201; Bronwen Douglas, *Science, Voyages, and Encounters in Oceania, 1511-1850* (London: Palgrave Macmillan, 2014), 252–86.

23. Johann Benjamin von Fischer, *Dissertatio Osteologica de Modo, Quo Ossa Se Vicinis Accommodant Partibus* (Leiden, 1743), 23–24.

24. Samuel Lamsveld, *Privilegien, willekeuren en ordonnantien, betreffende het collegium chirurgicum, Amstelœdamense* (Amsterdam, 1755); Petrus Camper, *Verhandeling over het natuurlijk verschil der wezenstrekken in menschen van onderscheiden landaart en ouderdom* (Utrecht, 1791), viii.

25. Camper, *Verhandeling*, 21, 23, 37.

26. Meijer, *Race and Aesthetics*, 101–78.

27. Camper, *Verhandeling*, iv, 91; Martial Guédron, "Panel and Sequence: Classifications and Associations in Scientific Illustrations of the Human Races (1770-1830)," in *The Invention of Race: Scientific and Popular Representations*, ed. N. Bancel, T. David, and D. Thomas (London: Routledge, 2014), 60–67; Paul van den Akker, "Petrus Camper on Natural Design and the Beauty of Apollo's Profile," in *Petrus Camper in Context: Science, the Arts, and Society in the Eighteenth-Century Dutch Republic*, ed. K. van Berkel and B. Ramakers (Hilversum: Uitgeverij Verloren, 2015), 243–55.

28. Camper, *Verhandeling*, 90, see 18, 91–92; see also Meijer, *Race and Aesthetics*, 160–66; van den Akker, "Petrus Camper on Natural Design," 243–55.

29. Pogliano, *Brain and Race*, 15–20.

30. Albrecht von Haller, *De partibus corporis humani sensilibus et irritabilibus* (Göttingen, 1752); Pogliano, *Brain and Race*, 23–24.

31. Johann Gottfried Herder, *Ideen zur Philosophie der Geschichte der Menschheit*, Band 1/2 (Berlin and Weimar, 1965), 90; Frank W. Stahnisch, " 'Dieu et cerveau, rien que Dieu et cerveau!'—Johann Gottfried von Herder (1744–1803) und die Neurowissenschaften seiner Zeit," *Wuerzburger medizinhistorische Mitteilungen* 26 (2007): 124–65; Zammito, *The Gestation of German Biology*, 144–49.

32. Sigrid Oehler-Klein, " 'Der "Mohr" auf der niedrigeren Staffel am Throne der Menschheit'? Georg Forsters Rezeption der Anthropologie Soemmerrings," in *Georg-Forster- Studien III*, ed. H. Dippel and H. Scheuer (Kassel: Kassel University

Press, 1999), 119–66; Sigfrid Oehler-Klein and Samuel Thomas Sömmerring, *Anthropologie: Über die körperliche Verschiedenheit des Negers vom Europäer (1785)*, ed. Sigfrid Oehler-Klein (Stuttgart: Gustav Fischer Verlag, 1998), 7–142.

33. Sömmerring, *Über die körperliche Verschiedenheit* (1784), 5.

34. Sömmerring, *Über die körperliche Verschiedenheit* (1784), 22, 16–20, 24, 32.

35. Sömmerring, *Über die körperliche Verschiedenheit* (1784), 4, 32.

36. Elliott W. Hoffmann, "Black Hessians: American Blacks as German Soldiers," *Negro History Bulletin* 44, no. 4 (1981): 81–82, 91; George Fenwick Jones, "The Black Hessians: Negroes Recruited by the Hessians in South Carolina and Other Colonies," *South Carolina Historical Magazine* 83, no. 4 (1982), 287–302; Wolfram Schäfer, "Von 'Kammer-Mohren', 'Mohren-Tambouren' und 'Ost-Indianern'. Anmerkungen zu Existenzbedingungen und Lebensformen einer Minderheit im 18. Jahrhundert unter besonderer Berücksichtigung der Residenzstadt Kassel," *Hessische Blätter für Volks-und Kulturforschung* 23 (1988): 35–79; Maria Diedrich, "From American Slaves to Hessian Subjects: Silenced Black Narratives of the American Revolution," in *Germany and the Black Diaspora: Points of Contact, 1250–1914*, ed. M. Honeck, M. Klimke, and A. Kuhlmann (New York: Berghahn, 2016), 92–111; Jeanette Eileen Jones, " 'On the Brain of the Negro': Race, Abolitionism, and Friedrich Tiedemann's Scientific Discourse on the African Diaspora," in *Germany and the Black Diaspora: Points of Contact, 1250–1914*, ed. M. Honeck, M. Klimke, and A. Kuhlmann (New York: Berghahn, 2016), 134–52; Friederike Baer, *Hessians: German Soldiers in the American Revolutionary War* (Oxford: Oxford University Press, 2022), 95–99, 370–86.

37. Jones, " 'On the Brain of the Negro,' " 138–41; Baer, *Hessians*, 370–86.

38. Schäfer, "Von 'Kammer-Mohren,' " 35–37, 44–45.

39. Diedrich, "From American Slaves to Hessian Subjects," 93.

40. Schäfer, "Von 'Kammer-Mohren,' " 48; Diedrich, "From American Slaves to Hessian Subjects," 97.

41. Schäfer, "Von 'Kammer-Mohren,' " 53–58nn218–19.

42. Schäfer, "Von 'Kammer-Mohren,' " 46; Oehler-Klein and Sömmerring, *Anthropologie*, 33–66.

43. Ulrike Enke, " 'Leichen für die Anatomie'—Samuel Thomas Soemmerrings Arbeitsbedingungen in Kassel," *Philippia* 14, no. 3 (2010): 248; Schäfer, "Von 'Kammer-Mohren,' " 69n125.

44. S. T. Sömmerring to P. Camper, November 29, 1785, in Franz Dumont, ed., *Samuel Thomas Soemmerring Briefwechsel November 1784–Dezember 1786* (Stuttgart: Gustav Fischer, 1996), 256.

45. Johann Friedrich Blumenbach, "Rezension: Sömmerring (1784)," *Göttingische Anzeigen von gelehrten Sachen* 12 (1785): 108–11; see also J. F. Blumenbach to S. T. Sömmerring, January 16, 1785, in Frank W. P. Dougherty, *The Correspondence of Johann Friedrich Blumenbach*, vol. 2: *1783-1785*, ed. Norbert Klatt (Klatt Verlag: Göttingen, 2007), 237.

46. Johann Friedrich Blumenbach, *De generis humani varietate nativa liber* (Göttingen, 1776), 7–9; Zammito, *The Gestation of German biology*, 186–214.

47. Johann Friedrich Blumenbach, *De generis humani varietate nativa* (Göttingen, 1775), 63; Böker, "Blumenbach's Collection of Human Skulls," 81.

48. Zammito, *The Gestation of German Biology*, 204–5; Böker, "Blumenbach's Collection of Human Skulls," 84–85.

49. Antje Kühnast, "Johann Friedrich Blumenbach's 'New Hollander,'" *Zeitschrift für Australienstudien/Australian Studies Journal* 33/34 (2020): 31–56.

50. Zammito, *The Gestation of German Biology*, 204–6; Böker, "Blumenbach's Collection of Human Skulls," 81–82.

51. Thomas Nutz, *Varietäten des Menschengeschlechts: Die Wissenschaft vom Menschen in der Zeit der Aufklärung* (Böhlau Verlag: Köln, 2009), 263.

52. Frank W. P. Dougherty, "Johann Friedrich Blumenbach und Samuel Thomas Soemmerring: Eine Auseinandersetzung in anthropologischer Hinsicht?," in *Samuel Thomas Soemmerring und die Gelehrten der Goethezeit*, ed. G. Mann and F. Dumont (Stuttgart: Gustav Fischer Verlag, 1985), 35–56.

53. Böker, "Blumenbach's Collection of Human Skulls," 81.

54. Johann Friedrich Blumenbach, "Einige naturhistorische Bemerkungen bey Gelegenheit einer Schweizerreise," *Magazin für das Neueste aus der Physik und Naturgeschichte* 4, no. 3 (1787): 1–12.

55. Böker, "Blumenbach's Collection of Human Skulls," 81.

56. J. F. Blumenbach to P. Camper, September 9, 1784, in Dougherty, *The Correspondence of Johann Friedrich Blumenbach*, 2:189–92.

57. Christian Friedrich Michaelis, "Briefe von D. Michaelis aus Neuyork," *D. August Gottlieb Richter's Chirurgische Bibliothek* 6, no. 1 (1782): 119–20; Christian Friedrich Michaelis, *Über die Regeneration der Nerven: Ein Brief an Herrn Peter Camper* (Cassel, 1785); Nicolas Philbert Adelon, "Chrétien Frédéric Michaelis," in *Dictionaire des sciences médicales* (Paris, 1812–1822), 271–72; Whitfield J. Bell, "A Box of Old Bones: A Note on the Identification of the Mastodon, 1766–1806," *Proceedings of the American Philosophical Society* 93, no. 2 (1949): 169–77.

58. Franz Dumont, ed., *Samuel Thomas Soemmerring Briefwechsel 1784-1792*, Part 1, 38n142; Therese Heyne to S. T. Sömmerring, June 18, 1784, in Franz Dumont, *Samuel Thomas Soemmerring Briefwechsel 1761/65-1784* (Gustav Fischer: Stuttgart, 1996), 525–26.

59. Böker, "Blumenbach's Collection of Human Skulls," 81–82n7; Sömmerring, *Über die körperliche Verschiedenheit* (1784), 18; William Snow Miller, "Abraham Chovet: An Early Teacher of Anatomy in Philadelphia," *Anatomical Record* 5, no. 4 (1911): 147–72.

60. Blumenbach, "Rezension: Sömmerring (1784)," 108–11.

61. J. F. Blumenbach to P. Camper, August 25, 1779, January 24, 1781, May 22, 1781, in Frank W. P. Dougherty, *The Correspondence of Johann Friedrich Blumenbach*, vol. 1: *1773-1782*, ed. Norbert Klatt (Klatt Verlag: Göttingen, 2006), 175–77, 220–21, 244–46; J. F. Blumenbach to S. T. Sömmerring, January 26, 1781, March 24, 1781, May 5, 1781, September 20, 1782, November 30, 1782, in Dougherty, *The Correspondence of Johann Friedrich Blumenbach*, 1:222, 226–29, 240–42, 348–52; J. F. Blumenbach to P. Camper, March 10, 1783, December 3, 1783, September 9, 1784, in Dougherty, *The Correspondence of Johann Friedrich Blumenbach*, 2:22–24, 102–6, 189–92; P. Camper to J. F. Blumenbach, August 8, 1784, September 24, 1784, in Dougherty, *The Correspondence of Johann Friedrich Blumenbach*, 2:182–86, 192–96; J. F. Blumenbach to

S. T. Sömmerring, May 14, 1784, in Dougherty, *The Correspondence of Johann Friedrich Blumenbach*, 2:150–51; S. T. Sömmerring to J. F. Blumenbach, after May 14, 1784, n.d. August 1784, in Dougherty, *The Correspondence of Johann Friedrich Blumenbach*, 2:151–52, 182.

62. S. T. Sömmerring to P. Camper, May 24, 1784, in Dumont, *Samuel Thomas Soemmerring Briefwechsel 1761/65–Oktober 1784*, 479.

63. R. H. Popkin, *Isaac La Peyrère (1596-1676): His Life, Work, and Influence* (Brill, 1987); David N. Livingstone, *Adam's Ancestors: Race, Religion, and the Politics of Human Origins* (Baltimore, MD: Johns Hopkins University Press, 2008), 26–51; T. D. Keel, "Religion, Polygenism, and the Early Science of Human Origins," *History of the Human Sciences* 26, no. 2 (2013): 3–32.

64. Keel, "Religion, Polygenism, and the Early Science of Human Origins," 7.

65. Lamsveld, *Privilegien, willekeuren en ordonnantien*; S. Ude-Koeller, W. Knauer, and C. Viebahn, "Anatomical Practice at Göttingen University Since the Age of Enlightenment and the Fate of Victims from Wolfenbüttel Prison Under Nazi rule," *Annals of Anatomy-Anatomischer Anzeiger* 194, no. 3 (2012): 304–13; Enke, "'Leichen für die Anatomie," 241–56; E. T. Hurren, *Dissecting the Criminal Corpse: Staging Postexecution Punishment in Early Modern England* (Springer Nature, 2016).

66. Suman Seth, "Materialism, Slavery, and the History of Jamaica," *Isis* 105, no. 4 (2014): 764–72; Suman Seth, *Difference and Disease: Medicine, Race, and the Eighteenth-Century British Empire* (Cambridge: Cambridge University Press, 2018); Suman Seth, "Performing Polygenism: Science, Religion, and Race in the Enlightenment," in *Critical Approaches to Science and Religion*, ed. M. Perez Sheldon, A. Ragab, and T. Keel (New York: Columbia University Press, 2023), 229–54.

67. Christoph Meiners, *Grundriß der Geschichte der Menschheit* (Lemgo, 1785), 43–44; Michael C. Carhart, "Polynesia and Polygenism: The Scientific Use of Travel Literature in the Early 19th Century," *History of the Human Sciences* 22, no. 2 (2009): 58–86; Britta Rupp-Eisenreich, "Christoph Meiners' 'New Science' (1747–1810)," in *The Invention of Race: Scientific and Popular Representations*, ed. N. Bancel, T. David, and D. Thomas (London: Routledge, 2014), 68–83; John S. Michael, "The Race Supremacist Anthropology of Christoph Meiners, Its Origins and Reception," unpublished manuscript, 2021.

68. Seth, "Materialism, Slavery, and the History of Jamaica," 771.

69. Olaudah Equiano, *The Interesting Narrative of the Life of Olaudah Equiano, or Gustavus Vassa, the African* (1789), 31.

70. Nell Irvin Painter, *The History of White People* (New York: Norton, 2010), 59–71.

71. Seymour Drescher, "The Ending of the Slave Trade and the Evolution of European Scientific Racism," *Social Science History* 14, no. 3 (1990): 415–50; Keel, "Religion, Polygenism, and the Early Science of Human Origins," 3–32.

72. P. Camper to S. T. Sömmerring, June 12, 1784, in Dumont, *Samuel Thomas Soemmerring Briefwechsel 1761/65–Oktober 1784*, 512; see Nutz, *Varietäten des Menschengeschlechts*, 126–41.

73. Blumenbach, "Rezension: Sömmerring (1784)," 110–11; Sömmerring, *Über die körperliche Verschiedenheit* (1785), 79, section 73.

74. Sömmerring, *Über die körperliche Verschiedenheit* (1785), xix–xx.

75. Johann Friedrich Blumenbach, "Rezension: Sömmerring (1785)," *Göttingische Anzeigen von gelehrten Sachen* 31 (1786): 302–3.

76. Dougherty, "Johann Friedrich Blumenbach und Samuel Thomas Soemmerring," 36–37; Oehler-Klein, " 'Der "Mohr" auf der niedrigeren Staffel am Throne der Menschheit?,' " 150n74; Oehler-Klein and Sömmerring, *Anthropologie*, 126–42.

77. Dougherty, "Johann Friedrich Blumenbach und Samuel Thomas Soemmerring," 47–48; Oehler-Klein, " 'Der "Mohr" auf der niedrigeren Staffel am Throne der Menschheit?,' " 119–66.

78. Goethe Universität, Frankfurt am Main, Universitätsbibliothek, Sonstige Nachlässe und Autographen, Soe2, Samuel Thomas Sömmerring, *Über die körperliche Verschiedenheit des Negers vom Europäer*, 1785/1813.

79. Böker, "Blumenbach's Collection of Human Skulls," 81–82.

80. Johann Friedrich Blumenbach, "Blumenbach an 'die erste Decade seiner Sammlung von Schädeln,' " *Göttingische Anzeigen von gelehrten Sachen* 3 (1790): 25–29; Böker, "Blumenbach's Collection of Human Skulls," 81–82.

81. Blumenbach, "Blumenbach an 'die erste Decade seiner Sammlung von Schädeln,' " 26–29.

82. J. E. Doornik, *Wijsgeerig-natuurkundig Onderzoek aangaande den oorspronglijken Mensch en de oorspronglijke Stammen van deszelfs Geslacht* (Amsterdam, 1808); S. T. Sömmerring, *Göttingischen gelehrten Anzeigen* 2, no. 147 (1812): 1457; Oehler-Klein and Sömmerring, *Anthropologie*, 47–50n110.

83. Nigel Aston, *Christianity and Revolutionary Europe, 1750–1830* (Cambridge: Cambridge University Press, 2002).

84. Drescher, "The Ending of the Slave Trade," 415–50; J. M. Charles, "The Slave Revolt That Changed the World and the Conspiracy Against It: The Haitian Revolution and the Birth of Scientific Racism," *Journal of Black Studies* 51, no. 4 (2020), 275–94.

85. White, *An Account of the Regular Gradation in Man*, cxxxix–clxvi.

86. Richard Saumarez, *A New System of Physiology* (London, 1798), 1:159–61.

87. Georges Cuvier, *Extrait d'observations faite sur le cadavre d'une femme connue à Paris et à Londres sous le nom de Vénus Hottentotte* (Paris, 1817), 271.

88. J. J. Virey, *Recherches médico- philosophiques sur la nature et les facultés de l'homme* (Paris, 1817), 12; J. J. Virey, *Histoire naturelle de genre humain* (Paris, 1824), 2:39; J. H. Guenebault, *Natural History of the Negro Race* (Charleston, 1837); Roger B. Taney, John H. van Evrie, and Samuel Cartwright, *Opinion of Chief Justice Taney, with An Introduction by Dr. J. H. Van Evrie, also an Appendix Containing an Essay on the Natural History of the Prognathous Race of Mankind by Dr. Samuel Cartwright* (New York, 1860).

89. Gunter Mann, "Franz Joseph Gall und Samuel Thomas Soemmering: Kranioskopie und Gehirnforschung der Goethezeit," in *Gehirn-Nerven-Seele: Anatomie und Physiologie im Umfeld Soemmerings*, ed. G. Mann and F. Dumont (Stuttgart: Gustav Fischer Verlag, 1988), 149–89; Michael Hagner, "The Soul and the Brain Between Anatomy and *Naturphilosophie* in the Early Nineteenth Century," *Medical History* 36, no. 1 (1992): 1–33.

90. Brit Rusert, *Fugitive Science: Empiricism and Freedom in Early African American Culture* (New York: New York University Press, 2017); James Poskett, *Materials of the Mind: Phrenology, Race, and the Global History of Science, 1815–1920* (Chicago: University of Chicago Press, 2019); Rebecca Walker, "Facing Race: Popular Science and

Black Intellectual Thought in Antebellum America," *Early American Studies: An Interdisciplinary Journal* 19, no. 3 (2021): 601–40.

91. William Hamilton, in Alexander Monro, *The Anatomy of the Brain: With Some Observations of Its Functions* (Edinburgh, 1831), 1–8; Friedrich Tiedemann, "On the Brain of the Negro, Compared with That of the European and the Orang-outang," *Philosophical Transactions of the Royal Society of London* 126 (1836): 497–527; Friedrich Tiedemann, *Das Hirn des Negers mit dem des Europäers und Orang-Outangs verglichen* (Heidelberg: K. Winter Verlag, 1837).

92. Fabian, *The Skull Collectors*, 41–43.

93. Mitchell, "The Fault in His Seeds."

94. J. C. Nott and G. R. Gliddon, *Types of Mankind* (Philadelphia: Lippincott, Grambo & Company, 1854); Frederick Douglass, *The Claims of the Negro, Ethnologically Considered: An Address Before the Literary Societies of Western Reserve College, at Commencement, July 12, 1854* (Rochester: Lee, Mann & Company, Daily American Office, 1854); Alexander Hamilton Stephens, "Alexander H. Stephens, in Public and Private: with Letters and Speeches, Before, During, and Since the War by Henry Cleveland," 717–729, March 21, 1861, courtesy of the Indiana State Library, https://iowaculture.gov/history/education/educator-resources/primary-source-sets/civil-war/cornerstone-speech-alexander; see also Christopher Willoughby, *Masters of Health: Racial Science and Slavery in US Medical Schools* (Chapel Hill: University of North Carolina Press, 2022).

95. Thomas Henry Huxley, *Evidence as to Man's Place in Nature* (London, 1863), 93–95.

96. A. S. Winston, "Why Mainstream Research will Not End Scientific Racism in Psychology," *Theory & Psychology* 30, no. 3 (2020): 425–30; Quinn Slobodian, "The Unequal Mind: How Charles Murray and Neoliberal Think Tanks Revived IQ," *Capitalism: A Journal of History and Economics* 4, no. 1 (2023): 73–108.

97. R. Wagner, *Samuel Thomas von Sömmerring's Leben und Verkehr mit seinen Zeitgenossen: Briefe berühmter Zeitgenossen an Sömmering* (Voss, 1844), 89–90; Thomas Brehm, Theo Horst-Werner Korf, Udo Benzenhöfer, Christof Schomerus, and Helmut Wicht, "Notes on the History of the Dr. Senckenbergische Anatomie in Frankfurt/Main. Part I. Development of Student Numbers, Body Procurement, and Gross Anatomy Courses from 1914 to 2013," *Annals of Anatomy-Anatomischer Anzeiger* 201 (2015): 103–5.

98. Sömmerring to Johann Friedrich Merck, May 8, 1784, Camper to Sömmerring, June 12, 1784, and Johann Heinrich Merck to Sömmerring, August 13, 1784, in Franz Dumont, *Samuel Thomas Soemmerring Briefwechsel 1761/65–Oktober 1784* (Gustav Fischer: Stuttgart, 1996), 453–57, 511–17, 561–63; Sömmerring to Petrus Camper, April 1785, in Franz Dumont, *Samuel Thomas Soemmerring Briefwechsel Teil 1 1784–Dezember 1786* (Gustav Fischer: Stuttgart, 1996), 169–71; Johann Friedrich Blumenbach, *Decas collectionis suae craniorum diversarum gentium illustrata* (Göttingen, 1790), 21.

TWO

Unbecoming Subjects

Psychiatry, Race, and Disordering the Human

ERIC REINHART

I AM RESEARCH material. I woke up yesterday with that in my head: *Pamela, you • are • re- • search • ma- • ter- • i- • al*—she says these words with a metronomic spacing between each syllable, emphasizing the deliberateness and authority of the formulation, as if it were a commandment.

Just like that. It finally dawned on me all of a sudden; it be like this clarity came down upon me from out my sleep and told me what I'm supposed to be in the world. I've always wanted to be useful for something—my mother raised us that way—and research might be the only thing I'm good for at this point. Fifty years of these medications, crack, lithium until it was killing me, all the antipsychosis pills I can't even pronounce. I haven't even always known what it is they're telling me to swallow every day. Most of the time, since they can't fix nothing real in my life, all they're trying to do is just knock me out, make me disappear, put me to sleep and not even be able to wake up or get up off the floor.

But I've figured out how to manage over these years: what to really take and what's actually worse for me than the crack. So, after all my shit—years of killing myself with drugs and life on the street and medications—maybe I could be good for research. . . . Not too many people like me still alive, you know, so I was just thinking maybe they could study me for all the damage from my life, right? Make some use of me? Like when doctors research the heart and what happens to it long-term, making notes on the scars and trau-matics and how long it keeps pumping before it gives up? 'Cause it could help somebody else if research can learn something from me, my body, but especially my mind—because I'm telling you, Eric, I have been falling apart. I almost can't believe it, what's happening to me. I . . . I . . . I am falling apart! I can't speak right. The words don't come. They just don't come, can't think right . . . oh, but I'm sorry: Did I wake you? I been wanting to call you all night to ask you about this, but my mother taught us you don't call no one after 10, just have to wait 'til morning. So I been waiting, and maybe it's too early to be getting you outta bed. I'm sorry . . . but do you know of any studies at the university I could be a part of? Is there some place people can use me to make some knowledge?

*　　*　　*

You are a Negro. There is only one thing *you have got to know:* you are a *Negro* in this world. I was a *Negro* when the King riots happened, and I'm still a *Negro* now while they're tearing up the 'hood after another *Negro* been murdered by the police.

These words come out of Pamela within a rhythm that feels as if it preceded her voice—a historical rhythm in which the Negro serves as anchor for each measure, the downbeat or gravitational center around which the words circuit in a weighted force field of violence.

I been looking at pictures of him, George Floyd. What they did to him . . . we all seen it before, but it's sick. It's just sickness everywhere: the police, shootings, pushers on my corner, politicians. . . . And they give me the antipsychosis? It almost funny, isn't it? But it's getting real bad here. Already ain't got no place to shop in a food desert—you know what it's like in the 'hood. All we got is the Arab store on the corner and the Dollar Store. And now those be all smashed up too. No place left to get something to eat. The truth is that most the white folks and the city, even with a Black mayor of Chicago who talk a big game but don't do nothing, they'll starve you before they let you be anything but a Negro. All kids coming up now gotta know, and you *need to know it: you are a Negro.* It's that simple. From my mama's time to my time to your time, it don't change. A Negro is a Negro, and that makes the world we know—this *sick* world—keep slipping into tomorrows.

*　　*　　*

Pamela and I have been regular writing companions and have spoken with each other almost every day for over seven years. In her writing and our conversations, she often invokes the ways in which power and history have placed her in the world and arranged her possibilities as an object of scientific knowledge and racial control.

She grew up and has spent most of her life in racialized Chicago neighborhoods that for a century have been made into a "laboratory" for manufacturing academic knowledge. The multiple waves of the Chicago School of sociology and related psychiatric and medical researchers have flooded through these neighborhoods time and time again. For Pamela, the terms of the ghetto, deviance and delinquency, cultures of poverty, race relations, Black families as tangles of pathology, welfare mothers, the underclass and inner city, drugs and policing, and urban violence—that is, the research-making languages of social science—have suffused the spaces of her life from her childhood in the 1950s to the present. Pamela necessarily works within these overdeterminations of her experiences and ways of being in the world. In the process, she simultaneously accepts and refuses the uses of her body and psyche to serve the knowledge-desires of others.

When she writes—and Pamela is always writing, if not on paper then always in her mind's eye—she does so both within and beyond the restrictive sanctioning of her self-articulations through the position of the raced, sexed subject of oppression and deviance. To make sense of herself, for herself and for others, Pamela has been compelled speak in relation to predetermined positions. She has been identified and identifies herself as a Negro and crazy—a Black woman productive of medical and psychiatric knowledge, first as a young girl and then adult overwritten by sexual imaginaries, pathologies, and desires forcibly projected upon and into her from without.

When sexually violated as a child and then controlled through street drugs and shame while subjected to years of traumatic violence via the manipulations of an older relative engaged in sex and drug economies, the response of her mother when the painful reality could no longer be fully ignored—though it remained unspoken—was to take her to a psychiatrist and ask for a medication. The doctor obliged: lithium and a diagnosis of bipolar disorder. As Pamela narrates it, she loved her mother, and I know she loved me. We had a real special bond—she was everything to me—before he did everything he could to break it, to break her trust in me, like stealing her money and hiding it in my drawer so she would find it there. I think she just didn't know what else to do, struggling by herself to feed all us kids and not knowing what to do with what was happening to me. It was either the psychiatrist or the police. I guess the psychiatrist seemed like a little better option for Black folks back then.

This is how Pamela, as a teenager, first became mentally ill.

Through psychiatric diagnoses and treatment, Pamela could be reabsorbed into the family and community. Violence and violation were transformed into disease, and disease offered a qualified form of acceptance and support. A continued silence with respect to incest and trauma was enabled by psychiatric interdiction: a seizure of an unspeakable reality and its transformation into the abstract terms of mental disorder.

But this didn't stop the violence. It now acquired multiplying forms that even more deeply shaped Pamela's circumstances and perception of herself: I was just a crazy Black girl. To the doctors and really to everyone then—even to myself—I was just a loose woman without morals. Out of control. Another crazy Negro. They couldn't see—they didn't want to see—what was going on.

Pamela soon left home and tried to make a life for herself on her own terms. But the terms have never been her own. One of her first acts of independence at age seventeen was to procure her own supply of drugs, to which she had become addicted by thirteen via the coercions of her abuser. Over the half-century since, by consenting to inhabit the positions assigned to her—mentally ill, disabled, battered woman, sex worker, criminal, addict, research material—and identifying herself through them, she has obtained essential resources like housing, health care, food, employment, and substances with which to endure the persistent wounds of trauma and loss. In the process, Pamela's life—literally staying alive—has both been made possible and, by the same gesture, foreclosed.

The foreclosure, however, isn't total. As Pamela said to me, in an effort to encourage and care for me, during one of our writing sessions after she had read passages from the book she is working on, *Prose from a Ho*: you can't escape the past, Eric. You not gonna get free from the way it make you feel and the way it make you think, but you can shake loose of yo'self sometimes. That's what you gotta do: you *have got* to shake loose of yo'self. That's what this writing discipline is all about, isn't it? Put things down and then, someday, something different comes out, something you don't even know about. That's what we're doing, and damn it, Eric . . . she laughs . . . we are going to keep doing it, so don't you tell me you haven't been writing.

*   *   *

Pamela's entrapment within intersecting racial and psychiatric discourses evokes the problematic at the center of Gayatri Spivak's critique in "Can the Subaltern Speak?"[1] There, Spivak takes aim at the easy abstractions and failure to confront Eurocentric assumptions in poststructuralist political thought. In particular, she criticizes Michel Foucault and Gilles Deleuze for their eagerness to decenter the Subject and announce its death without ever taking into account its differential enduring force in relation to colonization, racialization, sexuation, and the global division of labor.

As Spivak notes, not all are positioned in the same way in relation to the demand to be a Subject in the mold of Immanuel Kant's transcendental philosophy. As she argues in A Critique of Postcolonial Reason, the Enlightenment paradigm within which we remain submersed advances a particular European model of being that declares itself universal while simultaneously defining itself against the non-European "raw man" and via exclusion of "the primitive" from the realm of reason.[2]

As a consequence, not all share the possibility of simply declaring the Subject dead and embracing in its place a life of multiplicity, contradiction, and disregard for historical-material constraints. The subaltern, or what Spivak refers to as the third-world woman, is compelled to speak through the hegemonic Western frames of the Subject, its humanism, and corresponding concepts of human rights in order to be heard or known, including by herself. The subaltern, as such, thus cannot speak and can only ever remain a figure—that is, an abstraction rather than an anthropological reality. This figure is always beyond the reach of our necessarily ideologically overdetermined ears and eyes, which have been shaped by Enlightenment sciences of Man, the human, and their dissemination via colonialism and racial capitalism.[3]

To register in the world and to resist the power to which they are subjected, the subordinated are forced to adopt and work on the very terms of their own subordination within self-reproducing structures of historical-epistemological violence. Resistance, then, becomes a means of violation's ever-tightening hold over the violated. Crudely glossed, there is no outside of subjectifying power through which one comes to know oneself and others. There is no possibility of return to an imagined state of precolonial authenticity. And there is no horizon of freedom that the pursuit of which is not at the same time a reinscription of the impossible, enslaving command to be a self-determining, transparent Subject.

If one accepts these illiberal, pessimist arguments that are incompatible with ideologies of history as progress and that can at first appear disabling, then of what might struggle against one's own domination consist? What might it mean to get free? What is it from which one must get free, and might it include the very notion of freedom itself and of the human to which it is tethered? If so, what are the tools through which the concepts of freedom and the human have been built, which, by extension, also support the categories of the subhuman? Can we identify and break these apart so as to support other possibilities for knowing, relating, and being—or, as Pamela puts it, shaking loose of one self?

My hope in pursuing such questions through a mash-up of theoretical, historical, and ethnographic modes is that this effort might operate as an adjunctive alongside the lived practices of invention that orient my relationship with Pamela and with our broader community of writing partners on Chicago's South Side. Motivated by these relationships, this chapter is an expression of abolitional ambition that by rejecting and dismantling the institutions and concepts that have engendered our contemporary world, we might collectively invent new ways of being together.

* * *

Parallel ambitions animate strands of postcolonial and critical race theory,[4] Afropessimism,[5] Black feminist and posthuman studies,[6] and queer and trans theory[7] that seek to identify and deconstruct the epistemological and phenomenological frames of modern (i.e., racializing-rationalizing-colonizing) subjectivity and their supporting sciences. They also resonate with the ethical-political promise of psychoanalysis—that misfit science that has been rejected by the proper sciences because unfaithful to the demands of empiricism, experimental method and generalizability, and conventional materiality.

In particular, the return to Freud initiated by Jacques Lacan that, in turn, influenced what Azeen Khan identifies as the subaltern clinic of Frantz Fanon pushes us to consider that subjective alienation and the double bind of language (i.e., that language is both bondage and condition of being) are core to the particular subjection of the colonized and also universal.[8] As Fanon writes—in an echo of Lacan's early work, with which Fanon was familiar[9]—on the first page of Black Skin, White Masks: "To speak is to exist absolutely for the other . . . To speak means being able to use a certain syntax and possessing the morphology of such and such a language, but it means above all assuming a culture and bearing the weight of a civilization."[10] And as he later writes in *The Wretched of the Earth*: "It is the settler who has brought the native into existence and who perpetuates his existence. The settler owes the fact of his very existence, that is to say, his property, to the colonial system."[11]

Because the existence of both the colonized and the colonizer—the construction of both Black and white—is predicated on a structure of constitutive exclusion, decolonization cannot simply be a project of progress or development toward the model of subjective freedom inherited from this colonial order. It must instead, Fanon writes, "set out to change the order of the world" through a "program of complete disorder" by which is introduced a new "rhythm into existence, introduced by new men, and with it a new language and a new humanity."[12]

What psychoanalytic thought and practice, at its most subversive, offers to thinking and being is in line with Fanon's call for a programmatic *disordering of the human*—that is, of the regime of the human that is enforced by colonial systems of knowledge. This cannot be a conservative work of therapeutic repair and of shoring up of the Subject and his ego, as in the goals of psychiatry, psychology, and much of what has passed for psychoanalysis in the American traditions aligned with relational psychoanalysis and ego psychology. It is instead a labor of *subjective destitution*—a mortification of the colonial-Enlightenment self, its supporting discourses, and its relation to the colonial Other who governs the symbolic—so as to open space for inventing new modes of being, making, and of being with others and otherness.

For psychoanalysis, this task is not exclusive to the colonized or racialized; it is universal. Alienation in language and the failure to be a transparent, unified Subject is constitutive of all speaking beings—that is, beings who ask after the meaning of their being and necessarily do so through inherited languages that are never fully one's own and never adequate to one's sense of life, relations, and history. But, in line with Spivak's critique of Foucault and Deleuze, this universal fact of alienation always manifests within and is shaped by particular, historically determined positions of power and possibility. Demands to identify oneself and to satisfactorily appear as Subject (i.e., law-abiding, rational, transparent to knowledge and meaning, undivided by the unconscious, etc.), and the punishment for failing to present oneself as such, are not evenly distributed.[13]

The racialized other is subjected to this demand in particularly insistent and violent forms. Consider, for example, the racially variable interactions with police in the United States, where Black Americans are far more likely to be stopped, harassed, arrested, and killed across all class categories.[14] And for every such surface manifestation of power's racialized demand to identify oneself and to perform the Subject, countless more pernicious instantiations of racializing subjectification permeate everyday life, often requiring no external audience and no policeman. We are all interpellated by authority, as Louis Althusser argued in his example of being hailed by the police and instantly made to feel as if a criminal who must give an account of oneself, but few are subjected to this more frequently and deeply than what Pamela referred to as "the Negro" in America.[15] Over time, this external authority (e.g., the colonizer, policeman, judge, psychiatrist, priest, etc.) that demands one give an account of oneself is internalized both at individual and collective levels, passing from generation to generation. This oppressive Other becomes interwoven into the fabric of perception, language, and identity.

While Fanon acknowledges the role of language in racism and colonial domination, he is most noted for his emphasis in *Black Skin, White Masks* on the gaze and the operation of racism—the internalization of the white Other—at the level of the image. Fanon's portrait of the colonized is of one who is reduced by racialization to imaginary identifications imposed by the colonizer. Captured by the imaginary, Fanon's colonized figure suffers from a dearth of access to the symbolic—that is, to language and the possibility of signifying one's world. The symbolic, then, might appear to offer the possibility of freedom, except that, as Fanon noted, it too—like the imaginary—has been installed by the colonizer. To put it to use is to submit oneself to colonizing authority and its ways of knowing. To give an account of oneself is always to give an account of oneself on the terms of the Other and for the Other; it is to subordinate oneself once again.

It is from these cornered positions of Fanon's colonized Algerian comrades or of those in positions like Pamela's in the United States today that the urgency of disordering inherited structures of identification and signification is especially real. It is toward this project that this chapter and the writing that Pamela and I do together are oriented.

Central to the political-ethical interjection of psychoanalysis and why I find it especially useful for ethnographically motivated work is that, alongside the *universal* nature of alienation as beings caught by language and the historically *particular* forms of its instantiations, there is another scale upon which psychoanalytic theory insists: the irreducible *singularity* for each one of what it is to be—which is always to fail to be—the Subject that modern power demands. It is here that Lacan points to the third register of psychic life alongside the symbolic and the imaginary—the real: that which is beyond symbolization and in which each one's absolute difference, unique mode of enjoying, and fullest life possibility inheres. It is this refusal to reduce an interlocutor (i.e., one who i̶s̶ between speech) to the group or to a system of knowledge that makes psychoanalytic thought so subversive to the scientific enterprise and its totalizing, nomothetic demands. It is also what makes it especially useful for critiques of racist epistemologies that seek to avoid reinscribing racial essentialisms in inverted forms.

By drawing on these intersecting scales of the universal, particular, and singular that correspond to my uses of theory, history, and ethnography, I attempt to bring each to bear on my friendship with Pamela in pursuit of a shared effort to shake loose—each in our own way—of ourselves together. What follows here is thus a brief history of the present, motivated by friendship, that might cut against inherited ways of knowing and representing the human so as to open new ways of being alongside one another. To this end, the rest of this essay pulls on the threads of two intertwined scientific discourses of the self through which Pamela frequently finds voice and that have shaped her life with promises of both emancipation and domination: psychiatric science and racial reason.

*   *   *

In eighteenth-century Europe, with the ascent of the twinned ages of Enlightenment and colonialism, there arose a new scientific aspiration: the development of a "science of human nature" based on empirical observation.[16] Prominent figures such as David Hume, Charles de Montesquieu, and Immanuel Kant pursued, with varying emphases, this task of delineating types of human beings and their behaviors, perceptions, cognitions, sentiments, and capacities for judgment and morality. While the natural sciences concerned the external world of material bodies, forces, and laws, the emergent domain of human science took the primarily internal world of "Man" as its object of investigation.

The science of the mind was the cornerstone of this endeavor to systematically define the human, and the empirical study of the mind and its variations depended from its outset—in the cases of each Hume, Montesquieu, and Kant, for example—on references to racial difference, often invoked via comparative "national character" elaborated by quickly developing fields of geography and anthropology.[17] If the biomedical sciences "drew their terms from contemporary discourses about the human condition fed by Europe's encounter with the non-European world," then this dynamic was all the more pronounced and explicit in the domain of the human sciences, which both shaped and reflected the epistemological desires of contemporary audiences.[18] In this regard, it is worth nothing that Kant's most popular lectures were not on the metaphysics for which he is now most regarded but rather those he delivered on anthropology, in which he elaborated human types alongside a theory of mental disorder.[19]

Up to this point, race science from the seventeenth century forward along the lines of François Bernier, Carl Linnaeus, Comte de Buffon, and Johan Blumenbach, among others, had primarily relied upon biological classifications and physiognomy, debating matters such as monogenism versus polygenism (i.e., whether all races share a single origin), degeneration theory, and the effects of climate on physical development.[20] One could say that this science was organized around *biocentric* principles of racial differentiation. But with the advent of the human sciences, concepts of racial difference and the justifications for racial domination acquired a second, parallel paradigm: *psychocentric* definitions of the human. Alongside contemporary formulations of the Subject and subjectivity organized around a new concept of reason, the sciences of the mind offered new terms for demarcating the borders of the properly human, carrying extensive implications for the life of modern science, racial ideas and governance, and subject formation.

These two lines of racial science, the biocentric and psychocentric, are not mutually exclusive and have often overlapped in ongoing debates concerning, for example, whether psychic pathology is always attributable to a localizable physical lesion. But while biological race science soon began to decline—albeit not without resistance and persistent vestiges—and has become progressively discredited over the nineteenth and twentieth centuries, psychocentric racism manifested in more subtle forms that have enabled its intensification and dissemination through dissimulation.[21]

Biological race science was a field of empirical and theoretical investigation. Although tied to political discourses that provided moral justifications for colonial violence and extraction, it was not an applied science soldered to missions to treat, convert, or improve. Psychocentric conceptions of the human, on the other hand, took shape contemporaneously with the first systematic European medical study of mental diseases and associated treatments pursued on the principles of Enlightenment science and its humanistic imperatives. This coincidence was mutually constitutive and supplied the most important mechanism by which psychocentric racial subjection has acquired its enduring power: attachment to and diffusion through applied practices of humane care for the other of reason—a civilizing mission not primarily targeting culture or religion but rather the infinite recesses of the mind.

Indeed, the late-eighteenth-century protopsychiatric movement for the "moral treatment" of those suffering from mental disorders—care based on principles of unshackled freedom of movement, salubrious environments, and humanizing treatment rather than simply custodial confinement—led by Philippe Pinel in France and William Tuke in England indirectly laid the groundwork for various moral and moralizing applications of psychocentric definitions of the human, racial difference, and the civilizing mission of our still-colonial world.[22] Over the last two centuries, the humanistic pursuit of care practiced in service of supposedly universal principles of reason, science, and human dignity—that is, care for and of the human that ostensibly aims to reclaim for humanity its pathological deviants—has provided one of most effective tools by which racial ideologies have come to permeate life across the globe.[23] This therapeutic imperative has provided a vector for restrictive notions of the human and attendant politics of racial order and control, facilitating their far-reaching operation under other names.

*   *   *

To think the fundamental interrelation between psychiatric and racial thought in the ethnographic present in which Pamela and I are writing in Chicago requires accounting for its distinctive genealogy in the United States against the backdrop of slavery and its afterlives.[24] The figure of "the Negro" has always been a problem for the American medical science of human reason and its perversions. From its inception, U.S. psychiatry has been entwined with race, emerging as a core epistemological framework through which racial difference and hierarchies have been constructed, justified, and enforced. But the place of the racialized other in this scheme was for a long time—and, in some ways, still lingers as—uncertain. The racial other was both foundational to early psychiatric definitions of rationality, insanity, and normality and also simultaneously external to their reach and operations: a constitutive outside.

American psychiatry's "founding father" and prominent advocate of the Enlightenment, Benjamin Rush, whose image appeared until 2016 on the seal of the American Psychiatric Association as a homage to the man credited with the first systematic study of mental disorders in the United States, argued that black skin was a form of leprosy, curable only by becoming white. Blackness itself was a pathology beyond any known cure. And in his *Medical Inquiries and Observations, Upon the Disease of the Mind* (1812), the first American textbook on the medical treatment of the mind, Rush noted that slaves generally failed to benefit from his treatments due to noncooperation of a will diseased by lack of exercise.[25]

At the same time that he reflected many of the dominant racist presuppositions of his era and carried them into his medical-scientific ideas, Rush was also a noted abolitionist and advocate for the rights and care of enslaved Africans. At the conjunction of his study of mental disorders and his abolitionist activities, he wrote a brief report in 1789 in which he described "a wonderful talent for arithmetical calculation in an African slave living in Virginia," seventy-year-old Thomas Fuller, that "merits a place in the records of the human mind." Rush presented an account of Fuller's exceptional calculating abilities before the Pennsylvania Society for the Abolition of Slavery:

On being asked, how many seconds a man has lived, who is seventy years, seventeen days and twelve hours old, he answered, in a minute and a half, 2,210,500,800. One of the gentlemen [examining Fuller], who employed himself with his pen in making these calculations, told him he was wrong, and that the sum was not so great as he had said—upon which the old man hastily replied, "top, massa, you forget de leap year." On adding the seconds of the leap years to the others, the amount of the whole in both their sums agreed exactly.[26]

One might expect that Rush presented such examples of Fuller's "extraordinary powers in arithmetic" to his abolitionist colleagues as evidence of

Black peoples' capacity for reason to oppose dominant dehumanizing portraits of Africans as mentally deficient. (Hume, for example, had famously
declared Africans to be mentally inferior to Europeans and specifically cited
the absence of genius among them as support for this claim.)[27] Instead, Rush
presented Fuller as an object of pathological curiosity, describing him as a
slave "of such limited intelligence who could comprehend scarcely anything
either theoretical or practical, more complex than counting."[28] Fuller has
since been catalogued among the "lightning calculators" that supplied the
anecdotal basis for savant syndrome—"a rare, but extraordinary, condition
in which persons with serious mental disabilities . . . have some 'island of
genius' which stands in marked, incongruous contrast to overall
handicap."[29]

Although Rush told his colleagues that Fuller's talent should be entered
into "the records of the human mind," nineteenth-century literature on
mental diseases suggests that neither Rush nor his scientific contemporaries
regarded the mind of the Negro as fully, or simply, human in the way they
considered that of the European. Fuller, and the enslaved people taken from
Africa in general, represented a simultaneous surplus and deficit for the
early scientists who sought to categorize the human mind and its
aberrations.

On the one hand, in early-nineteenth-century American medical science,
with its close ties to German, Scottish, English, and French medical thought
and to the European human sciences of the preceding half-century, the
racialized African served as the constituting figure of insanity: the index
for the opposite of the normal, rational, and self-authorizing Subject of reason. As Achille Mbembe and Spivak—among many others—have noted, by
studying "the savage," Europeans constructed a supposed mirror in which
it could find an inverted portrait of itself. The non-European provided the
"raw" material against which civilized Man could emerge.[30] This rational
Man relied for assurance of himself—indeed, constructed the very notion
of the reason and of the European—upon appeal to his inverse, of which there
was no more paradigmatic instantiation than the figure of the African.[31]

But, on the other hand, to be insane meant to have lost the capacity for
reason. And to lose reason, one must have had it. The animal without reason,
for example, could not be said to be insane, never having had a rational
capacity to have lost. To be declared insane, then, implies an affirmation of
one's humanity, if only through its withdrawal. Did the African or enslaved

Negro ever possess reason such that they could lose it? Was he immune to insanity or especially susceptible to it?

If the Black American did have reason and then lost it or suffered from its underdevelopment, this might implicate the conditions in which he lived—a world of systematic denigration and disadvantage in the "free" North or of routinized violation and the flagrant brutality of slavery in the South. The ideological stakes of these questions were thus considerable on several levels. To acknowledge the presence and then loss of reason in the Negro would present challenges for psychocentric racial distinctions that supported justifications for European colonialism and American slavery on the premise of reason's right over nature and history's progress toward the fuller realization of reason.

Against this conflicted backdrop, the question of whether the condition of insanity properly applied to the African or American Negro remained a subject of debate for the much of the first half of the nineteenth century. The Scottish physician Sir Andrew Halliday, an expert on mental disorders, voiced a commonly held view in an 1828 book: "We seldom meet with insanity among the savage tribes of men. Among the slaves in the West Indies it very rarely occurs." The reason for this, Halliday claimed, was that insanity was an affliction only of those whose "organs of the mind" were further developed and had undergone "better cultivation."[32]

Similar views were common among American physicians, who drew on their European counterparts' writings to exclude those of non-European origin from scientific discussions of insanity. As reported in the *American Journal of Insanity* in a brief entry on the "Exemption of the Cherokee Indians and Africans from Insanity" (1844), no case of "decided insanity" had ever been observed among the Cherokee in decades of observation. Among Africans too, the article went on to note, insanity was "very rare in their native country." The application of the medical sciences of mental disorders to non-Europeans, then, the report concludes, was in most instances inappropriate and likely to misattribute delirium caused by physiological conditions to mental diseases that rarely, if ever, affected the less developed races.[33]

Insanity was widely believed a disease of civilized races—that is, those that subscribed to European social organization and Enlightenment norms.[34] As a matter of definition, insanity consisted of a deterioration or loss of the higher faculties of the mind, such as intellectual capacity, moral judgment, and control over animal impulses—faculties that only a civilized people possessed. An intrinsic unfreedom of primitive peoples, whose nonautonomous nature, in which reason was regarded as inadequate to overcome animal instincts and to manage the demands of civilization, was implicated in this immunity to insanity.

In 1835, Amariah Brigham—an American physician who was one of the founding members of the Association of Medical Superintendents of American Institutions for the Insane, which later became the American Psychiatric Association—wrote of association between insanity and liberty. As Brigham explained, freedom and participation in a civil society appeared to be, if not an absolute precondition, at least a major determinant of susceptibility to insanity: "Insanity is a disease that always prevails most in countries where the people enjoy civil and religious freedom, and where all are induced, or at liberty to engage in the strife for wealth, and for the highest honors and distinctions of society. We need therefore to be exceedingly careful not to add other causes to those already existing, of this most deplorable disease."[35]

Brigham's concerns—primarily directed in this passage toward women but resonant with contemporary justifications of slavery—that too much freedom and indulgence in the passions, such as inflammatory religious feeling and sermonizing, and departure from the regular order of work would provoke insanity echo the earlier worries of Benjamin Rush after the Revolutionary War. Rush observed that some Americans had been unprepared for the resolution of the war and became affected by an excess of freedom:

> The excess of the passion for liberty inflamed by the successful issue of the war, produced, in many people, opinions and conduct which could not be removed by reason nor restrained by government. . . . The extensive influence which these opinions had upon the understandings, passions, and morals of many of the citizens of the United States, constituted a species of insanity, which I shall take the liberty of distinguishing by the name of *Anarchia*.[36]

The questions of race, freedom, social order/disorder, and madness were wrapped closely but confusedly together in the unfolding early meanings of insanity in the United States. It was in this context that the sixth census of the United States, conducted in 1840 and published in 1841, famously interjected itself into scientific and popular debates on the supposed nature of race and insanity, providing a mythical early impetus for a biopolitics of race and mental disorders that continues to exert significant influence over governance in the U.S. today.[37]

The sixth census recorded for the first time the number of "insane and idiots." The tabulated results were striking: the proportion of those classified as insane and idiots was approximately eleven times greater among the "free coloreds" in the North than among enslaved people in the South. The farther from slave territories, the greater the proportion of insane among Black Americans. Emerging statistical sciences merged with nascent nosology of mental diseases to settle the debate of racial susceptibility to insanity and to inform a preeminent political debate of the time: the legitimacy of slavery and future of racial administration.

Like most, Edward Jarvis from Massachusetts—a physician with expertise in mental afflictions and an interest in anthropology who was also one of the founders of the American Statistical Association—initially concluded from the census returns that slavery must have "a wonderful influence upon the development of moral faculties and the intellectual powers" of Negroes. Conversely, Jarvis noted, the Negro's "false position" in the North must have a deleterious effect on his character.[38] Jarvis was among many Northern physicians publicly opposed to slavery who found their stance shaken by the young sciences of statistics and mental disease as they converged on the issue of governance. For perhaps the first time in the Americas, in response to the 1840 census, there emerged a peculiar biopolitical system of value that interrelated and thereby reshaped the questions of racial difference, reason, the health of the population, and the function of government.

The reframing of slavery via the census transformed, albeit temporarily, the most pertinent ideological question into one concerning the health of the statistical aggregate rather than one of individual rights, interpersonal violence, or economics. But it did so within a notable difference relative to Foucault's genealogy of biopolitics in nineteenth-century Europe.[39] What was reflected in the rhetoric of American politicians, physicians, and periodicals at the time was a concept of the population that was not simply indexed to birth rates or life expectancy consistent with a paradigm of bare life. Instead, alongside this familiar ideology, a psychocentric biopolitics took form in which the task of government and justification for state policy hinged on the cultivation of a rational population against, as contemporary Southern publications put it, the threat of "diseased imaginations."

This supplement to biopolitics made the framework of reason—or what Jarvis referred to as "moral faculties" and "intellectual powers" so well protected by slavery—into a new scientific alibi for a racial politics of being in which the physical body is deemed secondary to, because derivative of, the rational-moral integrity of the mind.[40]

With the conjunction of state statistics and the science of insanity, government acquired a core function: the maximization of the ratio of rational to irrational—or the prevention, measurement, and disciplining of madness. In the decades ahead and rippling up to the present, the modalities of racial subjectification and domination have increasingly come to operate not just upon the body but through constructions of psyche and associated concepts of immorality, irrationality, and criminality.[41] In turn, the dominant ideological framing of racial inequality has become less a matter of bodily freedom or economic distribution and instead more one of interminable, essentializing projects of psychical remediation: moral uplift, education, and overcoming the trauma of history.[42]

Unsurprisingly, periodicals and politicians in the slave states quickly seized on the census returns as scientific validation of slavery as vital for the health and safety of the American public. The *Southern Literary Messenger* printed the insanity tabulations along with a long editorial, including this description of the dystopian future that the end of slavery would bring: "Let us then suppose a half of a million of free Negroes suddenly turned loose in Virginia, whose propensity it is, constantly to grow more vicious in a state of freedom. . . . Where should we find Penitentiaries for the thousands of felons? Where, lunatic asylums for the tens of thousands of maniacs? Would it be possible to live in a country where maniacs and felons met the traveler at every cross-road?"[43]

Not only did the census provide scientific evidence for white supremacy and the benevolence of white domination over the hundreds of thousands of Africans that had been taken captive, sold, beaten, raped, and killed; it also informed scientific thought on the nature of insanity. It appeared partially to answer the question of whether insanity was the result of "physical" or "moral" causes, or some interaction of the two. Moral causes must be predominant, if freedom so afflicted the Negro, but the disproportionate incidence of insanity and idiocy among the Negro population in the North relative to their white European counterparts must also implicate the physical constitution of the Negro. The primacy of moral causes alongside a hereditary predisposition to mental debility among Africans and their descendants in America was thus confirmed, the editors of the *Southern Literary Messenger* concluded.[44]

The 1840 census was widely heralded, even by many in the North, as affirmation of the rightness of racial subordination, the fragility of the Negro mind, and a cautionary counterpoint to abolitionist fervor.[45] Historians have largely focused on the fact that it was later discovered that statistical errors accounted for the racial disparity in insanity and idiocy in the census, producing unsuccessful efforts—led by Edward Jarvis—to persuade the government to correct the results. In the process, historians have often discounted how the influence of the census on popular and professional perception of the relation between race, madness, and social order remained significant independent of its statistical merits.[46]

The cutting edge of science appeared to have confirmed the unfitness of the Negro for freedom, paving the way for a psychocentric American biopolitics that has become a core mechanism of ongoing racialization and racial domination in its multiplying forms. By the same gesture with which the Negro was made subject to insanity, he was both inscribed within the realm of the human and written at its margin in a liminal position always subject to question and the threat of expulsion.

The notion that descendants of Africans belonged to a psychically inferior race not capable of full human freedom commensurate with that of the imaginary European Subject overrepresented, in Sylvia Wynter's terms, as universal "Man," remained prominent within the American medical and social sciences in the decades following formal emancipation.[47] Well into the first half of the twentieth century, physicians in both the North and South—paralleling colonial psychiatrists in Africa—continued to invoke the thesis that exposure to modern civilization and the demands of freedom threatened the underdeveloped African psyche, putting darker-skinned people at high risk of insanity.[48]

This trope of the Negro as psychically unfit for freedom found frequent repetitions both in the scholarly writings of eminent psychiatrists into the 1910s and '20s and also, in more subtle and implicit forms, in the newly formed field of American sociology, with its pathologizing preoccupations with the ghetto, deviants and delinquents, race relations, and criminality.[49] In a circular process, racial difference was—and still is—repeatedly invoked to identify and explain mental disorders. These have, in turn, supplied a central paradigm for enforcing racialization, the exclusion of the figure of the Black from status as a Subject of reason in supposed contrast to the European, and rationalizations of benevolent white governance to counteract a Black tendency toward individual pathology and collective disorder.

The reach of psychocentric racism is deep and enduring in both the medical and social sciences, and its many legacies could fill many volumes. For ready examples of its material consequences in U.S. governance today, one need only look to the ongoing American system of racial control via intertwined systems of incarceration and psychopathologization to see signs of its ongoing hold over minds, bodies, popular discourse, and official policy. For example, U.S. jails—into which Pamela has been booked thirty-eight different times and each time subjected to a mental health screening interview in which she is made to repeat her various diagnoses and treatment histories—today represent many of the largest psychiatric facilities in the world, with Chicago's Cook County Jail chief among them.[50]

*　　*　　*

At stake in the eighteenth-century emergence of the human sciences and the subsequent nineteenth-century delineations of the objects of the medical science of the human mind is the modern meaning and boundary of the human itself—a category forged together with colonialism and transatlantic slavery.[51] This claim leans on Sylvia Wynter's essay "Unsettling the Coloniality of Being/Power/Truth/Freedom: Towards the Human, After Man, Its Overrepresentation—an Argument."

Wynter seeks to "unsettle the coloniality of power" by, beginning from fifteenth- and sixteenth-century Spanish and Portuguese colonialism, tracing the rise of the modern "order of race" and its "space of Otherness" via a particular "descriptive statement" of the human as "Man." This Man overrepresents itself as if universal and encompassing the limits of the human species. For Wynter, who foregrounds the continued operationalization of this order of racial difference in the American prison-industrial complex and imperial ideologies of development, "the struggle of our times, one that has hitherto had no name, is the struggle against this overrepresentation."[52]

I will not summarize the whole of Wynter's complex and important intervention, which is now well known for how it extends and critiques Foucault's genealogy of the invention of Man as modern juridical-political Subject by emphasizing its fundamental imbrication with coloniality and racial difference.[53] I will only draw out a few points at which my interjection here draws on Wynter's emphases and seek to supplement her argument and its implications.

Wynter describes a process by which a particular European ethnoclass, Man, was written into a position as the ostensibly universal Subject, producing the "first 'degodded' (if still hybridly religio-secular) 'descriptive statement' of the human in history, as the descriptive statement that would be foundational to modernity." In this process by which the West reinvented its "True Christian Self in the transumed terms of the Rational Self of Man," Wynter writes,

> it was to be the peoples of the militarily exploited New World Territories (i.e., Indians), as well as the enslaved peoples of Black Africa (i.e., Negroes), that were made to reoccupy the matrix slot of Otherness [previously occupied on theological grounds by heretics and Enemies-of-Christ infidels]—to be made into the physical reference of the idea of the irrational/subrational Human Other.[54]

In this new paradigm, which Wynter designates as "Man1" and that reigned from the Renaissance to the eighteenth century, "Race was therefore to be, in effect, the non-supernatural but no less extrahuman ground ... of the answer that the secularizing West would now give to the Heideggerian question as to the who and the what we are." Man1 was principally a "ratiocentric" descriptive statement of the human—that is, it was a formulation of the human/subhuman distinction that relied upon claims concerning a capacity for rationality and thus morality. Accordingly, the discourses of knowledge functioned, Wynter writes,

> to construct all the non-Europeans ... as the physical referent of ... its irrational or subrational Human. ... While the "Indians" were portrayed as the very acme of the savage, irrational Other, the "Negroes" were assimilated to the former's category, represented as its most extreme form and as the ostensible missing link between rational humans and irrational animals.[55]

In the space of the eighteenth century and contemporaneous with the rise of the human sciences, Wynter marks the advent of "the West's second wave of imperial expansion" and with it a "reinvention of Man now in purely biologized terms." Wynter characterizes this as a shift from Man1 to a more thoroughly scientific descriptive statement of the human, Man2, as "purely secular and biocentric."[56] Although Wynter's epistemes of Man1 and Man2 appear to overlap synergistically to some degree rather operating via simple successive replacement (much like Foucault's epistemes and shifting paradigms of sovereignty), Wynter emphasizes a progressive shift toward biology and away from the ratiocentric, opening the possibility for what she views as a more properly universal science.

In the process, Wynter endorses the promise of science to overcome the overrepresentation of Man and to create, in its place, a genuine universalism of the human and of humanity. Importantly, Wynter defines this through by bracketing any concrete particular, as she seeks to avoid simply replacing one faux-universal human with another. Still, in this recuperative gesture, as both Katherine McKittrick and Denise Ferreira da Silva note,[57] Wynter invests a surprising and cruel optimism in the capacity of science to overcome its constitutive particularisms.

My own narrative, by arguing that the science of man is inseparable from its racializing foundations, which persist into our present psychocentric regimes of governance, takes a more pessimistic stance on the status of science and the possibility—and desirability—of a new humanism or universalism. Science, tethered as it is to psychocentric conceptualizations of reason and subjectivity, is inseparable from the racializing regimes of the human and freedom that constitute the post-Enlightenment subjectivity in which we all remain caught. Scientific rationality depends upon the exclusion of its opposite, and it must constantly produce its outside in order to legitimate itself.

It is to the excluded outside, then, that we must look for possibility. Science cannot lead us out of itself. It cannot be reformed out of its constitutive exclusions of difference. Nor can it simply be overthrown via a mythical dissolution of the past and purified reformulation of the lexicon of knowledge through which we have come to think, speak, resist, and be.

What an ethical confrontation—one that welcomes and multiplies difference rather than instrumentalizing and constraining it—with the history of science compels, then, is a practice of subversion. It demands not a reform of the old but the abolitional invention of the new. To shake loose of ourselves and to realize Fanon's decolonizing vision of a project of complete disorder, we must together make a new language, sense of being, and mode of relation by which we might unbecome what we have long told ourselves we must be.

From within the death of subjectivity lies the promise of subversivity as a way of life.

Notes

This chapter is drawn from a forthcoming monograph, *Subversivity: Race, Psychiatry, and Aesthetic Anteriority*, that is centrally concerned with the interrelated overdetermination of language and subjectivity and what it means to write when the language through which we articulate both ourselves and others is never our own. In that text, I endeavor to refuse conventions of superficial attribution, such as the use of quotation marks to punctuate dialogue, which often function to silence ongoing problems of power, extraction, debt, ownership, responsibility, boundaries between self and other, and the origins and ends of language. This is intended as a formal means by which to discomfit the experience of reading and to leave permanently unresolved questions of authorship, the ethics of representation, and "proper"--meant in its etymological resonance with property and ownership--form. Consistent with that broader project, I omit quotation marks when invoking spoken conversation in this chapter. Of note, the uses of Pamela's voice have been developed in ongoing dialogue with Pamela, who has reviewed this chapter and consented to its publication.

1. Gayatri Chakravorty Spivak, "Can the Subaltern Speak?," in *Marxism and the Interpretation of Culture*, ed. Cary Nelson and Lawrence Grossberg (London: Macmillan, 1988).
2. Spivak, "Can the Subaltern Speak?"; Gayatri Chakravorty Spivak, "Chapter 1: Philosophy," in *A Critique of Postcolonial Reason: Toward a History of the Vanishing Present* (Cambridge, MA: Harvard University Press, 1999).
3. Spivak's arguments resonate through the work of Sylvia Wynter and Denise Ferreira da Silva, albeit with substantive differences and disagreements. The terms I have invoked here are thus not only drawn from Spivak but also anticipate the intersection of these three thinkers' contributions, to which I will come later in this chapter.
4. Denise Ferreira da Silva, *Toward a Global Idea of Race* (Minneapolis: University of Minnesota Press, 2007); Achille Mbembe, *On the Postcolony* (Berkeley: University of California Press, 2001); Achille Mbembe, *Critique of Black Reason*, trans. Laurent Dubois (Durham, NC: Duke University Press, 2017); Dipesh Chakrabarty, *Provincializing Europe: Postcolonial Thought and Historical Difference* (Princeton, NJ: Princeton University Press, 2000); Spivak, *A Critique of Postcolonial Reason*; Homi K. Bhabha, *The Location of Culture* (London: Routledge, 1994); Edouard Glissant, *The Poetics of Relation*, trans. Betsy Wing (Ann Arbor: University of Michigan Press,

1997); Antonio Viego, *Dead Subjects: Towards a Politics of Loss in Latino Studies* (Durham, NC: Duke University Press, 2007).

5. David Marriott, *Lacan Noir: Lacan and Afro-Pessimism* (London: Palgrave, 2021); Frank B. Wilderson III, *Afropessimism* (London: Liveright, 2020); Christina Sharpe, *In the Wake: On Blackness and Being* (Durham, NC: Duke University Press, 2016).

6. Zakiyyah Iman Jackson, *Becoming Human: Matter and Meaning in an Antiblack World* (New York: New York University Press, 2020); Saidiya Hartman, *Wayward Lives, Beautiful Experiments: Intimate Histories of Riotous Black Girls, Troublesome Women, and Queer Radicals* (New York: Norton, 2019); Sadiya Hartman, *Scenes of Subjection: Terror, Slavery, and Self-Making in Nineteenth-Century America* (New York: Oxford University Press, 1997); Saidiya Hartman, *Lose Your Mother: A Journey Along the Atlantic Slave Route* (New York: Farrar, Strauss and Giroux, 2008); Hortense J. Spillers, "Mama's Baby, Papa's Maybe: An American Grammar Book," *Diacritics* 17, no. 2 (1987): 63–81.

7. Judith Butler, *Gender Trouble: Feminism and the Subversion of Identity* (New York: Routledge, 1990); Paul B. Preciado, *Countersexual Manifesto*, trans. Kevin Gerry Dunn (New York: Columbia University Press, 2018).

8. Azeen Khan, "The Subaltern Clinic," *boundary 2* 46, no. 4 (2019): 181–217. See also Azeen Khan, "Race and Lacan," in *After Lacan*, ed. Ankhi Mukjerkjee (Cambridge: Cambridge University Press, 2018). Additionally, see Khan's forthcoming monograph on the intersection of psychoanalysis, deconstruction, and postcoloniality, provisionally titled *The Subaltern Clinic*.

9. Jean Khalfa, "Fanon and Psychiatry," *Nottingham French Studies* 54, no. 1 (2015): 52–71.

10. Frantz Fanon, *Black Skin, White Masks*, trans. Richard Philcox (1952; New York: Grove, 2008), 1.

11. Frantz Fanon, *The Wretched of the Earth*, trans. Constance Farrington (1961; New York: Grove, 1963), 36.

12. Fanon, *The Wretched of the Earth*, 36.

13. Antonio Viego, *Dead Subjects: Towards a Politics of Loss in Latino Studies* (Durham, NC: Duke University Press, 2007).

14. Frank Edwards, Hedwig Lee, and Michael Esposito, "Risk of Being Killed by Police Use of Force in the United States by Age, Race-Ethnicity, and Sex," *Proceedings of the National Academy of Sciences* 116, no. 34 (2019): 16793–98.

15. Louis Althusser, "Ideology and Ideological State Apparatuses," in *Lenin and Philosophy and Other Essays*, trans. Ben Brewster (London: New Left Books, 1971), 121–76.

16. Christopher Fox, Roy Porter, and Robert Wokler, eds., *Inventing Human Science: Eighteenth-Century Domains* (Berkeley: University of California Press, 1995); Esther Engels Kroeker, "The Science of Human Nature in the Scottish Enlightenment," *Journal of Scottish Philosophy* 18, no. 3 (2001): 227–32; John H. Zammito, *Kant, Herder, and the Birth of Anthropology* (Chicago: University of Chicago Press, 2002); Alix Cohen, *Kant and the Human Sciences: Biology, Anthropology, and History* (London: Palgrave Macmillan, 2009); Thomas Sturm, *Kant und die Wissenschaften vom Menschen* (Paderborn: Mentis Verlag, 2009); Michel Foucault, *The Order of Things: An Archaeology of the Human Sciences* (New York: Vintage, 1994).

17. Charles de Montesquieu, *Spirit of the Laws* (Cambridge: Cambridge University Press, 1989); David Hume, *Treatise of Human Nature* (1740; Oxford: Oxford University Press, 2000); David Hume, "On National Characters" (1748), in *Political Essays* (Cambridge: Cambridge University Press, 1985); Immanuel Kant, *Anthropology from a Pragmatic Point of View* (1798; Cambridge: Cambridge University Press, 2006).

18. Jean Comaroff and John Comaroff, "Medicine, Colonialism, and the Black Body," in *Ethnography and the Historical Imagination* (Boulder, CO: Westview, 1992), 218.

19. Michel Foucault, *Introduction to Kant's Anthropology* (New York: Semiotext(e), 2008); Patrick Frierson, "Kant on Mental Disorder. Part 1: An Overview," *History of Psychiatry* 20, no. 3 (2009): 267–89; Patrick Frierson, "Kant on Mental Disorder. Part 2: Philosophical Implications of Kant's Account," *History of Psychiatry* 20, no. 3 (2009): 290–310.

20. Robert Bernasconi, ed., *Concepts of Race in the Eighteenth Century* (Bristol: Thoemmes, 2001); Sara Eigen and Mark Larrimore, eds. *The German Invention of Race* (Albany: SUNY Press, 2006).

21. Robert Bernasconi and Tommy Lee Lott, eds., *The Idea of Race* (Indianapolis, IN: Hackett, 2000).

22. Michel Foucault, *The History of Madness* (New York: Routledge, 2006); Richard Keller, *Colonial Madness: Psychiatry in French North Africa* (Chicago: University of Chicago Press, 2007); Leonard Smith, *Insanity, Race, and Colonialism: Managing Mental Disorder in the Post-Emancipation British Caribbean, 1838-1914* (London: Palgrave Macmillan, 2014); Jonathan Sadowsky, *Imperial Bedlam: Institutions of Madness in Colonial Southwest Nigeria* (Berkeley: University of California Press, 2009); Mab Segrest, *Administrations of Lunacy: Racism and the Haunting of American Psychiatry at the Milledgeville Asylum* (New York: The New Press, 2020); Marin Summers, " 'Suitable Care of the African When Afflicted with Insanity': Race, Madness, and Social Order in Comparative Perspective," *Bulletin of the History of Medicine* 84, no. 1 (2010): 58–91; Jock McCulloch, *Colonial Psychiatry and "The African Mind"* (Cambridge: Cambridge University Press, 1995); Megan Vaughan, *Curing Their Ills: Colonial Power and African Illness* (Stanford, CA: Stanford University Press, 1991).

23. Although their respective foci are slightly to the side of the argument I make here, for two important critiques on the history of humanitarian care in relation to European colonial power, see Vaughan, *Curing Their Ills*; and Didier Fassin, *Humanitarian Reason: A Moral History of the Present* (Berkeley: University of California Press, 2011).

24. Peter McCandless, *Moonlight, Magnolias, and Madness: Insanity in South Carolina from the Colonial Period to the Progressive Era* (Chapel Hill: University of North Carolina Press, 2013). On the "afterlife of slavery," see Sadiya Hartman, *Lose Your Mother: A Journey Along the Atlantic Slave Route* (New York: Farrar Strauss & Giroux, 2006), 6: "If slavery persists as an issue in the political life of black America, it is not because of an antiquarian obsession with bygone days or the burden of a too-long memory, but because black lives are still imperiled and devalued by a racial calculus and a political arithmetic that were entrenched centuries ago. This is the *afterlife of slavery*—skewed life chances, limited access to health and education, premature death, incarceration, and impoverishment. I, too, am the afterlife of slavery."

25. Benjamin Rush, *Medical Inquiries and Observations Upon the Diseases of the Mind* (Philadelphia: Thomas Dobson, 1794), 1:267.

26. Benjamin Rush, "Account of a Wonderful Talent for Arithmetical Calculation, in an African Slave Living in Virginia," *American Museum*, January 1789, 61–62.

27. David Hume, "On National Characters" (1748), in *Political Essays* (Cambridge: Cambridge University Press, 1985).

28. Edward Needles, *An Historical Memoir of the Pennsylvania Society for Promoting the Abolition of Slavery; the Relief of Free Negroes Unlawfully Held in Bondage, and for Improving the Condition of the African Race* (Philadelphia: Merrihew & Thompson, 1848), 32.

29. Darold A. Treffert, "The Savant Syndrome: An Extraordinary Condition. A Synopsis: Past, Present, Future," *Philosophical Transactions of the Royal Society of London B* 364, no. 1522 (May 27, 2009): 1351–57.

30. Achille Mbembe, *Critique of Black Reason*. (Durham, NC: Duke University Press, 2017); Spivak, *A Critique of Postcolonial Reason: Toward a History of the Vanishing Present*.

31. Robert C. Young, *White Mythologies* (London: Routledge, 2004).

32. Andrew Halliday, *A General View of the Present State of Lunatics and Lunatic Asylums in Great Britain and Ireland, and in Some Other Kingdoms* (London: Thomas & George Underwood, 1828), 80.

33. "Exemption of the Cherokee Indians and Africans from Insanity," *American Journal of Insanity* 1 (1845): 287–88.

34. Daniel H. Tuke, "Does Civilization Favour the Generation of Mental Diseases," *Journal of Mental Science* 4 (1857): 94–110. Tuke, an English physician and expert on insanity, observed that "the liability to mental disease is greater (other things being equal) in a civilized and thinking people than in nomadic tribes" (94). For a historical overview of this idea and its legacies into the twentieth century, see Ana Maria G. Raimundo Oda, Claudio Eduardo M. Banzato, and Paulo Dalgalarrondo, "Some Origins of Cross-Cultural Psychiatry," *History of Psychiatry* 16, no. 2 (2005): 155–69.

35. Amariah Brigham, *Observations on the Influence of Religion Upon the Health and Physical Welfare of Mankind* (Boston: Marsh, Capen & Lyon, 1835), 275.

36. Benjamin Rush, *Medical Inquiries and Observations Upon the Diseases of the Mind* (Philadelphia: Thomas Dobson, 1794), 277.

37. Gerald N. Grob, "Edward Jarvis and the Federal Census: A Chapter in the History of Nineteenth-Century American Medicine," *Bulletin of the History of Medicine* 50, no. 1 (1976): 4–27; Albert Deutsch, "The First U.S. Census of the Insane (1840) and Its Use as Pro-Slavery Propaganda," *Bulletin of the History of Medicine* 15, no. 5 (1944): 469–87; Leon F. Litwack, "The Federal Government and the Free Negro, 1790–1860," *Journal of Negro History* 43, no. 4 (1958): 261–78; Seymour Leventman, "Race and Mental Illness in Mass Society," *Social Problems* 16, no. 1 (1968): 73–78.

38. Edward Jarvis, "Statistics of Insanity in the United States," *Boston Medical and Surgical Journal* 27 (1842): 116–21. See also William Stanton's discussion of Jarvis's initial reaction to the census result in *The Leopard's Spots* (Chicago: University of Chicago Press, 1960), 58.

39. Michel Foucault, *Territory, Security, Population: Lectures at the Collège de France, 1977–1978*, trans. G. Burchell (New York: Picador, 2009); Michel Foucault, *The Birth of Biopolitics: Lectures at the Collège de France, 1978-1979*, trans. G. Burchell (London: Palgrave Macmillan, 2008); Michel Foucault, *The History of Sexuality*, vol. 1: *An Introduction*, trans. Robert Hurley (New York: Vintage, 1978); Michel Foucault, *"Society Must Be Defended": Lectures at the Collège de France, 1975-1976*, trans. D. Macey (New York: Picador, 1997).

40. The origins of psychiatric epidemiology are tightly connected to the 1840 census and its social-epistemological consequences, producing a set of persistent problems for psychiatric epidemiology still to this day in which descriptive statistics consistently fuel problematic etiological claims while often distracting from root political causes of psychic distress and racial inequality. See Gerald N. Grob, "The Origins of American Psychiatric Epidemiology," *American Journal of Public Health* 75 (1985): 229–36.

41. Khalil Gibran Muhammad, *The Condemnation of Blackness: Race, Crime, and the Making of Modern Urban America* (Cambridge, MA: Harvard University Press, 2010).

42. Mical Raz, *What's Wrong with the Poor? Psychiatry, Race, and the War on Poverty* (Chapel Hill: University of North Carolina Press, 2016). For influential examples of this tradition focused on the social and psychic rehabilitation of the Black American, see Orlando Patterson, *Rituals of Blood: The Consequences of Slavery in Two American Centuries* (New York: Basic Books, 1999). See also the voluminous literature on "the underclass" following William Julius Wilson's interjections in the 1980s, which themselves were extensions of work from the 1960s, such as Kenneth Clark's *Dark Ghetto* and Daniel Patrick Moynihan's *The Negro Family*.

43. "Reflections on the Census of 1840," *Southern Literary Messenger* 9, no. 6 (June 1843): 340–52.

44. "Reflections on the Census of 1840."

45. F. Leon, "The Federal Government and the Free Negro, 1790–1860," *Journal of Negro History* 43, no. 4 (1958): 261–78.

46. Edward Jarvis, "Insanity Among the Coloured Population of the Free States," *American Journal of Medical* Science 7 (1844): 71–83; Grob, "Edward Jarvis and the Federal Census"; Deutsch, "The First U.S. Census of the Insane"; Litwack, "The Federal Government and the Free Negro."

47. Sylvia Wynter, "Unsettling the Coloniality of Being/Power/Truth/Freedom: Towards the Human, After Man, Its Overrepresentation—an Argument," *New Centennial Review* 3, no. 3 (2003): 257–337.

48. Summers, " 'Suitable Care of the African When Afflicted with Insanity.' "

49. With respect to twentieth-century psychiatric expressions of psychocentric racism, see, for example, J. E. Lind, "The Color Complex in the Negro," *Psychoanalytic Review* 1 (1914): 404–14; A. B. Evarts, "Dementia Praecox in the Colored Race," *Psychoanalytic Review* 1 (1914): 388–403; W. M. Bevis, "Psychological Traits of the Southern Negro with Observations as to Some of His Psychoses," *American Journal of Psychiatry* 78 (1921): 69–78. In the field of sociology, the Chicago School is a key reference point for this diffusion of psychocentric racism. I am not aware of any historical text that has as of yet thoroughly elaborated the Chicago School's

imbrication with early psychological, psychiatric, and psychoanalytic ideas alongside its racial foci, but its incorporation of medico-scientific theories of mind is not difficult to discern in its basic thematic orientations. From at least Robert Park, who began at the University of Chicago in 1913 with a first course entitled "Negroes in America," and his theories of social disorganization and social pathology onward, concepts of psychic constitution and development have been key to the sociological paradigm and the pathologizing frameworks with respect to Black Americans. For a historical overview of the Chicago School's organizing frameworks, see Jean-Michel Chapoulie, *Chicago Sociology* (New York: Columbia University Press, 2020). Another important reference point for the diffusion of psychocentric racism into sociology is found in W. E. B. Du Bois's pioneering work *The Philadelphia Negro*. Even as Du Bois takes great pains to emphasize the primacy of socioeconomic causes of the so-called problem of the Negro and to oppose the racist presumptions of his scholarly peers, the study nonetheless reflects the early sociological presumption of psychic deficiency and pathology in its attention to family structures and psychic development. Some historians have thus criticized Du Bois's early work for laying a foundation for later framings of the Black family as a "tangle of pathology" (as Daniel Patrick Moynihan put it, borrowing from the Black psychologist Kenneth Clark and sociological literature, in his 1965 report *The Negro Family: The Case for National Action*) and as fertile ground for psychiatric disorder and criminal tendency. For historicizations of Du Bois in relation to these criticisms, see Tera Hunter, " 'The Brotherly Love' for Which This City Is Proverbial Should Extend to All: The Everyday Lives of Working-Class Women in Philadelphia and Atlanta in the 1890s"; Jacqueline Jones, " 'Lifework' and Its Limits: The Problem of Labor in *The Philadelphia Negro*"; and Antonio McDaniel, "*The Philadelphia Negro* Then and Now: Implications for Empirical Research," all in *W. E. B. Du Bois, Race, and the City*, ed. Michael Katz and Thomas Sugrue (Philadelphia: University of Pennsylvania Press, 1998): 126–51, 102–25, 154–93. For a much fuller account of Du Bois's contributions to sociology and the later shifts in his intellectual orientation, see Aldon Morris, *The Scholar Denied: W. E. B. Du Bois and the Birth of Modern Sociology* (Berkeley: University of California Press, 2015).

50. Matt Ford, "America's Largest Mental Hospital Is a Jail," *Atlantic*, June 8, 2015.
51. Young, *White Mythologies*; Spivak, "Chapter 1: Philosophy"; Jackson, *Becoming Human*.
52. Wynter, "Unsettling the Coloniality of Being/Power/Truth/Freedom," 262.
53. Foucault, *The Order of Things*. For an excellent set of commentaries on Wynter's contribution, see Katherine McKittrick, ed., *Sylvia Wynter: Being Human as Praxis* (Durham, NC: Duke University Press, 2015). In particular, Denise Ferreira da Silva's contribution, "Before Man: Sylvia Wynter's Rewriting of the Modern Episteme," gives an especially cogent summary and critical reading of Wynter's "Unsettling the Coloniality of Being/Power/Truth/Freedom."
54. Wynter, "Unsettling the Coloniality of Being/Power/Truth/Freedom," 266.
55. Wynter, "Unsettling the Coloniality of Being/Power/Truth/Freedom," 264.
56. Wynter, "Unsettling the Coloniality of Being/Power/Truth/Freedom."
57. See their respective essays in McKittrick, ed., *Sylvia Wynter*.

THREE

Locating the Child in Racial Science

Scenes from Latin America

SEBASTIÁN GIL-RIAÑO AND JULIA E. RODRIGUEZ

IN 1915, THE distinguished Argentine pediatrician and public health official Gregorio Aráoz Alfaro gave a speech entitled "For the health and vigor of the race" in which he equated children's health with the strength and progress of the nation. He argued that assistance for mothers and small children was crucial to achieving the nation's goals. He exhorted his audience to pay attention to child health and feeding because "the efficient protection of mother and infant, [and] hygiene and medicine of school age children . . . [will] create stronger, healthier, more intelligent, and better educated generations capable later of victoriously resisting the harmful causes of disease, vice, and crime."[1] Like a growing number of physicians, public health officials, social scientists, and reformers in Latin America, Aráoz Alfaro urged his peers to take children seriously. These activist doctors proposed a wide range of reforms that extended from social welfare to labor regulation to public health. They were concerned with children's well-being for its own sake and for the sake of larger society. Only by creating the ideal environment for growing children could the nation be saved from the myriad physical and social ills it faced. In taking this broader view, they tapped into an increasingly common belief that the child was the future citizen and therefore a natural resource.[2] Such views were reflected in Latin America, where physicians, politicians, and reformers elevated child well-being in new organizations and conferences, such as the Pan American Child Congresses, regularly held after 1916.[3]

As professional and social reform movements in Latin America coalesced in the first few decades of the twentieth century, debates about what to do with the nation's youngest members were fraught with unsettling racial tensions. Latin American experts interpreted children's well-being (both at the individual and population level) as having profound consequences for the evolutionary trajectory of their societies. For example, in Argentina in the early 1900s, medical experts grappled with how to balance concerns about avoiding pathology in developing children while strengthening the chances for normative growth. That said, the state focused more resources on deviant, ill, and potentially dangerous youth than on building universal programs for child well-being. Moreover, most elites in Argentina ignored the nation's sizable Afro-, Indigenous, and mestizo populations and forwarded a Eurocentric model of the model child and future citizen. Fears of racial degeneration, of society's backward slide into primitivism, rested on the nation's youth.

Pathology was not the only lens through which reformers looked at children. During the interwar period in Brazil, practitioners of eugenics and mental hygiene advocated for an environmentalist approach to studying children's mental health that displaced the racial and hereditarian frameworks of fields like criminal anthropology. Yet, like their Argentinean counterparts, Brazilian reformers fretted over the potential impact of racialized and impoverished children and ultimately retained hierarchical beliefs that privileged whiteness. During a period of nation building when ruling elites sought to assert a distinctively Brazilian identity, mental hygienists clumsily attempted to replace categories drawn from race science like "abnormal" and "mentally retarded" with notions like "backwardness" and "maladjusted."

The new and increasingly specialized focus on children in medicine and mental hygiene raised a number of questions: How should societies deal with specific problems facing children, such as malnutrition, disease, disability, poor environment, and weak or defective moral development? Who was responsible for children's healthy development—parents, doctors, the state, or the children themselves? Who should take the blame should a child, or their society at large, fail to thrive? Many Latin American physicians, especially the first female doctors at the turn of the century, informed by long-standing pronatalist and maternalist assumptions, proposed "positive" programs like milk stations and improved living conditions for families. In addition, some of them were motivated by leftist social reform movements that pushed for improvements in workers' wages and living conditions, and even (in a limited sense) women's rights. Thus, a focus on child pathology coexisted with a focus on normative child health, often in the same people, publications, and proposals advocating for answers to the era's questions about children.

In addition to the tension between pathology and normative growth, a second important scientific approach shaped the new theories of human development: racial science, especially eugenics. These approaches focused not only on children in their family and social contexts but were also informed by hereditarian assumptions, albeit diverse and sometimes opposing interpretations thereof. Studies of racial science have largely overlooked a genealogy of the child; often the focus is on aspects of the body such as crania, skin pigment, hair pattern, blood samples, and genes as empirical focal points for investigating the shifting history of racial thought. In this chapter, we explore what happens to the history of racial science if we focus not on the science of static body parts but on how race experts conceptualized the whole human being—in particular, children and the earliest stages of human development. While seldom explicitly stated, reproduction and its products—children—are central to racial science. While reproduction ("to improve the race") has implicated a variety of actors, at its center is the mother-child dyad. In these contexts, we investigate the medical gaze as it rested specifically on children *qua* children.

The child is a useful lens through which to better understand two powerful and racialized medical projects in early-twentieth-century Latin America: *puericultura* and mental hygiene. These two important projects of child-focused and eugenics-adjacent science were populated by powerful figures in their scientific communities, and both were concerned with the quality of "the race" in their nations. In Argentina and Brazil, the racial sciences contributed to making children visible. Professionals saw youngsters not just as preadults (i.e., future citizens or criminals) but as individuals who were also the offspring and charges of adults, especially women. For the first time perhaps, beginning around 1900, physicians saw children as a distinct group, and saw them through a scientific and medical lens tinted by urgent social goals. In both cases discussed in this chapter, scientists simultaneously ascribed to and deviated from eugenics and evolutionary models of human development. With our analysis of these cases, we hope to draw attention to the complexities inherent in attempts to address social problems via children and the entanglements with racial ideologies as shaped by specific social conditions.

Racialized studies of childhood were particularly intense, we argue, in Latin American nations set on progress. In the last decades of the nineteenth century, many Latin American governments made significant investments in medical and scientific institutions, along with modern state apparatus building. Influenced by European (especially French) trends in medicine, Latin Americans incorporated new thinking from Europe into their infant institutions. The result was the early elaboration and nurturing of public health, hygiene, eugenics, and the psych fields, as has been explored in a substantial historiography.[4] There was also a significant state and social investment in practices shaped by these fields. Here we show how racial evidence was mobilized in the interest of child development, but with other motives (e.g., national growth) and outcomes (e.g., both positive and negative practices with effects on children's health and well-being).

Thus, from the vantage of development (in its multiple senses) Latin American children and childhoods emerge as central objects and themes in the global history of racial science. At the height of scientific racism in the late nineteenth century, evolutionary theorists conceptualized childhood as a racialized placeholder symbolizing a state of primitiveness and inferiority on the path toward an adulthood (and modernity) that only European men could achieve. In the early twentieth century, evolutionary anxieties and concerns over racial fitness also prompted eugenicists and proponents of puericulture throughout the Americas and elsewhere to devote significant attention to the surveillance and cultivation of babies and children as a political imperative in which the health and well-being of the nation-state was at stake. Eugenicists in Latin America, in particular, made children and human development a central part of their programs. During the interwar period, researchers in the human sciences working under rubrics such as the mental hygiene movement, race psychology, the culture and personality school, and the settlement house movement homed in on families and children as places where attitudes, values, and ways of life were transmitted across generations. And they often brought normative assumptions with them that privileged white and heteronormative families as an aspirational model and often led to the conceptualization of non-European families as disorganized and dysfunctional that perpetuate cultures of poverty and stagnation.

Puericultura, Maternalism, and the Politics of Care in Argentina

Early-twentieth-century discourse on child health and welfare in Argentina emerged from and overlapped with eugenics and evolutionary theory.[5] *Puericultura* (literally, the cultivation of children) was one of a number of newly coined terms in the human sciences of the late nineteenth century. The term originated in France in the 1890s, coined by the obstetrician Adolphe Pinard, who sought a remedy for the French population "crisis," which encompassed concerns about the potential degeneration of the French "race."[6] Pinard advocated medical fixes for social problems such as disease among the poor and dangerous work conditions for women. The main focal point of French puericulture was reproduction and heredity. They saw the cultivation of healthy "germ plasm" as the best way to guarantee healthy outcomes. Direct and indirect exposure to toxins such as alcohol, tuberculosis, and venereal infections would damage the fetus and young child. Pinard advocated that "the task of the obstetrician and the pediatrician was to reduce all the adverse factors threatening health in reproduction, through sexual education, aid to families, and the new obstetric techniques."[7] He explicitly linked such policies with the survival of the nation.

In Argentina, as Rodriguez has argued, scientific and medical fields like *puericultura* assumed a central role in expanding state power and in the service of the national dream of progress.[8] Moreover, according to the historian Nancy Leys Stepan, "Children especially were thought of as biological-political resources of the nation, and the state was regarded as having an obligation to regulate their health."[9] Argentine doctors were enthralled with the promising, optimistic science of *puericultura* as the key to improving their population across the lifespan. In 1915, Genaro Sisto, a prominent professor of pediatrics at the University of Buenos Aires, called *puericultura* the "science of the moment."[10] The focus on environment and flexibility in heredity, as well as an emphasis on practical measures such as public health campaigns and hygienic education, merged with Latin American physicians' self-perceptions, as Stepan has argued.[11] The youngest members of society were the starting point for advancement of the national body as a whole. The preventative dimensions of *puericultura*, that is, its promise to reduce juvenile deviance and disability, appealed to Argentine medical reformers, even as such strategies would ultimately fail because of a lack of political consensus on how to address underlying social and economic inequalities. But they also expressed hope that growing a strong and healthy normative youth population would also combat racial degeneration. As in applications of eugenics more broadly, in the Argentine context, *puericultura* addressed individual children's well-being at the same time it projected a racialized vision of a healthy national body.

Argentine puericulturists' ideas were shaped by eugenic assumptions, and they viewed both normal and abnormal children through racialized evolutionary models of human development. Central concepts within the evolutionary framework that resonated in Argentina at a time of intense national self-identification and postcolonial ambition included the idea of progress, degeneration and regeneration, heredity and the need for remedial methods (both environmental and medical), and moral education as a key ingredient to child development and therefore to society's survival and progress. Historians have debated the emphasis of Argentine eugenics and the extent to which "positive" (e.g., prenatal care) versus "negative" (sterilization) approaches existed.[12] Of course, eugenics as a worldview could (and still does) encompass both punitive and nurturing attitudes and policies. Both expressions, in fact, are important parts of social history in twentieth-century Latin America. Nonetheless, by focusing primarily on the institutional impacts of the ideology, historians have neglected the centrality of the importance of maternalist ideology and *puericultura*'s accentuation of the mother-child dyad.

Argentina was fertile ground for *puericultura*; the new specialization fed a national mythology that cast it as a "young" country full of youngsters. Elites believed that the nation's growth and development depended on the healthy growth and development of its youth. Young people would supply healthy labor and, ideally, an educated citizenry who could continue to lead Argentina on its path to progress. As Argentina sought to maintain its economic momentum (built on thirty years of phenomenal profits from cattle and grain), the large number of children in the population from immigration and a high birth rate represented both the promise of continued profit generation as well as a potential threat to that prosperity, should the nation's offspring prove to be deficient. Even though Argentina had one of the highest birthrates in the world in 1914, reaching nearly 37 births per 1,000 inhabitants, the census figures revealed that in ten years the percentage of women without children had increased from 11 to 14.8. Officials worried that infertility was rising.[13] They were also extremely concerned about the still-high infant mortality rate.

Argentine doctors and officials were intent on increasing the population *and* improving the "quality" of children reproduced. In their view, mothers were on the front lines, as signs of "degeneration of the race" could be caused by faulty or ignorant mothering. The prominent pediatrician Enrique Feinmann wrote in the introduction to his 1915 textbook *Ciencia del niño* (The science of the child), "Woman will be the good fairy of the new era. Her nursery of human beings will be an immense blooming garden, and the children, instead of going to heaven as angels, will populate the earth as men, to make it better and more beautiful."[14]

In this sense, Argentines' embrace of *puericultura* reflected the ideologies of maternalism (a longstanding and widespread belief that held the primacy of motherhood as a universal ideal) and pronatalism (a newer concern, focused on increasing the nation's labor force). The historian Marcela Nari documented how maternalism, always a strong ideology in Argentina, transformed over time. In the early twentieth century, economic, social, and political forces shaped the consolidation of women's main, if not in most cases exclusive, role as mother.[15] Virtually all members of Argentine society agreed that the nurturing of children was women's most natural calling and noble labor. Women who rejected motherhood were regarded as barbaric and immoral; proper women understood their maternal role well. Women were said to have a "civilizing and beneficial" effect as mothers, with a responsibility to fulfill their natural duties. By 1900, the woman's role as mother was a key component of liberal reformers' visions of a modern and progressive society (based on North Atlantic norms).

In Argentina as elsewhere at this time, children's health, well-being, and development became a central social and political concern not just because of the professionalization of medicine and economic growth but also as the result of the increasing presence of women in medicine. The historian Anne-Emanuelle Birn has pointed out that in many Latin American countries, the rise of *puericultura* and modern pediatrics in the region emerged from "a mix of feminist-maternalist ideas, indigenous cultural practices, nationalist concerns, and physician advocacy."[16] Feminist and socialist reformers, among them some of Argentina's first female physicians, began to devote themselves to improving the living conditions of women and children. They championed issues like prenatal leave for working women and mandatory breaks for nursing mothers, proposed protectionist legislation to reduce infant mortality and debilitating disease, and worked to ensure equal wages for women. They sought improved conditions in factories. One socialist document of 1894 called exploitative female and child labor in factories "the source of many ills in the family." Buenos Aires factories, by all accounts, were poorly lit and ventilated and rarely clean, with women and children workers forced to sit doing repetitive work for hours on end and denied bathroom breaks. Some industries, such as match manufacturing, used noxious chemicals. Gabriela Laperriere de Coni, an active social reformer and wife of the hygienist Emilio Coni, warned that the substandard environment of Argentina's factories would destroy the reproductive health of women and the health of their future children.[17]

As the idea of *puericultura* caught on in Argentine medical circles, Feinmann, along with the physicians Cecilia Grierson, Elvira Rawson de Dellepiane, Aráoz Alfaro, and others, began to issue chemical analyses of breast milk and formula and instructions for proper swaddling and dressing of infants. They intended puericulture to be incorporated into girls' curriculum in public schools. In 1913, the First National Congress of the Child was held in Buenos Aires, one of the first scientific gatherings in Argentina where women professionals such as Cecilia Grierson could participate. The state welfare apparatus, or Asistencia Pública, in 1908, had created a department devoted to the protection of children under two years old. The Protección de la Primera Infancia, as it was called, established and oversaw milk dispensaries, visited nursing mothers and provided them with food, and, after 1919, examined wet nurses. The number of children helped by the agency expanded from 232 in 1908 to nearly 12,000 by 1920; even more numerous were the women who had contact with the Protección de la Primera Infancia: more than 500,000 consultations or inspections in 1916.[18]

These programs built on the early work of Aráoz Alfaro, who in 1899, seven years after graduating from medical school, published *El libro de las madres. Manual práctico de higiene del niño* (The mother's book. Practical manual for child hygiene), a book intended to popularize scientific knowledge about mothering, especially for those of the "many poor families who do not have access to a doctor at all times."[19] In his view, women's ignorance led them to make terrible errors in childrearing. His goal in this book was to provide the most up-to-date techniques of child development according to the latest hygienic principles. In the preface, Aráoz Alfaro made it clear that the stakes of childrearing were high:

> They—who are today the depositories of this dark and latent force, visible and powerful tomorrow, that is called the child—are destined to mold this soft mass that generally depends on the momentum of its early years, in its physical organization as well as in its moral texture. [The mothers] alone, taught and led by the doctor, who is now . . . the authorized mentor of the family and of society, can give us healthy and vital seeds from which school and state will make men who are physically strong with healthy souls, and flexible and open intelligence.[20]

Aráoz Alfaro's manual focused largely on feeding, care, and hygiene, with later chapters on normal growth and development, needs of the home environment, and education of the child.[21] He also called for state action to bring about ideal inputs for the developing child. It was in the interest of the state to guarantee healthy environments for children since they would affect the evolutionary path of Argentine society. The 1922 edition also included a section on eugenics, a "new science destined to guarantee the health and beauty of coming generations." He called on Argentines "to establish and propagate this new science in our new nation [because] it is so necessary for a healthy and vigorous population." His call for attention to children's health and development were intended to "increase the vigor and energy of the new race we are creating in Argentina."[22]

Scientists' acceptance of the evolutionary model of child development tightened the Argentine elite's embrace of pronatalism and maternalism. While scientists recognized that fathers contributed healthy or damaged "germ plasm," they uniformly believed that mothers had additional abilities to abate or abet degeneration. A 1915 article in the *Revista de Criminología, Psiquiatría, y Medicina Legal* (Review of criminology, psychiatry, and legal medicine) on "Women's growing impotence to nurse their children," warned of the decline in breastfeeding as contributing to the degeneration of the race. A writer for the Argentine journal, a Swiss professor of chemical physiology, attributed to lower rates of breastfeeding "symptoms" of social pathology.

Because a mother's responsibility to nurse was so great, "men of good health, desiring of healthy descendants," should marry only a woman who had herself been breastfed as an infant.[23]

There was a growing consensus that addressing and preventing childhood pathology was necessary for national strength. Physicians and officials saw pathological childhoods as a threat to national, if not species, survival. Just one example of many in which the links between the child and the nation were drawn is found in the criminologist M. A. Lancelotti's 1905 article in the *Revista Penitenciaría* on "Education and crime." In Lancelotti's view, criminals began their lives in dysfunctional homes and if reared without proper guidance would turn to criminal activity. The desire to survive was driven further by inborn "instincts" that humans shared with animals. Children, according to Lancelotti, were similar to "primitives." He concluded that "the greatest modern psychological studies concur that the child's indolence in general is wicked, and that his psychology is similar to primitive man, to those of the modern savages, [and] to those of criminals. In light of the laws of heredity, there is in the child a powerful tendency to reproduce in miniature the history of humanity."[24]

After over two decades of promoting state interventions in juvenile institutions, advocates of puericulture were able to push through new legislation. In 1919, the Argentine Congress enacted the National Law of Protection of Abandoned and Criminal Minors, known as the "Agote Law" after its main sponsor, the physician-legislator Luis Agote. The law gave the state expanded power over delinquent children.[25] It funded the establishment of four homes dedicated to the "teaching and regeneration of the minors" and created a codified procedure in which judges would send children to the designated institution, according to their mental status and/or behavior.[26] With its emphasis on deviant youth, the Agote Law revealed the state's views about which children were in most urgent need of attention. The medical literature focused on the child as the future citizen but was inextricably linked to anxieties about abnormal children dragging society backward to an earlier state of barbarism. In an era of intense state social engineering, including attempts to harness the labor and civic power of children, the concept of protection versus segregation was not fully worked out. While medical studies of children's individual suffering and social danger led to protective (e.g., antipoverty, hygiene, nutrition) proposals, approaches to children's health just as often in this period sought to protect society *from* children. Thus, the legacy of *puericultura* children was mixed.

The focus on pathology notwithstanding, physicians never fully lost sight of the normative child. As liberal democratic governments gave way to the populism and nationalism of the 1930s and beyond, the health of children remained a national priority. After 1930, in the context of a regional mobilization for health care reform, the Argentine state increased its investment in normative child development.[27] Moreover, in the second half of the twentieth century a strong movement for children's rights coalesced in Argentina and Uruguay, advocating for the right to health and, in this sense, informed by the more socialist, optimistic aspects of *puericultura*.[28] The punitive, segregating practices of the early twentieth century were eventually, if incompletely, balanced with policies informed by social medicine. With its roots in both racial science and sociomedical perspectives, the new specialization of *puericultura* had encapsulated the complexities of scientific approaches to the crisis in child health.

Race and Mental Hygiene in Brazil

Like Argentina, during the 1930s the Brazilian state identified the investment in and care of children and families as a crucial element of nation building and planning for the future. Yet in Brazil this occurred during a period of political polarization and turmoil that gave rise to the authoritarian and populist regime of Getulio Vargas, who ruled with military support and through the suppression of free speech and democratic opposition from 1930 to 1945. After taking control of Brazil's government in 1930, Vargas faced intense opposition from the right-wing Brazilian Integralist Action Party and from the Communist Party, both of whom mobilized supporters and mounted protests across the country in 1935 and 1936. In response, during the lead-up to the 1937 elections Vargas, with support from the military, used the pretext of an alleged communist plot to assassinate members of government to suspend parliament, ban political parties, and imprison or exile political opponents. His regime also announced a new constitution and declared the foundation of an Estado Novo, or New State, that prized "modernity and progress" as its core principles.[29]

As several recent works have shown, two cornerstones of the nationalist ideology of the Vargas regime were a valorization of race mixing as a tool of whitening and a strategic investment in early childhood education.[30] In both cases, Vargas's vision for Brazilian society was informed by positivist and eugenic philosophies that identified uplifting Brazil's population as a means of modernizing the nation. Vargas also fashioned himself as a populist and champion of the poor and identified the care of children and investment in early childhood education as a central strategy for modernizing Brazil.[31] Just three days after taking power in November 1930, Vargas created the federal Ministry of Education and Public Health (MES). During Vargas's tenure, the MES initiated a series of educational reforms that gave white educators, physicians, and social scientists the opportunity to transform the education system in a way that valorized whiteness and pathologized Blackness. By coupling education and public health, the MES also paved the way for an infusion of eugenic themes and concepts, such as nutrition, physical education, and "mental hygiene," into the education system.

One of the key sites for this eugenic-themed educational reform was the Rio de Janeiro school system. In 1931, under the leadership of the Columbia-trained Anisio Teixeira, officials from Rio's school system created an Institute for Educational Research (IPE) modeled on the Institute for Education Research at the Columbia University Teachers College in New York. Under Teixeira's leadership, the IPE became a hub of experimental scientific research that used Rio's racially diverse schoolchildren as an experimental population and data mine.[32] At the IPE, Teixeira, who had trained with the pragmatist philosopher and educator John Dewey, recruited some of the nation's brightest human scientists—including the physicians Edgar Roquette-Pinto and Arthur Ramos (both also practiced anthropology)—to head four research units, or "sections": Tests and Measures, Educational Radio and Cinema, Orthophrenology and Mental Hygiene, and Anthropometry. As Jerry Davila has demonstrated, the experts recruited to lead this research section framed their work in eugenic terms and envisioned their task as an anticipatory one concerned with identifying the causes for the "degeneration" of Rio's population and with implementing preventive programs. Yet in keeping with the Lamarckian orientation of most eugenic projects from South America and southern Europe, these experts tended to reject the notion that the causes of degeneration and social problems were caused by inborn hereditary traits and instead saw them as stemming from disorders and deficiencies in the social and cultural environment.

One of the key programs that the reformers of Rio's school system implemented during the Vargas era was an ambitious system of monitoring and data gathering for the purpose of maximizing the psychological and physical development of children. Although each research unit focused on a different aspect of the IPE's eugenic program, all IPE researchers worked from a standardized data set composed of *fichas*, or records, that they gathered for each child, in some cases without knowledge of the parents. For each student, IPE researchers generated two fichas—a *ficha antropometrica* and a *ficha de higiene mental*—that they then used to track each student's physical and psychological development and for their ongoing research. In a context of significant race mixing that researchers likened to a melting pot, researchers attempted to use data from these *fichas* to discern patterns of racial formation. For instance, Bastos D'Avila, a biometrician who headed the IPE's anthropometry section, used data from the *fichas antropometricas* in an attempt to fine tune an anthropometric measure called the Lapicque index, which he believed could be used to identify latent "African" traits in those who were phenotypically white. D'Avila also used the *fichas* to compare patterns of growth between Black and white children and reasoned that the faster growth of Black children could be explained by the fact that upperclass whites did not send their children to public schools and thus the white children in the school system tended to be of a "physically deficient" type.[33]

While the work done in the anthropometry section of the IPE was firmly in line with standard practices of racial science, the research conducted by Arthur Ramos in the orthophrenology and mental hygiene section drew from social scientific disciplines that emerged during the interwar period and sought to break from the biological and evolutionary orientation of late-nineteenth-century racial science. Although he fit seamlessly in the IPE's eugenic agenda, Ramos's research was also crucial for the institutionalization of Afro-Brazilian studies in Brazilian universities, especially in the northeastern states of Bahia and Recife. Ramos trained in clinical psychiatry and forensic medicine at the prestigious Faculty of Medicine of Salvador de Bahia, where he graduated in 1926 after defending a dissertation on a topic that would stay with him for the rest of his career—"Primitiveness and Madness."[34] In medical school, Ramos became enamored with the work of

Nina Rodrigues—a late-nineteenth-century practitioner of "legal medicine" who studied Afro-Brazilian religious practices yet conjectured that non-Europeans were innately disposed to criminality and that Brazil's pattern of race mixing could lead to racial degeneration. Later, during the 1930s, when the Recife-based historian Gilberto Freyre attempted to position himself as the leading interpreter of Afro-Brazilian history, Ramos began to fashion himself as Rodrigues's devoted disciple as a way to challenge Freyre's authority and to assert the centrality of Bahia within Afro-Brazilian studies.[35]

By identifying with the "Nina Rodrigues School," Ramos was forced to confront the racist conclusions of his teacher's work. Before the second Afro-Brazilian congress, Gilberto Freyre undermined the work of Bahian scholars by describing their interest in Afro-Brazilian folklore as unscientific and by questioning Nina Rodrigues's scientific contributions and their emphasis on what he called "biological pathology" and the "absolute inferiority of the black and the mulatto."[36] In response to Freyre's critiques, Ramos recognized that Rodrigues's evolutionist framework and hereditarian conceptions of race were no longer in line with contemporary discourse. Accordingly, in his major publications on Afro-Brazilians, Ramos systematically attempted to substitute Rodrigues's key theoretical concepts with ones from contemporary social science. "If, in the works of Nina Rodrigues," Ramos asserted, "we substitute the terms race for culture and miscegenation, for acculturation, for example, his concepts become completely and perfectly contemporary."[37]

In his research for the orthophrenology and mental hygiene section of the IPE, Ramos also steered away from narrow hereditarian explanations. At the IPE, Ramos made detailed observations of Rio schoolchildren's behavior and recorded them in *fichas de higiene mental*. Ramos viewed the *fichas*, which also listed results from psychometric and anthropometric tests, as tools for documenting a wide range of environmental conditions that could potentially shape the development of the child. According to Ramos, the *fichas de higiene mental* recorded information on "family data (parents, siblings, other relatives . . .), family environment (material and psychological conditions of the housing . . .), maternal obstetric history, development and habit formation (growth, general health, food, gait and language, physiological rhythms, discipline and school life), temperamental and characterological facade, psychological functions . . . medical examination, personality diagnosis."[38] From the frame of mental hygiene, Ramos thus used the *fichas* to cast a wide net in an effort to identify the causes of maladjustment in children; he also believed they could provide context for early therapeutic interventions.

By viewing children's psyches through an environmentalist lens, Ramos also challenged the psychiatric classifications and tools that figured prominently in early-twentieth-century race science. In *Criança problema: A higiene mental na escola primária* (1939), the book that Ramos published after his five years of service at the IPE, Ramos argued that approximately 90 percent of children classified as "abnormal" would be better classified as "problem children." While "abnormal" implied a physical defect, Ramos insisted that most children classified as such were in fact suffering from "maladjustments in their social and family environments."[39] According to Ramos, the "abnormal" view stemmed from the popularity of French psychometric approaches and from researchers' tendency to rely on tests (such as Binet and Simon's intelligence tests) that reduced complex phenomena to a simple quotient or scale.[40] By contrast, Ramos insisted that psychology should be conceptualized as a science that "studies the totality of the psyche, which has no limit and is therefore immeasurable" and reasoned that "the notion of mental retardation cannot be resolved through the application of simple tests."[41] Yet the thick data collected through the *fichas* and his use of the term "problem child" suggest that despite his objections to psychometric tests, Ramos still paternalistically viewed the psychiatrist as an authority figure who ultimately knew what was best for children and families.

By 1937, when Vargas declared the Estado Novo and tightened his control over Brazilian society, relations with childhood experts like Teixeira and Ramos began to unravel. After a failed communist uprising in November 1935, Teixeira and his team were forced to leave Rio's school system, and some were even jailed.[42] After Teixeira's departure, Ramos continued working at the IPE but eventually resigned from his position and closed the orthophrenology and mental hygiene section in 1938. Eleven years later, in a preface to the second edition of *O criança problema*, Ramos pointedly revealed that he had chosen to close the orthophrenology section because of political interference from the Vargas regime. In addition to Teixeira's dismissal, one of the main reasons Ramos gave for closing the unit was his distaste for the ideological direction imposed by the Vargas regime and the increased surveillance of his work. Vargas's declaration of the Estado Novo, Ramos explained, ushered in a new regime based on "a false nationalist education" that replaced the "humane and comprehensive [approach to] education" that he valued with a "classical pedagogy" based on "rigid discipline of interdictions and coercion." Although he attempted to keep his section going during this "period of shadows," Ramos explained that he eventually could not withstand the "holy inquisition of brave censors" who did not openly formulate any allegations against his work but nevertheless prevented it from functioning in an "atmosphere of freedom where science can really be done."[43]

In addition to denouncing the Vargas regime, Ramos also used his 1949 preface to distance himself from the eugenic framing of his work at the IPE. Indeed, Ramos used his preface to align his mental hygiene work with trends that had become fashionable in the social sciences in the Americas during the 1940s. "Children's problems can only be understood in today's world in relation to their family circles and to society," Ramos professed, arguing that once the "organicist" aspects of mental hygiene are removed, the field essentially becomes a "comparative social psychology." Ramos also extensively cited the work of psychologists and anthropologists from the "culture and personality" movement such as Ralph Linton, Cora Du Bois, Géza Roheim, and Margaret Mead and described how his own engagement with anthropology had served to deepen and sharpen his own longstanding "conceptions of human behavior as a function of [the] social and cultural environment." Instead of the capacious Lamarckian and eugenic framework that researchers at the IPE had favored, Ramos now attempted to reframe his work through the comparatively narrower approach of "culture and personality" studies that focused on the relation between culture and psyche. Anthropological research, Ramos asserted, had corrected the "primitive teachings of psychoanalysis" (and presumably eugenics as well) by showing that human instincts are "conditioned by the cultural environment."[44]

Ramos' strident views in this 1949 preface reflect not only a disciplinary reorientation but also a political awakening of sorts. During the 1940s, Ramos moved away from the psychological and psychiatric focus of his previous work and pivoted to institutionalizing anthropology in Brazil and positioning himself as its figurehead. As Vargas's control over Brazilian society tightened and war broke out in Europe, Ramos also strengthened his international connections and spent time teaching and lecturing in the United

States, where he also participated in academic campaigns against racism.[45] During this period, he published pamphlets in Brazil decrying Nazi racism and began voicing forceful critiques of Western imperialism. In English-language articles on Afro-Brazilians, which he published in the *Journal of Negro Education*, Ramos also diligently replaced notions of a "primitive" or "prelogical" cultural mentality with analyses of Afro-Brazilian religions that foregrounded the lasting effects of slavery and the precarious economic conditions it had created. He also began to celebrate what he viewed as a distinctly Brazilian approach to race relations—what he called "one of the purest racial democracies of the Western Hemisphere"—and suggested that Brazil's most urgent task ahead was to raise "the economic standards of all races" and specifically to raise these standards much higher than the one "prevailing throughout the vast territory in which a large colored population is to be found."[46]

Ramos's work thus serves as a potent illustration of how Latin American experts wrestled with the racial and eugenic logics they inherited from European science and attempted to break from their evolutionary schemas of progress. Yet as his strategic professionalism and self-fashioning demonstrates, this was ultimately an elitist project primarily invested in preserving and expanding the authority of scientific experts. In 1949, Ramos died suddenly a few months after serving as the director of UNESCO's social science department. At the department's first "meeting of race experts" that year, the sociologist Luiz Aguiar Costa-Pinto, Ramos's Brazilian colleague, paid tribute to Ramos's career and unwittingly highlighted the contradictions he had grappled with. "One of the outstanding characteristics of Dr. Ramos's work," Costa-Pinto explained to the committee of experts before him, "was his deep sympathy with all the backwards peoples and oppressed races." According to Costa-Pinto, Ramos's noble sentiments stemmed from a remarkable career spanning several decades during which he achieved fame as "an expert on African problems and on integrating Negroes in the culture of the New World." For Costa-Pinto, Ramos's tireless efforts on behalf of backward peoples made him "one of the greatest representatives of the new scientific humanism." Although he did not mention Ramos's work on "problem children," the paternalistic sentiments and logics that Ramos developed during his tenure at the IPE can also be discerned in the language of "backwards people and oppressed race" that Costa-Pinto deployed. Ramos's lasting legacy, as the tributes to his career make clear, was the articulation of an elitist alternative to scientific racism that ultimately continued to draw analogies between non-Europeans and children.[47]

Conclusion

Viewing the history of racial science through the lens of child health and mental hygiene clarifies these fields' social meaning in Latin America during the first half of the twentieth century. Ideas of progress and social evolution that dominated political and social discourse among the region's scientific elites imbued the crisis in child health and development with urgency. Tropes of racial superiority and inferiority also sharpened scientists' definitions of degenerate and regenerative traits in the population and informed the solutions proposed and sometimes carried out.

In Argentina, calls for treatment of both normal and pathological children spoke to an idealized (and racialized) vision of national progress. The education and health of children were crucial ingredients of modernization, continued cultural advances, and, perhaps above all, hopes for the nation's political economy. It was generally assumed, at least among scientific, literary, and policy figures, that citizens should be made healthy, strong, and beautiful in mind and body. The task was clearly complicated by material realities in Brazil and Argentina alike: incomplete postcolonial nation building, a volatile economy, and the need to educate and homogenize the immigrant and multiracial population, not to mention rural poverty, urban crowding, and political upheaval, all of which impinged on any straightforward child welfare project. In Argentina, the *puericultura* model promised that normal children would be the "natural" outcome if conditions were optimal. At first, scientists largely laid blame for failure, and responsibility for success, on the individual—above all, the child and its mother. In Brazil, the mental hygienist Arthur Ramos moved beyond the individualist framing of puericulture and instead identified childhood problems as arising from familial disorder and the persistence of "backward" cultural beliefs. Although he sought to distance the care of children from concerns about race and evolution, Ramos clung to the modernizing ambitions of puericulturalists and their notions of civilizational progress.

The proliferation of biomedical and cultural "fixes" to social "degeneration" via the production of healthy children created material benefits in some cases, for example, social welfare for families, protective legislation for women and children, and additional funding for educational institutions. Yet for many marginalized citizens, scientific scrutiny of children paradoxically resulted in the neglect and blame of families for their failure to transcend their imperfect genetic inheritances. This anxiety pervaded and ultimately undermined the optimism of childcentric health programs in the first decades of the twentieth century. It was not until the 1930s and 1940s that normative child development strategies emerged in any serious form.

By midcentury, throughout Latin America physicians and reformers increasingly looked to the state to improve children's living conditions and health. They aspired to a nation that did not condemn its poor children to ongoing misery—the pattern since colonial times—but that invested in social needs to create a healthier nation. These scientists also contributed to a larger sociomedical political movement, that of the "right to health." In 2015, an article in the *Lancet* reflected on its rich history, calling the sociomedical commitment (at least in theory) a "key and distinctive feature of Latin America."[48] At least in principle and to an extent in practice, the authors noted that nations in the region had "incorporated rights, principles and standards in constitutions and legislation, together with health policies and programmes . . . understanding governments as duty bearers and health system users as claims holders." That said, one could easily argue that to this day neither country—as in many places throughout the world—has fully invested in the health of its children, given ongoing inequalities in rates of poverty, malnutrition, education, and exploitation.[49] Despite a growing awareness of the rights of children and overall improvements in their basic living conditions and health care, persistent inequalities show us that, even in the twenty-first century, the central tension between protection and punishment of vulnerable and at-risk children remains sadly unresolved.

Puericulture and mental hygiene, which reached their heyday in early-twentieth-century Latin America, were cutting edge and progressive in their own time but ultimately failed to fulfill their vision of healthy children in a healthy society. Proponents of these eugenics-adjacent projects were locked into evolutionary and racialized frameworks. Moreover, health and welfare reforms aimed at children and families remained overshadowed and over-powered by individualist market-driven solutions and state policies that avoided fixing the longstanding structural inequalities and racism inherent in their societies.

Notes

1. Gregorio Aráoz Alfaro, *Dos conferencias en la Universidad de Tucumán, julio 1915* (Buenos Aires: Imprenta de Coni Hermanos, 1915), 58.
2. See Julia Rodriguez, *Civilizing Argentina: Science, Medicine, and the Modern State* (Chapel Hill: University of North Carolina Press, 2006), 112–20. See also Marcela M. Nari, *Politicas de maternidad y maternalismo político. Buenos Aires 1890-1940* (Buenos Aires: Editorial Biblos, 2004); Ann S. Blum, *Domestic Economies: Family, Work, and Welfare in Mexico City, 1884-1943* (Lincoln: University of Nebraska Press, 2010), 123–28.
3. See Donna Guy, "The Pan American Child Congress, 1916 to 1942: Pan Americanism, Child Reform, and the Welfare State," *Journal of Family History* 23, no. 3 (July 1998): 272–92.
4. For an overview of the history of medicine and psych fields in Latin America, see Diego Armus, ed., *Disease in the History of Latin America: From Malaria to AIDS* (Durham, NC: Duke University Press, 2003).
5. On the history of eugenics in Latin America after Stepan, see Natalia Milanesio, "Redefining Men's Sexuality, Resignifying Male Bodies: The Argentine Law of Anti-Venereal Prophylaxis, 1936," *Gender & History* 17, no. 2 (August 2005); Yolanda Eraso, "Biotypology, Endocrinology, and Sterilization: The Practice of Eugenics in the Treatment of Argentinean Women During the 1930s," *Bulletin of the History of Medicine* 81, no. 4 (2007); Andrés H. Reggiani, "Depopulation, Fascism, and Eugenics in 1930s Argentina," *Hispanic American Historical Review* 90, no. 2 (2010).
6. The Cuban physicians Eusebio Hernández and Domingo F. Ramos had coined a similar term, *homicultura*, in a 1911 book; see Nancy Leys Stepan, *"The Hour of Eugenics": Race, Gender, and Nation in Latin America* (Ithaca, NY: Cornell University Press, 1991), 76–77, 79–80.
7. Cited in Stepan, *"The Hour of Eugenics,"* 78.
8. On national science and the state in Argentina, see Rodriguez, *Civilizing Argentina*.
9. Stepan, *"The Hour of Eugenics,"* 78.
10. Genaro Sisto, "Prologo," in Enrique Feinmann, *Ciencia del niño* (Buenos Aires: Cabaut y Compañía, 1915), v.
11. Stepan, *"The Hour of Eugenics,"* 17.
12. Diego Armus, "Eugenics in Buenos Aires: Discourses, Practices, and Historiography," *História, Ciências, Saúde—Manguinhos* 23 (December 2016): 1–20.
13. Among the 606,174 women surveyed for the third census, native-born Argentines had the highest fertility rates, Italian women the largest families. The average

number of women per children was about 4.3 in 1914, with Spaniards having the lowest number at 3.7, Italians the highest at 4.9. Census officials were nonetheless concerned that birthrates were declining. For census figures on marriage rates, birth rates, and fertility, see *Tercer censo nacional*, 10:283–300.

14. Feinmann, *Ciencia del niño*, 5.

15. Marcela Nari, *Políticas de maternidad y maternalismo político. Buenos Aires (1890-1940)* (Buenos Aires: Biblos, 2005).

16. Anne-Emanuelle Birn, "Child Health in Latin America: Historiographic Perspectives and Challenges," *História, Ciências, Saúde—Manguinhos* 14, no. 3 (2007): 688.

17. Cecilia Grierson, the first woman physician in Argentina and a feminist activist, took on the health and well-being of women and children as her primary concern. Elvira Rawson de Dellepiane, a physician colleague, also at the Congress in 1910, urged better nutrition and medical care for poor children.

18. Victoria Mazzeo, *Mortalidad infantil en la ciudad de Buenos Aires, 1856-1986* (Buenos Aires: Centro Editor de América, 1993), 70–71.

19. Gregorio Aráoz Alfaro, *El libro de las madres. Manual práctico de higiene del niño* (1899; Buenos Aires: Cabaut y Cía, 1922), vii.

20. Aráoz Alfaro, *El libro de las madres*, x.

21. In 1936, Aráoz Alfaro wrote a new book, *Por nuestros niños y por las madres. Protección, higiene, y asistencia social* (Buenos Aires: Cabaut, 1936); the study was framed by eugenic notions of strength but also had chapters on protection of the child and children's rights.

22. Aráoz Alfaro, *El libro de las madres*, xi.

23. G. von Bunge, "Las fuentes de la degeneración," *APCCA* 12 (1913): 48; G. von Bunge, "De la impotencia creciente de las mujeres para amamantar a sus hijos," *RCPML* 2 (1916): 80–81.

24. M. A. Lancelotti, "Educación y delito," *Revista Penitenciaria* 2, no. 2 (1905): 34.

25. Donna Guy, "Parents Before the Tribunals: The Legal Construction of Patriarchy in Argentina," in *White Slavery and Mothers Alive and Dead: The Troubled Meeting of Sex, Gender, Public Health, and Progress in Latin America* (Lincoln: University of Nebraska Press, 2000), 175.

26. Luis Figueroa Alcorta, "Luis Agote: El sabio, el legislador, el escritor, el hombre," *Rev. Arg. Transf.* 15, no. 1 (1989): 8.

27. Donna Guy, "The Pan-American Child Congresses," in *White Slavery and Mothers Alive and Dead*, 49.

28. Anne-Emanuelle Birn, "The National-International Nexus in Public Health: Uruguay and the Circulation of Child Health and Welfare Policies, 1890–1940," *História, Ciências, Saúde* 13, no. 3 (2006).

29. C. Lucia and M. Valladares de Oliveira, "Psychoanalysis in Brazil During Vargas' Time," in *Psychoanalysis and Politics: Histories of Psychoanalysis Under Conditions of Restricted Political Freedom*, ed. Joy Damousi and Mariano Ben Plotkin (Oxford: Oxford University Press, 2012), 113–34.

30. Jerry Davila, *Diploma of Whiteness: Race and Social Policy in Brazil, 1917-1945* (Durham, NC: Duke University Press, 2003); Anadelia A. Romo, *Brazil's Living Museum: Race, Reform, and Tradition in Bahia* (Chapel Hill: University of North Carolina Press,

2010); Manuella Meyer, *Reasoning Against Madness: Psychiatry and the State in Rio de Janeiro, 1830-1944* (Boydell & Brewer, 2017).

31. Lucia and Valladares de Oliveira, "Psychoanalysis in Brazil," 115–116.

32. Davila, *Diploma of Whiteness*, 34–35.

33. Davila, *Diploma of Whiteness*, 38.

34. Brad Lange, "Importing Freud and Lamarck to the Tropics: Arthur Ramos and the Transformation of Brazilian Racial Thought, 1926–1939," *The Americas* 65, no. 1 (July 2008): 17.

35. Mariza Corrêa, *As ilusões da liberdade: A escola Nina Rodrigues e a antropologia no Brasil* (SciELO-Editora FIOCRUZ, 2013); Romo, *Brazil's Living Museum*; Mariana Ramos de Morais, "Race, Culture, and Religion: The Afro-Brazilian Congresses and Anthropology in 1930s Brazil," in *Bérose—Encyclopédie internationale des histoires de l'anthropologie* (Paris, 2020).

36. Romo, *Brazil's Living Museum*, 68.

37. Arthur Ramos, *A aculturaçao negra no Brasil* (São Paulo: Companhia Editora Nacional, 1942), 180.

38. "Dados da família (pais, irmãos, outros parentes . . .), ambiente familiar (condições materiais e psicológicas da habitação . . .), história obstétrica materna, desenvolvimento formação de hábitos (crescimento, saúde geral, alimentação, marcha e linguagem, ritmos fisiológicos, disciplina e vida na escola), fachada temperamental e caracterológica, funções psicológicas . . . exame médico, diagnóstico da personalidade." Arthur Ramos, *A criança problema: A higiene mental na escola primária* (Rio de Janeiro: Livraria Editoria da Cada do Estudante do Brasil, 1939), 11.

39. Ramos, *A criança problema*, 13.

40. Ramos described this psychometric approach as an "experimental period" that began with the introduction of Binet and Simon's intelligence tests in 1907. Ramos wrote that although French physicians and teachers had been concerned with the "glorious work of assistance and special teaching for the mentally weak" since the eighteenth century, Binet and Simon's tests offered a new diagnostic tool that gave scientists a way to quantify to different degrees what had previously been an undifferentiated clinical notion of "mental retardation." Yet Ramos did not celebrate these developments as a triumph of scientific reason. Instead, he wrote that they had produced an endless multiplication of intelligence and other psychometric tests that had led to a "testologization" of contemporary pedagogy.

41. Ramos, *A criança problema*, 13.

42. When Teixeira joined the IPE in 1931, he promptly sidelined conservative Catholics from his educational reforms and described the church's approach to education as dogmatic and superstitious. Teixeira thus became a lightning rod for Catholic conservatives who seized on his team's secular approach to education as evidence that they were communists who posed a threat to Brazil's social order.

43. Ramos, *A criança problema*, 8.

44. Ramos, *A criança problema*, 10.

45. Antonio Guimarães, "Africanism and Racial Democracy: The Correspondence Between Herskovits and Arthur Ramos (1935–1949)," *Estudios Interdisciplinarios de América Latina y el Caribe* 19, no.1 (2008).
46. Arthur Ramos, "The Negro in Brazil," *Journal of Negro Education* 10, no. 3 (1941): 522.
47. Ramos died shortly before this meeting, which he had convened. During this meeting, the invited race experts drafted UNESCO's 1950 Statement on Race and paid tribute to Ramos's career by describing him as a "scientific humanist" who held "deep sympathy with all the backwards people and oppressed races." For more on this meeting, see Sebastián Gil-Riaño, "Relocating Anti-Racist Science: The 1950 UNESCO Statement on Race and Economic Development in the Global South," *British Journal for the History of Science* 51, no. 2 (June 2018): 281–303.
48. Alicia Ely Yamin and Ariel Frisancho, "Human-Rights-Based Approaches to Health in Latin America," *Lancet* 385 (April 4, 2015): e26–30.
49. "Children's Rights: Argentina," Library of Congress, https://www.loc.gov/law/help/child-rights/argentina.

FOUR

Race and Sameness

On Ordering the Human and the Specificities of Us-ness and Other-ness

AMADE AOUATEF M'CHAREK

"TELL ME, THERE is no biological race, is there?"

You won't believe how often I get this question. The people posing it seem to be looking for reassurance. Looking for comfort that there is nothing in nature, nothing deep down in our bodies that supports the persistent racism in society. This question is understandable, but each time I start answering it, I stumble over my words: the question seems simple, yet it is full with ambiguities. Ambiguities that are to be found in the two words *biology* and *race*. Neither of these words can be reduced to one singular entity. Biology, for example, is not a matter of genes, bones, or hormones. And it is not a matter of adding up all these entities: genes, hormones, bones, skin color, and what have you. Biological differences are perhaps best viewed as configurations of scientific work, where theories, methods, and materials, such as samples (however biological), chemicals, and devices, are configured to present a "natural phenomenon."[1] This phenomenon is not less valid or less true, by contrast. The point is that it is irreducible to one thing. The power of its truth claim, its validity, is precisely the fact that it is this meticulous con-figuration. I cannot convince you of the existence of the melamine gene by making you stare at your skin. As Bruno Latour and many others have shown, we need a laboratory to do that. The gene is not simply in your body but exists in the connection between bodily material, books, journals, labs, samples, and data. This allows us to study it, act upon it, determine, for example, whether it is mutated and the cause of cancerous cells, etc. The gene is con-figured, even if it seems independent and self-contained. Are you noticing the detours I am taking? If what I have just said offers some space for maneuvering and for navigating the biological differently, it is the juxtaposition of biology and race that takes my breath away. This juxtaposition is precisely suffocating because it has a long and encrusted history.

Race had long been the prime workhorse in studies of human diversity.[2] Since the late nineteenth century, the prospect of finding the racial type has driven a feverish collection of data. In the slipstream of colonial projects and equipped with novel statistical methods, scientists started to measure length, skin color, head shape, hair structure, iris color, lip thickness, ear form, fingerprint, the shape and print of hand and foot, and so on and so forth. More details, better methods, and larger data sets would determine once and for all what the human racial types were, or so the story went. But as the data accumulated it became clear: race was an illusion. It could not be pinpointed on real existing human bodies.

But the very idea of race was also doing work in society. Assumed hierarchies, where the white man figured as the crowning glory of evolution, were mobilized to justify injustice. To justify colonial extractions, killings, slavery, humiliations . . . Some aspects of this violence have been in the open, others silenced.

In the aftermath of World War II, in 1950, UNESCO issued the Statement on Race, indicating that there is no biological basis for race. Throwing this antiracist stone surely got the pond of science and society rippling.[3] However, this did not end the preoccupation with biological race. And while genetic research has produced ample evidence for the nonexistence of race, race is definitely making a comeback. First, while most genetic research teaches that differences are probabilistic and cannot be pinpointed to specific groups of individuals, as these results start to circulate social categories are mobilized to do precisely that.[4] This conflation of statistical distributions with social categories, which happens in science and society, renders the fluid genetic results rigid. Second, given the persistence of social problems, for example, crime or poverty, and the enormous interest in the life sciences, it is assumed that life science research will finally provide us with the answers, contributing to the biologization of social categories.[5]

So how can I simply answer the question that biological race does not exist, when it is constantly produced, even if never simply as biology but always as a configuration of nature-culture? This then means that we have to "stay with the trouble" of race.[6] But it also means that we cannot assume we know what race is and that we need to examine how it is produced and what it is made to be in practice.

In this chapter, I analyze a Dutch homicide case that evoked racist violence against asylum seekers. This case could be analyzed in terms of its effects on policy and legislation, both in the areas of forensic policing (for example, allowing for the use of race as a biological category in DNA research) and migration policy, and it could help us elaborate the policy implications of leaving race behind.[7] Here I mobilize it to engage in a conversation on what going beyond our usual take on race means for our understanding race and its capacity to order the human. I show how in practice different configurations of race emerge and demonstrate their normative content, including their tendency to engender belonging or violence against others. My analyses show that race is not a clearly defined category but rather a mode of ordering diversity in practice, mobilizing intricate elements, such as the care for dairy cows.[8]

Tweeting

It is Monday morning, November 19, 2012. I woke up very early that morning and was listening to the radio when I heard the news. After twelve years of exhaustive criminal investigation, a suspect was found in the highly mediatized rape and murder case of Marianne Vaatstra. I could not believe my ears. In a population screening involving almost eight thousand men, the forensic examiners managed to find a match within one month. I knew that they were expecting at least some nine months of work before coming to a conclusion. It was even more surprising that they found a full match, because the screening was done in the context of familial searching. It was assumed that the suspect would not be among the participants but that a partial match between a participant and the profile of the perpetrator would indicate that the suspect is a relative of that particular participant. Earlier that morning, at 5:06 AM, the famous Dutch crime reporter Peter R. de Vries was the one to share the news via Twitter, stating: "Man gearresteerd. *Blanke* verdachte, *Fries*, woonde 2,5 km van plaats delict. 100 procent DNA-match! [Man arrested. White suspect. Frisian, lives 2.5 km away from crime scene. 100% DNA match!]."

This tweet indicates that I was not the only one who was surprised. The news was not merely the fact that the suspect was identified or that this was a result of a 100 percent DNA match but also the fact that the suspect was "white" and "Frisian." Following the media circus, I was indeed mesmerized by how the Frisian-ness of the suspect came to play a central role. Over the years in the Netherlands and beyond we have come to view the Marianne Vaatstra case as one that functioned to highlight xenophobia, racism, and violence vis-à-vis the migrant Other.[9] This changed radically once the suspect was found, giving way to a remarkable sense of consideration and care toward the community in the village where the suspect lived, toward his family, and even toward the suspect himself: "the poor guy who lived all those years with this secret." I was more than amazed by this response, especially in light of the racism that had always been attached to the case. This care for the community kept resonating in my head, and over the years it had translated into questions about the issue of *race and sameness*. I should add here that I have grown attentive to sameness because of the Dutch situation. Ever since the 1990s we have witnessed a growing xenophobia coupled with right-wing nationalism and a naturalization of Dutchness.[10] But the Marianne Vaatstra case in particular made me think about sameness in relation to race.

In this chapter I zoom in on this issue of sameness to tease out lessons about race. I wonder whether a focus on sameness rather than on difference could help us refine our analysis of race and how it is produced and sustained in practice. But also, what kind of ordering device does race become once attended to through the lens of sameness? Our collective habit of thinking race in relation to difference has contributed to the idea that differences are produced while sameness is given, a baseline. Indeed, an emphasis on difference seems to suggest that in the context of race, differences are political, while similarities, by contrast, are curiously apolitical. The production of sameness, resemblance, and equivalence has thus received little attention in critical analyses. Here I suggest switching focus from difference to sameness in order to better grasp the politics of different versions of racialization.

Over the years, I have produced different analyses of the Marianne Vaatstra case, because of my interest in forensic genetics and race.[11] It is a case that had provoked different forensic DNA legislations in the Netherland and one in which all possible investigative techniques, including forensic genetic technologies, have been tried out.[12] In this chapter I provide but a brief history of this case and of the forensic genetic technologies mobilized and focus on the family search through which the case was eventually solved. I will first situate my concern with race and sameness in a theoretical context, drawing on historical as well as philosophical work. I end with an analysis of how the care for the collective figured in the media and what it might teach us about race and sameness.

On Sameness and Race

Critical studies of race and genomics, and of race more generally, have for good reasons been concerned with a politics of difference.[13] Markers that have been mobilized to suggest innate differences between groups of people, from skulls to IQ, from genes to hormones, have received ample attention.[14] This work has helped us situate the work of producing differences in science and society as well as identify the very technologies that do that work. Importantly, this critical scholarship has convincingly shown that differences are not to be located in bodies but in the ways bodies and bodily markers are engaged and related to a variety of technologies as well as other (cultural) elements, including preestablished ideas about race. This has led me to conclude that race is a relational material semiotic object.[15] Yet, such an alertness to difference has sustained the idea that when it comes to race, differences are political while similarities are apolitical.[16] Although it is commonly acknowledged that race and racialization are based on configurations of differences *and* similarities,[17] similarities are usually taken for granted. This might in fact have deeper, more structural roots.

The historian and political scientist Siep Stuurman has written a seminal book, *The Invention of Humanity*, in which he shows how modernity and the modern state of justice are based on the sameness of humans as the norm and their equality before the law as its consequence.[18] He argues that historically, particularly during the eighteenth century and the emergence of what we now call the Enlightenment, there have been three crucial "modalities" that have helped invent this thing called humanity. First there is the acknowledgment of a common humanness, that is, that humans belong to the same species. The second modality is related to the anthropological turn through which cultural differences came to be understood as variations on a common theme, assuming a shared human culture. The third modality is a temporal regime that helped think civilization in terms of an evolutionary development, in the way that even if some peoples are not there yet, they are assumed to be undergoing similar development and will eventually arrive in modern times to come. Although the three modes that have helped establish the paradigm of humanity have been widely shared by different civilizations across the globe, so Stuurman argues, the coupling of equality and sameness has become pivotal in racial Europe during the Enlightenment. The dictum was: To become equal is to *become like those* who are already equal, that is, the European whites. Enlightenment thus became the obligatory point of passage for becoming equal. The crux of Stuurman's argument is that sameness had become a *normative baseline* in the modern equality paradigm. Stuurman's argument makes clear that becoming equal requires work and entails an ideological take on human relations, but since it has become the norm, its effect is that difference and deviance have typically come to attract attention, alarm, or dismay.

Rather than taking sameness for granted, I focus on what sameness is made to be, how it comes about and, also, to what effect. I am particularly interested in different versions of sameness and the different versions of race they bring about. In focusing on the politics of sameness, my aim is to specify different modes of racialization. I was encouraged by events related to the Marianne Vaatstra case but also by the work of Deleuze and Guattari. Stuurman's historical analysis resonates with a theoretical take on race and face developed by Deleuze and Guattari in their essay "Year Zero: Faciality," in which they argue that racism is fundamentally a device for doing sameness: "Racism never detects the particles of the other; it propagates waves of sameness until those who resist identification have been wiped out (or those who only allow themselves to be identified at a given degree of divergence). . . . There are only people who should be like us and whose crime it is not to be."[19]

In this essay Deleuze and Guattari focus on the politics of the face and develop the concept of facialization, facialization as the effect of what they call the "faciality machine." The faciality machine functions as a political device and creates faces by eroding diversity and reproducing sameness. While facialization operates though the abstract machine that makes sameness, those who do not move on the wave of sameness are bestialized and become killable, erasable. In Stuurman's sense, and relevant for the theme of this volume, they do not enter the realm of humanity. This version of sameness and race works from the center, from the position and perspective of, one could say, European whiteness. Philomena Essed and David Goldberg have termed this "cloning cultures," and Sarah Ahmed has analyzed this version of sameness through the politics of emotion.[20] I elaborate and specify race and sameness by introducing the notion of *us-ness* and by contrasting this to the production of sameness as a version of *other-ness*. The insights about facialization and racialization of sameness, as well as the radical othering of the nonsame, are strikingly relevant for the forensic setting. As we will see, the Vaatstra case will help illustrate that.

"A Non-Dutch Manner of Killing": On Sameness and Phenotypic Othering

Marianne Vaatstra was sixteen years old when she went out partying with friends on the night of April 30, 1999, but she would never return home. The next day, her body was found in a meadow in the village of Veenklooster, not far from the village where she lived. The villages are in the rural area of Friesland, a northern province of the Netherlands. Upon finding her body in the meadow, the coroner who had examined the crime scene and the manner of death concluded that she had been raped and then her throat had been slit. According to him, cutting a victim's throat was not a typically Dutch method of murder. The violent death of Marianne Vaatstra moved the local population in this rural area and beyond. And because it was but the latest of a series of violent crimes against young girls,[21] it led to large silent marches, heated debates about social safety, and much media coverage.[22] In addition, since the meadow where the victim was found was near a center for asylum seekers, the societal response to Marianne's death slowly changed into racist violence toward the inhabitants of the center, who were mostly from the Middle East. Suspicion and accusations were immediately directed toward what I have come to call the *phenotypic other*, the inhabitants of the center.[23] Local villagers threatened to tear down the center and threw stones through its windows, and in October 2000 a resident of the center was stabbed with a knife by two young men as he was going home from the train station. Upon this, the asylum seekers organized a protest, describing the ongoing violence and hatred they had been confronted with since the Vaatstra murder.[24] The racism and violence against refugees and volunteers was more widespread in this rural area of Friesland throughout the years of investigation[25] and grew beyond the rural area to become a national sentiment, especially after a vocal right-wing member of parliament, the late Pim Fortuyn, wrote a column in a weekly magazine in which he labeled the crime "a non-Dutch manner of killing."[26] In the media, the asylum seekers' center was described as "a hotbed of criminal activities."[27] The suspicion of the local population was further fueled by the late crime reporter Peter R. de Vries, who in 1999 claimed that the main suspects were two former center inhabitants. He broadcasted their pictures on his TV program. However, both assumed suspects, Ali Hassan and Mohammad Akbari, could be excluded based on DNA analysis. Yet the idea that this was an "un-Dutch" manner of death continued contributing to the phenotypic othering of people housed in the center and beyond. Marianne Vaatstra died because the perpetrator had slit her throat with a knife. This cutting of the throat became a central device of phenotypic othering and of race making. The scenario the locals had in mind was sketched as follows: "The perpetrator was well prepared. Like a predator looking for prey, he was waiting to attack Marianne from the bushes. After that, he killed her by slitting her throat. Given this *modus operandus*, the suspect cannot but be an inhabitant of the asylum seekers center."[28] Let me give another example in which a similar scenario was sketched in an opinion piece written in a reputable Belgian journal, concerning a murder that took place in Belgium, at Brussels's central train station.[29] "On video screens you can see them, like predators along the walls of the central station, waiting, alert and on the watch to find easy prey in the passing herds of passengers for them to kill. . . . The unlucky one will not stand a chance. The predators have knives. In childhood they have learned, during the annual sacrifice, how to slit the throat of warm-blooded herd animals."[30]

This quote helps us understand the framing of the knife. Relating knives to a religious custom indicates that the perpetrator is a non-Western, non-Dutch Other and, more specifically, a *Muslim* man. He is inclined to violence and killing with knives. The cutting of a throat, an animal-like mode of relating, thus contributes to the bestialization of the other. To put this in the words of Guattari and Deleuze, the other who resists or cannot be like us becomes killable.[31] Importantly here is that suspicion is typically directed not toward any specific individual but toward a whole group. This group is phenotypically othered through markers ascribed to the perpetrator, in this case, a Muslim male.

This process of othering comes with a specific notion of sameness. This sameness does not only racialize; it also leaves no space for differentiation. A group is lumped together. An individual cannot but stand for the whole group. This version of sameness indeed reduces a group of people to one specific quality, in this case violence. While the version of sameness that is connected to otherness produces a homogenous racialized group, in what follows we will see that sameness can also open up the category to various differentiations. We will consider sameness not as related to other-ness but to *us-ness*.

A Farmer from Here: On Sameness and Us

Although the quality of the biological traces left by the unknown suspect were praised by forensic experts, and although the investigating police had mobilized every possible resource to find leads, over more than a decade of almost constant forensic policing, the question of who he was remained on the table.[32] Suspicion was directed toward the inhabitants of the asylum seekers' center. The grim atmosphere and the repeated violence against the center and its residents had prompted the police to attempt forensic genetic research into the biogeographic ancestry of the unknown suspect, using the Y chromosome. At that time, June 2000, such tests were explicitly prohibited by the Dutch law.[33] The results, officially unusable because illegal, were widely publicized. They indicated that, in contrast to the commonly held idea among the public that the suspect was from the Middle East (as were most of the inhabitants of the center), DNA tests suggested a suspect of northwestern European, probably Dutch, ancestry and descent. Although these results did make people think, or blink, they did not remove all suspicion against the phenotypically othered.

This suspicion would dissolve overnight when a DNA match was found. Almost thirteen years after Marianne's death, on November 19, 2012, the suspect was identified using newly introduced DNA technology, namely, familial searching. Familial searching is routinely conducted through a search in DNA databanks where partial matches might suggest that the suspect is a relative of the person in the databank. Since DNA databank searches had not

FIGURE 4.1 Explaining the DNA research, indicating the range of the dragnet, a radius of 15 km.
Source: This diagram was used in the media and produced by the Dutch News Agency ANP.

found any exact matches, it was decided to look for relatives of the suspect in the community.

After a small study into the mobility of people in that particular area of Friesland, and while finding that the inhabitants of those villages tend to stay put, 8,080 men were invited to donate DNA for the search for a relative of the suspect. In the end 7,581 men participated in the study. Working with

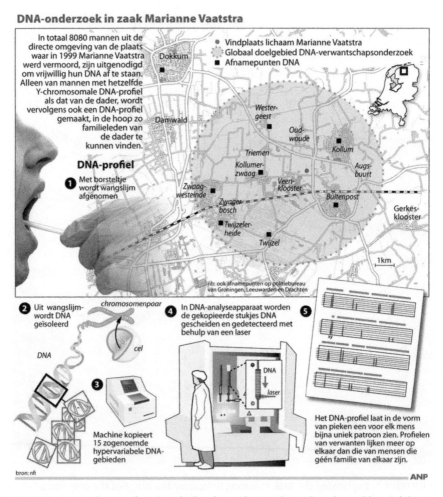

FIGURE 4.2 Brochure informing the local population about the why and how of the DNA familial searching to take place (September 2012).
Source: Openbaar Ministerie Nederland / Public Prosecution Service Netherlands.

such a large number of samples, it was expected that it would take more than nine months to produce all the DNA profiles and process the data. However, the forensic researchers were lucky, and in the first batch two samples were found in which the Y-chromosomal DNA profiles matched that of the DNA found at the crime scene.[34] This helped focus the rest of the analysis via genealogical research based on family names.[35] Within a month of genealogical mapping and targeting relevant samples from other batches, a full DNA match was found, leading the investigators to the suspect, the Frisian farmer Jasper S. This find came as a major surprise. One of the forensic investigators had put it as follows: "Jasper was just about the last person on whose door you would be knocking, with his farm and little family and all. Because you tend to presuppose a usual criminal." Upon hearing this news, the community in the Frisian village was in dismay. "This cannot be true. It must have been a mistake. If this is true, I lose my faith in humanity," one person reacted. A woman responded: "When everybody here believed that the perpetrator was an inhabitant of the asylum seeker center, at least that gave us some peace. You don't want it to be a father from here."[36]

What was remarkable about the months of preparation for the dragnet and familial searching was the sense of community that emerged. Everybody wanted to help solve the crime. But to make all the men in the area participate a carefully designed campaign was developed. For example, the primary leaflet through which the local population was informed about DNA familial searching did not only explain to readers what the research was about and how the DNA will be handled and then destroyed; it also included statements from local villagers expressing their hope that this murder case would finally be solved and that every man will show up to donate DNA. The cover was cleverly designed. It displays a typical Dutch landscape (a dyke), a typical Dutch activity (cycling), and a typical Dutch scenario (a family enjoying leisure time), as if to say that contributing to this DNA dragnet is a normal, Dutch thing to do. The family is connected and held together by DNA, the DNA double helix. In addition, it displays a gesture of care and safety. The young girl is cycling next to her mother. They seem to be cycling behind the father of this family. During one of the presentations he gave in the aftermath of the case, one of the police investigators explained that it was important that the father was cycling in front of them and not behind them, because that would diminish the feeling of safety. A man cycling behind them might be read as not belonging to the family and as chasing and threatening the mother and the child.[37] DNA research is contributing to a sense of safety and community, or such is the message that is projected.

I was even more struck by how this sense of community became stronger once the identity of the suspect was revealed. It became not the community of violence and aggression vis-à-vis the (phenotypic) other that had been articulated in the years before but a community of *care* vis-à-vis "us" and those who belong to us. Care was not merely directed toward the family of the victim, who lived in an adjacent village, but toward the family of the suspect and even the suspect himself. In the turmoil of media attention, this sense of *us-ness* never let go of me again, and it provoked my thinking about race and sameness. In what follows I unravel this version of sameness and its relation to race. I will do so based on a selection of responses in the media.[38]

The very first tweet disclosing the identity of the suspect, sent around 5 AM by the crime reporter Peter R. de Vries, already contains the ingredients of my analysis and illustrates how race came to matter. The suspect was "Frisian," "white," a "local," and a "farmer." The link between Frisian whiteness and his traditional occupation alerted me to race. While any of these markers by themselves do not necessarily enact race, together they become a potent technology of racialization.[39] Upon hearing the news, the father of Marianne Vaatstra mused: "So it is someone from our midst [*van ons*], a *farmer, a white man.*" A fellow villager of the suspect was quoted saying: " 'Well, DNA doesn't lie,' mumbles Nycklo de Vries (19). But it remains hard to believe. He knew the arrested man. Just like everyone else here in Oudwoude. A very normal, social man. *With a lot of land and a livestock farm.* Married, a son and daughter in her twenties."[40] In another account of what the villagers were going through, we read: "Yesterday people in Oudwoude responded with dismay to the arrest of the *friendly fellow townsman,* who was always in for a chat with everyone. His family was quickly relocated to a quiet area. *His nearly 100 dairy cows are being looked after.*"[41]

It is in fact remarkable that the suspect is addressed as a white man. Since whiteness is the norm, it hardly ever gets articulated. However, in this case, despite the huge investment in familial searching and thus the possibility that the suspect would be related to the local population, the whiteness of the suspect still sparked disbelief and surprise. It thus marked the persistent suspicion that was placed on migrants and refugees, a group that was phenotypically othered. But whiteness was also related to the occupation of the suspect, a farmer who takes care of his dairy cows and owns lots of land. On the evening of the rape and murder of Marianne Vaatstra, the suspect and his father went out at 11 PM to milk the cows, we learn from his statement.[42] He is one of us, "from our midst," as Vaatstra's father said. This coupling between whiteness, land, and relation to the land, as well as activity (caring for his cows) or occupation (being a farmer) is a classic way of racializing a community.[43] However, though the suspect was made a member of a community of us-ness through his skin color, occupation, and relation to land, the accounts described here also make space for him as an individual. He is somebody everybody knows, he is kind, normal, a social man, and, as we will see, has a friendly word for everybody. Also, in a long and calm interview with his lawyer the viewer is presented a portrait not of a monster or beast but of a torn person, full of remorse and shame over his uncontrolled behavior on that night thirteen years before.[44] In an interview about this TV appearance, his lawyer Jan Vlug said: "There I have tried to portray Jasper as a human being, as a nice man who had done something horrible."[45] This room for individuality, I want to suggest, is a key element of this version of sameness in relation to us-ness. Whereas the coupling between sameness and otherness takes away all individuality and reduces individuals to a homogenous and othered group, by contrast *sameness in relation to us-ness makes space for individuality*. In this case, the suspect-ness of the suspect came as a surprise because he was so normal and kind, which explains why this room for individuality in other cases leads to the proverbial "rotten apples" that do not spoil the identity of the whole group.[46]

Sameness in relation to us-ness makes room for individuality, but it does more. The suspect was referred to as a family man; he is married with a son and a daughter in her twenties. The family figured prominently in the care articulated by the local villagers. The municipality organized a meeting for the villagers at which the mayor spoke: "About 300 residents showed up in the village hall. Bilker [the mayor] speaks after a 'modest and heartwarming' meeting. The village will not let the family down, he says. The mayor knows the parents of the arrested man. 'They are overloaded with cards, phone calls and best wishes expressing support. That gives a good feeling.'"[47]

This excerpt makes clear that the family aspect of sameness is not only the fact that the suspect has children but that he himself is a child of parents who are also part of the same community. The mayor of the village of Oudwoude continued:

Everyone knows the parents; they are very well known in the village. Imagine: you lead a very normal life and then suddenly something like this happens. It was my pleasure to convey the commonly shared feeling among the inhabitants of Oudwoude. The feeling of: "You belong here, you belong." [*Jullie horen hier, jullie horen erbij.*] The parents were very happy with that. They responded very emotionally, in tears. They are doing reasonably well under the circumstances. . . . I did expect that something like "we stand by and around the family" [*we staan om de familie heen*] would arise, but I am pleasantly surprised that it is so strong.[48]

It took me some time to understand how this care for the *parents* of the suspect was relevant to my analysis and to see that it signals a particular family relation. Here the suspect is not merely a family man with his own household and children.[49] Crucially, he is addressed as a child, the child of someone. By caring for the parents, the suspect becomes a child. This obviously evokes a sense of innocence; even if the child is a forty-five-year-old man, he is still addressed as the object of care and concern for his parents. In addition, the attention to the parents puts the suspect in a genealogical relation, a relation of kinship. In this way we come to realize that not only does the suspect have a family and children of his own who deserve care and attention but that he has parents, and probably grandparents, and thus a history in that place. The continuation of kinship produces a long durée and a historical connection to the place, to Friesland and the village of Oudwoude. "You belong *here*," says the mayor. A version of "autochtony" and "nativism."[50]

In fact, this belonging to the place was the very reason that this large-scale, dragnet familial searching could take place. It was determined that in this region of the Netherlands there was not much mobility and that people tend to stay put. Moreover, after finding the two Y-chromosomal DNA matches, genealogical research was crucial. The Dutch Central Bureau for Genealogy was approached for help in searching their database and constructing family trees. The two Y-chromosomal profiles represented two families with one common ancestor, a man called Jasper Jans, who was an innkeeper in 1748 and lived in the nearby village of Westergeest. The investigating team then worked their way from this shared ancestor back into the present day looking for relatives and determining not just their Y-chromosomal profile but also their autosomal DNA profiles.[51] And on

November 14, 2012, a full match could be reported by the forensic geneticists to the investigating police team.[52]

Based on this account, I want to suggest that the coupling of sameness
and *us-ness* entails three specific elements: the individual, the family, and
the place. In the case I discuss here, sameness racialized the community
through a mobilization of markers such as whiteness, tradition, and occupation, as well as rootedness, a nativism of sorts. While this version of sameness was racialized, it still allowed for individuality, humanness, or goodness and thus for space within the community.

Staying with the Trouble: On Sameness, Race, and Ordering the Human

In this chapter we have encountered two different versions of sameness that
have produced different versions of race. First, sameness was related to *otherness* and the phenotypic othering of the unknown suspect. Therein, the
coupling between sameness and othering produces a racialized category, one
that subsumes differences and lumps of people together by reducing them
to one or a small number of markers that are deemed relevant. Here it was
crime and Muslim background. In the second part of this chapter we have
identified another version of sameness; sameness was there coupled to *us-
ness*. This configuration has also produced a racialized category, but one that
leaves space for differences within the group. While belonging to a racialized category, there remained space for *individuality* (he is a normal and social
guy), for *family* (he is a family man and the child of a well-known family),
and for tradition and *belonging to the place* (he is a farmer from here), all contributing to the creation of a sense of the normality of the suspect.

Related to this we can conclude that there are different machineries of
sameness at work. There is a version of sameness that came along with the
familial searching technology and produced a seemingly unmarked collective of "us." But upon looking closely it actually produced a white, Dutch,
farming community. This version of sameness did not readily translate into
individuals, immuring or firmly prescribing what they are. It did not reduce
all individuals within this collective to one characteristic or criminal behavior or keep mobilizing additional "evidence" to understand or support

"their" behavior. It was *not generative* of suspicion against all of them by lumping Dutch men together and assuming that they have a tendency for crime or a taste for raping and murdering young girls. Jasper S. was and is viewed as a singular case and his crime as that of one specific individual, with a specific biography.

By contrast, the version of sameness based on phenotypic othering produced the excludables. This version was much more *virulent* and *generative*: the "phenotypes" translated into individuals who allegedly belonged to that group, and it kept on mobilizing support (cultural elements) as evidence for the link between the homicide and the phenotypically othered.[53] This virulent version of sameness was generative of racist violence against individuals that belonged to the group of excludables, leading to the eventual shutting down of the asylum seekers' center and contributing to the emerging anti-Muslim racism fueled by right-wing politicians such as Geert Wilder.[54] It is remarkable that while the violence was directed toward the asylum seekers, their criminalization and unjust suspicion still looms in political and public debates about migration and the criminalization of migrants in the Netherlands.[55]

In her *Staying with the Trouble*, Donna Haraway explains this as follows:

> In urgent times, many of us are tempted to address trouble in terms of making an imagined future safe, of stopping something from happening that looms in the future, of clearing away the present and the past in order to make futures for coming generations. Staying with the trouble does not require such a relationship to times called the future. In fact, staying with the trouble requires learning to be truly present.[56]

Being truly present is precisely about attending to troubling things in practices, about making space for their various manifestations and politics. Given its tainted histories, *race* is par excellence a trouble.[57] Race as a word and concept should make us nervous. Making ourselves nervous about race could lead to the conclusion that since it has done so much harm, we need to get rid of it as quickly as possible. My purpose here was to convince the reader that we do precisely the opposite. While being nervous about race should indeed make us wary of mobilizing race as a social classification, even if it is used as self-identification, a practice common to, for example, the U.S. context but fairly unusual in continental Europe and other parts of the world.

For as Ian Hacking has taught us, categories do not represent but "make up people," thus contributing to ordering the human and to the reification of race.[58] At the same time, being nervous about race should invite us to keep it within the view of our studies and on the table of our conversations. Although race has often been declared dead,[59] as if race has done its work and we are done analyzing its manifestations, the world around us keeps manifesting its violent politics. We might thus rather embrace Haraway's suggestion to stay with the trouble, to critically follow how race keeps manifesting itself in practices and wonder how it orders humans in sometimes unexpected ways, what political work it does, and how its manifestation here becomes something else there.

Notes

I am grateful to the editors Eram Alam, Dorothy Roberts, and Natalie Shibley for their generosity and patience and for inviting me to be part of this beautiful volume. This chapter has benefited from Eram's feedback and suggestions to an earlier draft, for which I am very grateful. Also, since a version of this chapter has been published as a paper in *Critical Migration Studies*, I would like to thank the editors of Beyond Race Commentaries, Sawitri Saharso and Tabea Scharrer, for their guidance and help. I am indebted to Peter de Knijff (head of the Forensic Laboratory for DNA Research, Leiden) for collegiality and friendship and for over the years keeping me up to date with developments in the field while critically engaging with my work. The analyses presented here are the result of years of collaborations with members of the RaceFaceID team (https://race-face-id .eu/), whom I thank for thinking together hard about this wild object called race. I thank Bobby Witte for assisting with the collection of some of this material. Finally, I thank the European Research Council for supporting my research through an ERC Consolidator Grant (FP7-617451-RaceFaceID-Race Matter: On the Absent Presence of Race in Forensic Identification).

1. See, e.g., Amade M'charek, *The Human Genome Diversity Project: An Ethnography of Scientific Practice* (Cambridge: Cambridge University Press, 2005).

2. George W. Stockings, *Race, Culture, and Evolution: Essays in the History of Anthropology* (Chicago: University of Chicago Press, 1982).

3. See Perrin Selcer, "Beyond the Cephalic Index," *Current Anthropology* 53, no. 5 (2012): 173–84.

4. See, e.g., Amade M'charek, "Tentacular Faces: Race and the Return of the Phenotype in Forensic Identification," *American Anthropologist* 122, no. 2 (2020): 369–80.

5. See, e.g., Troy Duster, *Backdoor to Eugenics* (London: Routledge, 2003); Katya Gibel Azoulay, "Reflections on Race and the Biologization of Difference," *Patterns of Prejudice* 40, no. 4–5 (2006): 353–79.

6. Donna Haraway, *Staying with the Trouble: Making Kin in the Chthulucene* (Durham, NC: Duke University Press, 2016).

7. Amade M'charek, "Silent Witness, Articulate Collective: DNA Evidence and the Inference of Visible Traits," *Bioethics* 22, no. 9 (2008): 519–28; Amade M'charek, Victor Toom, and Lisette Jong, "The Trouble with Race in Forensic Identification," *Science, Technology, and Human Values* 45, no. 5 (2020): 804–28; Martijn de Koning, "Een Nederlander Snijdt Geen Keel Door," *Volkskrant*, November 24, 2012; Sawirtri Saharso and Tabea Scharrer, "Beyond Race?," *Comparative Migration Studies* 10, no. 1 (2022).

8. Michel Foucault, *"Society Must Be Defended"* (1970; London: Penguin, 2004).

9. E.g., de Koning, "Een Nederlander Snijdt Geen Keel Door"; Lisette Jong and Amade M'charek, "The High-Profile Case as 'Fire Object': Following the Marianne Vaatstra Murder Case Through the Media," *Crime, Media, Culture* 14, no. 3 (2018): 347–63; M'charek, Toom, and Jong, "The Trouble with Race in Forensic Identification."

10. Peter Geschiere, *The Perils of Belonging: Autochthony, Citizenship, and Exclusion in Africa and Europe* (Chicago: University of Chicago Press, 2009); Paul Mepschen, Jan Willem Duyvendak, and Evelien H. Tonkens, "Sexual Politics, Orientalism, and Multicultural Citizenship in the Netherlands," *Sociology* 44, no. 5 (2010): 962–79; Amade M'charek, "Fragile Differences, Relational Effects: Stories About the Materiality of Race and Sex," *European Journal of Women's Studies* 17, no. 4 (2010): 307–22; Rogier van Reekum, *Out of Character: Debating Dutchness, Narrating Citizenship* (Amsterdam: University of Amsterdam, 2014).

11. See, e.g. Jong and M'charek, "The High-Profile Case as 'Fire Object'"; Amade M'charek, "Performative Circulations: On Flows and Stops in Forensic DNA Practices," *Tecnoscienza: Italian Journal of Science & Technology Studies* 7, no. 2 (2017): 9–34; M'charek, Toom, and Jong, "The Trouble with Race in Forensic Identification."

12. See M'charek, Toom, and Jong, "The Trouble with Race in Forensic Identification."

13. See, e.g., Steven Epstein, *Inclusion: The Politics of Difference in Medical Research* (Chicago: University of Chicago Press, 2008); Ian Whitmarsh and David S. Jones, *What's the Use of Race? Modern Governance and the Biology of Difference* (Cambridge, MA: MIT Press, 2010).

14. See, e.g. Stockings, *Race, Culture, and Evolution*; Richard C. Lewontin, Steven Rose, and Leon J. Kamin, *Not in Our Genes: Biology, Ideology, and Human Nature* (New York: Pantheon, 1984); Joan H. Fujimura and Ramya Rajagopalan, "Different Differences: The Use of 'Genetic Ancestry' Versus Race in Biomedical Human Genetic Research," *Social Studies of Science* 41, no.1 (2011): 5–30; Dorothy Roberts, *Fatal Invention: How Science, Politics, and Big Business Re-create Race in the Twenty-First Century* (New York: New Press, 2012); Katharina Schramm, David Skinner, and Richard Rottenburg, eds., *Identity Politics After DNA: Re/creating Categories of Difference and Belonging* (Oxford: Berghahn, 2012); Rebecca Jordan-Young and Katrina Karkazis, *Testosterone: An Unauthorized Biography* (Cambridge, MA: Harvard University Press, 2019).

15. Amade M'charek, "Beyond Fact or Fiction: On the Materiality of Race in Practice," *Cultural Anthropology* 28, no. 3 (2013): 420–42.

16. An exception is a master's student thesis by Roos Metselaar, in which she analyses donor matching for reproductive purposes in terms of doing resemblance, which mostly is about doing race. The thesis is rich and does a lot; for example, she shows beautifully how resemblance is about the proximity with the sperm donor, which is managed against the distance making him invisible and the family normal. See Roos Metselaar, " 'Not Too Different': Doing Resemblance, Enacting Boundaries in Sperm Donor Matching Practices," Thesis, Research Master Social Sciences, University of Amsterdam, 2023.

17. See, e.g. Epstein, *Inclusion*; Amade M'charek, "Technologies of Population: Forensic DNA Testing Practices and the Making of Differences and Similarities," *Configurations* 8, no. 1 (2000): 121–58.

18. Siep Stuurman, *The Invention of Humanity: Equality and Cultural Difference in World History* (Cambridge, MA: Harvard University Press, 2017).

19. Gilles Deleuze and Félix Guattari, *A Thousand Plateaus: Capitalism and Schizophrenia* (Minneapolis: University of Minnesota Press, 1987).

20. Philomena Essed and David Goldberg, "Cloning Cultures: The Social Injustices of Sameness," *Ethnic and Racial Studies* 25, no. 6 (2002): 1066–82; Sara Ahmed, *The Cultural Politics of Emotion* (Edinburgh: Edinburgh University Press, 2004).

21. Jong and M'charek, "The High-Profile Case as 'Fire Object.' "

22. The Vaatstra murder case became the most mediatized case in the Netherlands; Jong and M'charek, "The High-Profile Case as 'Fire Object.' " Also, because of the exhaustive forensic technologies that were applied over the years, it became an internationally famous case; see M'charek, Toom, and Jong, "The Trouble with Race in Forensic Identification."

23. See, e.g., Amade M'charek, Katharina Schramm, and David Skinner, "Topologies of Race: Doing Territory, Population, and Identity in Europe," *Science, Technologies, and Human Values* 39, no. 4 (2014): 468–87; M'charek, Toom, and Jong, "The Trouble with Race in Forensic Identification."

24. M'charek, Toom, and Jong, "The Trouble with Race in Forensic Identification."

25. See the Dutch journal *Trouw*, December 8, 2012.

26. See Pim Fortuyn, "Kollumerstront," *Elsevier*, https://www.pimfortuyn.com/pim-fortuyn/archief-columns/165-kollumerstront; de Koning, "Een Nederlander Snijdt Geen Keel Door."

27. For a thorough media analysis and the racism toward the inhabitants of the asylum seekers' center, see Jong and M'charek, "The High-Profile Case as 'Fire Object.' "

28. This so-called *modus operandus* was articulated by the crime reporter Peter R. de Vries in a show dedicated to the Marianne Vaatstra case. To be sure, in that moment in time de Vries did not support this scenario but was merely reporting on a dominant view among the local population before disclosing the latest scenario put forward by the investigating police, namely, that Marianne was not killed by the slitting of her throat but first through strangulation with her bra, after which the suspect cut her throat. Given this method of murder, it is more likely that the suspect was to be found among/through the local population. In this show, broadcast on May 20, 2012, de Vries shifted position and was by then collaborating with the investigative team in a concerted effort to help persuade

the local population into participating in DNA familial searching research by donating DNA.

29. See for an analysis of this case, M'charek, "Silent Witness."

30. Paul Belien, *De Standaard*, April 25, 2006.

31. See also Michel Foucault, *Society Must Be Defended* (London: Penguin, 2004).

32. See, for a more detailed course of events, M'charek, Toom, and Jong, "The Trouble with Race in Forensic Identification." In this case, every novel forensic genetic technology, including DNA phenotyping and familial searching based on a population screening, was put to use. To do that in the Dutch context, national DNA legislation had to be revised.

33. See Peter de Knijff, *Meehuilen met de Wolven*, inaugural lecture (Leiden: Leiden University, 2006).

34. To be sure, a Y-chromosomal match does not imply a full match. It only indicates that these individuals are somehow related through their paternal lineage, which is not uncommon in a rural and traditional community.

35. Lex Meulenbroek and Paul Poley, *Kroongetuige DNA: Onzichtbaar Spoor in Spraakmakende Zaken* (Amsterdam: Uitgeverij de Bezige Bij, 2014).

36. In the Dutch newspaper *Trouw*, November 21, 2012.

37. Jelle Tsjalsma and Ron Rintjema, talk delivered at the "Publieksmiddag Wijzer over DNA: Op zoek naar de 'dader' met behulp van bevolkingsscreening," Rijksmuseum Boerhaave Leiden, November 4, 2018.

38. For a thorough media analysis of the case in the years before the suspect was identified, see Jong and M'charek, "The High-Profile Case as 'Fire Object.'"

39. Moreover, the racialization of the Frisian identity has a vested history in Dutch physical anthropology; see Rob van Ginkel, "Antropologie van Nederland," *Sociologische Gids* 42, no. 1 (1995): 7–59.

40. *Trouw*, November 20, 2012, emphasis added.

41. *Dagblad van het Noorden*, November 20, 2012, emphasis added.

42. See, e.g., the news website of Dutch national television: https://nos.nl/artikel /489559-jasper-s-doet-huilend-z-n-verhaal.html.

43. Other well-known examples of groups that have historically been racialized and classified through occupation are the Roma people and the Jewish people; see Mihai Surdu, *Those Who Count: Expert Practices of Roma Classification* (Budapest: Central European University Press, 2016).

44. See https://www.youtube.com/watch?v=HpoU5e8EuTI.

45. Meulenbroek and Poley, *Kroongetuige DNA*, 452.

46. And it is also the mechanism through which right-wing terrorism immediately leads to a psychologization of the suspects (think of Andres Breivik) rather than the default mobilization of culture, background, or religion as the explanation in cases of Muslim terrorism. This holds, of course, for crime in general.

47. *Trouw*, November 21, 2012.

48. *Dagblad van het Noorden*, November 22, 2012.

49. On the crucial and complicated politics of innocence, see Miriam Ticktin, "A World Without Innocence," *American Ethnologist* 44, no. 4 (2017): 577–90.

50. Geschiere, *Perils of Belonging*.

51. To be sure, while the police did have all the identifying information about those who participated in the DNA research, the Forensic Institute only received the registration numbers that accompanied the swabs. So, this part of the process required a fair amount of communication back and forth between the investigative team and the forensic geneticists.

52. Meulenbroek and Poley, *Kroongetuige DNA*, 446.

53. It is important to observe that sameness in forensics is an important technology, or *operator*, through which the profile of the suspect is shaped. Racializing the category of sameness and therewith the profile contributes to its value for criminal investigation. The category of *white* in the Dutch context was not of much value to solving the case, as it included too many individuals.

54. See, e.g., de Koning, "Een Nederlander Snijdt Geen Keel Door."

55. See Irene van der Linde, "Volledige Transparantie over Assielzoekers Móet," *De Groene Amsterdammer* 21 (May 22, 2019).

56. Haraway, *Staying with the Trouble*, 1.

57. See, e.g., Stockings, *Race, Culture, and Evolution*; M'charek, "Tentacular Faces."

58. Ian Hacking, "Making Up People," in *The Science Study Reader*, ed. M. Biagioli (New York: Routledge, 1999), 161–72.

59. See, e.g., David Goldberg, *Are We All Postracial Yet?* (Cambridge: Polity, 2015).

FIVE

The Racial Calculus

Security and Policy During the COVID-19 Global Pandemic

DENISE FERREIRA DA SILVA

> If slavery persists as an issue in the political life of black America, it is not because of an antiquarian obsession with bygone days or the burden of a too-long memory, but because black lives are still imperiled and devalued by a racial calculus and a political arithmetic that were entrenched centuries ago. This is the afterlife of slavery—skewed life chances, limited access to health and education, premature death, incarceration, and impoverishment. I, too, am the afterlife of slavery.
>
> —SAIDIYA HARTMAN, *LOSE YOUR MOTHER: A JOURNEY ALONG*
> *THE ATLANTIC SLAVE ROUTE*

EMERGENCY RESPONSES TO public health crises result from—and lead to—decisions by state authorities that affect transportation, education, housing, and other aspects of the everyday existence of its population.[1] Insofar as they are proposed as means for managing the spread of a disease, emergency health measures are biopolitical tools and, as such, provide the opportunity for mapping a given political conjuncture. In this chapter, the emergency responses to COVID-19 provide an avenue to examine how the colonial, racial, cisheteropatriarchal matrix operates in the capitalist political architecture, that is, how governments today enact their role as embodiments of the liberal state. Soon after the mantra "we are all in it together" resonated in the airwaves across the planet, we quickly learned that if indeed this was the case, we have never been *in* it in the same way:

the new disease was just as racially mapped as the global context on which it was rapidly spreading. Put differently, even before the numbers showing that nonwhite persons in the global north and the populations of the former colonies of the global south would be more dramatically affected by the new disease, everyone expected that that would be so.[2] Thus, precisely because there is nothing unexpected or unprecedented regarding the racial distribution of Sars-CoV-2 infections, it offers an occasion for moving beyond the explanatory mode of approaching racial subjugation with either conventional (sociological) or critical (sociohistorical, historical materialist, or poststructuralist) approaches that consistently treat the racial as a datum (racial difference as a natural trait that unduly impacts social processes) or residuum (racial classification or racial strife as a leftover of premodern times) and matter of ideology or belief but not as truth.

When tracking the emergency responses to the COVID-19 global pandemic, I explore the operations of the racial in the abstract tools of models, maps, and curves upon which governments devised their biopolitical strategies. This analysis reflects an approach that considers racial subjugation—as well as cisheteropatriarchal subjugation—as resulting from mechanisms of truth embedded within the liberal political architecture, in the interplay of the juridic, the ethic, and economic moments. Beyond showing that racial differentials do exist, one can also ask how racial differentials come to be; in other words, instead of proving once again that, one can try and expose how raciality played out in the pandemic. Of interest here, to be more precise, is how racial differentials are produced and reproduced in public policy. To approach racial subjugation in this manner requires an analysis of how scientific constructions of the Human and their related modes of formalization—calculation and categorization—are at work in our present post-Enlightenment governance regimes.

In what follows, I discuss how raciality accounts for the afterlife of slavery—that is, how total violence and generalized indifference have marked the existence of Black and Indigenous people and populations across the globe. This, I find, has to do with how, like coloniality (a mode of governance) and slavery (a mode of labor appropriation), raciality allows for a mobilization of the ethic and the symbolic, which attaches necessity to decisions that support the juridic and economic needs of the capitalist polity.

I

Bio-power [covers] the set of mechanisms through which the basic biological features of the human species became the object of a political strategy . . . how, starting from the eighteenth century, modern Western societies took on board the fundamental biological fact that human beings are a species.

—Michel Foucault, *Security, Territory, Population*

Nearly a year after the official announcement of the COVID-19 pandemic, the *Guardian* reported on UK officials' favorable views on easing mitigation measures such as "restrictions on social mixing."[3] Though the statements in the piece are not explicitly about how the decision would affect different sectors of the population, by that time it was evident that any harm resulting from easing mitigation measures would have disastrous consequences for the country's Black and Asian people. The article included a commentary by the former UK prime minister Boris Johnson, who enthusiastically supported these measures. Johnson is quoted as saying he was "optimistic, I won't hide it from you. I'm optimistic, but we have to be cautious. Our children's education is our number one priority, but then working forward, getting non-essential retail open as well and then, in due course as and when we can prudently, cautiously, of course we want to be opening hospitality as well." The reason for the optimism, he explained:

A new disease like this will take time for humanity to adapt to, but we are. The miracles of science are already making a huge difference, not just through vaccinations but therapies as well. New therapies are being discovered the whole time which are enabling us to reduce mortality, improve our treatments of the disease. I do think that in due time it will become something that we simply live with. Some people will be more vulnerable than others—that's inevitable.

Johnson's sentiment resonated with that of the Conservative MP David Davis, who, according to the same report, had told BBC Radio 4's *Today*: "We don't think for a second of locking down the country over flu," adding, "there will come a point where there is a death rate from Covid, but it is at normal level." The only dissonant voice in the piece is that of Steven Riley, a professor of infectious disease dynamics at Imperial College London and a member of the Scientific Pandemic Influenza Group on Modelling, which advises the UK government on emerging infectious disease threats. Professor Riley warned it was too early to begin making comparisons with influenza. "I'm not sure Covid settles down to look like flu so quickly," the newspaper quotes him saying to the *Today* program, "and there's quite a long time to go from now until next winter." "In some ways," he warns, "it's not the right time to make that decision, we need to see how low we can get the prevalence of Covid in the community." He also said that getting coronavirus vaccinations and treatments to the level of seasonal flu would be a "very good position to be in" and "certainly a reasonable scenario."[4]

In February 2021, when this report was published, we knew two things: first, decisions regarding when to ease mitigation measures were dependent on infection and hospitalization rates presented in curves, graphs, and other abstract mathematical representations; and second, in the United Kingdom, as in Canada and the United States, COVID community spread and deaths were highest among nonwhite persons. Thus, even if abstract mathematical presentations (models, curves, and graphs) depicting the temporal unfolding of the invisible killer suggested that it could affect anyone, anywhere, the spatial distribution of the numbers of infected and dead—in the United States, for instance—revealed a concentration of disease in urban and regional areas with a larger proportion of Black, Latinx, formerly colonized, and Indigenous residents.

Knowing how this new disease operates in the mapping of its occurrence requires more than collecting the data and exposing the racial consequences of the pandemic in terms of differentials (how nonwhite persons were worst affected). It requires interrogating how racial differentials prefigured the emerging pandemic data. At the theoretical level, this analysis recomposes Foucault's biopower by attending to something he would never consider, which is how racial categorization operates with/in calculation as a strategy of power. And at the analytical level, this approach finds (past) effects of previous racially determined decisions working from within, as categorization operates within the abstract tools on which biopolitical calculations, which are designed to have (future) effect, are based.

While the *Guardian* article conveyed the disagreements regarding the timing for the easing out of the United Kingdom's mitigation measures, it remained silent on the impact on the country's population. Instead, these decisions were based on predictive modeling visualized in maps, graphs, and curves. Therefore, easing mitigation measures was a biopolitical decision based on calculation, but how was it also a racial one? This question cannot be asked using Michel Foucault's formulation of biopower. For this reason, I begin by making a case for including raciality when mapping of modern bio-power. To do this, I reformulate biopower by retaining Foucault's tripartite analysis of power but considering what he ignored—and, heeding Saidiya Hartman's invitation in the epigraph, I track the workings of the "racial calculus."[5]

When describing his endeavors up to the 1977–1978 lectures delivered at the Collège de France entitled *Security, Territory, Population*, Foucault begins by recalling the mechanisms of power he had previously mapped in other texts: the juridical (the binary of permissions and prohibitions) and the disciplinary (surveillance, correction, and the culprit). He then introduces a third mechanism, the "apparatus of security," which "inserts the phenomenological in question . . . within a series of probable events." He explains, "The reactions of power to this phenomenon are inserted in a calculation of cost" and "establishes an average considered as optimal" as well as "a bandwidth of the acceptable that must not be exceeded."[6] And crucially, because biopolitical mechanisms govern through mathematical abstractions, the sovereign right to decide on life and death, which is always present, is not immediately evident. When describing the specificity of biopower as a mode of governing the inhabitants of a given territory, as the focus turned from the "individual" to the population, Foucault lists several consequences of this shift in the object of the decision. Among its several innovations, he highlights two of interest: first, the emergence of a notion of freedom that is now attached to movement and circulation, not to a person, and, second, that governing is done through abstract representations (calculation) of processes that unfold in time and inform the regulation of how circulation is managed in the territory.[7] Both of these aspects of biopower indicate its entanglement with the needs of the liberal capitalist state, namely, industrial capital's need for a dispossessed "multiplicity of individuals" concentrated in the cityscape.

According to Foucault, biopower was predicated on a rearrangement in the setting of the modern episteme in the late eighteenth century.[8] However, when describing this and other modern modalities of power, he does not attend to how, from the later nineteenth century on, the racial would operate within them. Instead, in line with the post–World War II rejection of the racial as a proper political concept, he treats race as a residuum of a previous epistemological and historical context whereby it operated by setting up an opposition between conflicting groups seeking the elimination of the other based on the principle of identity.[9]

My reading of the role of the racial in the modern episteme differs greatly from Foucault's and the post–World War II mandate. In my view, the political force of the racial results from how the nineteenth-century sciences of society and man deployed categories, which presume *time* in the framing of the efficient, formal, and final cause of the human forms and processes under investigation. This shift registers the underground operations of linearity, which was introduced by classical physics in the grammar of modern thinking and consolidated in Hegel's and Cuvier's formulations of the historic and the organic, respectively. Through this new mode of formalization, by the way of linearity, these projects of truth assembled determinative (in the nineteenth century) and interpretive (in the twentieth century) strategies that fundamentally changed preexisting notions, such as class and race.[10] That is, these knowledge projects assembled concepts (the racial and the social), which yielded categories that comprehended and distributed individual human beings into qualitatively different wholes (groups) that composed the ethical whole (humanity) along a linear temporality. In sum, these scientific projects recomposed the "multiplicity of individuals" through distinct racial, and later social, categories and reformulated the racial as a modern strategy of power—a scientific tool in the symbolic arsenal, which combines with the juridic, economic, and ethic moments in the post-Enlightenment liberal capitalist political architecture. The racial—more precisely, the knowledge arsenal I call *analytics of raciality*—performs its symbolic role in the post-Enlightenment liberal political architecture through two mechanisms, namely, the logic of obliteration and the logic of exclusion.[11] My argument is that the logic of obliteration, which prescribes the eventual but certain elimination of nonwhite populations, works within (or underneath) the mechanisms of exclusion, which prescribe that nonwhite persons should be kept at bay because they lack the moral and intellectual capacity to perform in modern social conditions. If one attends solely to the logic of exclusion, which remains the prevailing account of racial subjugation, the analysis stops at the point where racial difference is operative—missing how it is prescriptive—and that nonwhite persons and groups fare the worst.

When tracking the operations of the racial in mathematical and other abstract representations of the COVID-19 pandemic, one must look beyond the correlations that expose the operations of the logic of exclusion. Why? Because the former only captures a simple efficient causal chain: race (natural or social datum) accounts for the higher mortality rates in certain zip codes, neighborhoods, or towns where one observes the prevalence of preexisting conditions, which are in turn explained by low income, low rates of formal education, and kind of employment, which are explained by discrimination based on supposed racial (physical and mental) differences. However, when one attends to the logic of obliteration, the focus shifts from the chain of efficacy that begins with the natural or social datum of race to the prescriptive role of raciality, as a truth arsenal of the liberal political architecture, in the interplay of its ethic, economic, and juridic dimensions. The logic of obliteration captures exactly how raciality allows otherwise unacceptable (ethically) decisions (juridically) to be immediately justified on the basis of scientific necessity (symbolic-scientific or symbolic-historic), that is, as part of objective reality or natural processes. The analytical move I will outline in what follows explores the pandemic as an occasion to track precisely that which usually goes unattended, when one limits the examination of racial subjugation to the tasks of establishing or finding correlations or identifying causations: how the logic of obliteration, when raciality plays a prescriptive function with the force of necessity, which usually manifests in the moment of the *decision* (ethical or juridical), has facilitated capital accumulation, as a crucial component of the biopolitical modality of power in the post-Enlightenment era.

II

The population is an abstraction if I leave out, for example, the classes of which it is composed. These classes in turn are an empty phrase if I am not familiar with the elements on which they rest. E.g. wage labour, capital, etc. These latter in turn presuppose exchange, division of labour, prices, etc. For example, capital is nothing without wage labour, without value, money, price etc. Thus, if I were to begin with the population, this would be a chaotic conception [*Vorstellung*] of the whole.

—Karl Marx, *Grundrisse*

After charging political economists with treating the population as an empty mathematical multiplicity, an abstraction but not a complex rational formation, Marx introduces what he calls the proper method with the famous statement: "The concrete is concrete because it is the concentration of many determinations, hence unity of the diverse. It appears in the process of thinking, therefore, as a process of concentration, as a result, not as a point of departure, even though it is the point of departure in reality and hence also the point of departure for observation [*Anschauung*] and conception." When offering the proper scientific approach, he delineates a moment of the political that escaped Foucault's account of power. To recall: according to Marx, "along the first path [the economists'] the full conception was evaporated to yield an abstract determination; along the second, the abstract determinations lead towards a reproduction of the concrete by way of thought."[12] Put differently, the proper approach consists in comprehending the "multiplicity of individuals" in a way that, instead of tracing the effects of what they do, seeks to determine the causes for how they appear. For the most part, my effort here, which is not that different from Marx's method for the political economy, is to treat the decisions regarding the mitigation of the COVID-19 pandemic with an approach to racial subjugation that acknowledges how it results from determinative moves that account for its prescriptive effects.

We know *that* (operative function) raciality is at play in the pandemic given the outcomes of mitigating measures, which resulted in nonwhite people in the global north and populations of former colonies in the global south among the largest numbers of the infected, the hospitalized, the killed, and those maimed by Sars CoV-2. What needs more treatment is the description of *how* (prescriptive function) raciality plays as a mechanism of modern power and what this reveals about the ways that the liberal political architecture of the global present attends to the needs of capital. To investigate this question and analyze the decision making regarding mitigation protocol, I propose an image of the political that besides the juridical, the mechanisms of disciplinary (identification) power, and apparatus of security (calculation) also considers how the weapons of necessity (determination) play out in what distinguishes the political per se, namely, the decision.

A great contribution of Foucault's analysis of security is that it provides a framework for examining the political role *formalization* played before the science of life reorganized the modern scientific field in the nineteenth century. One way to distinguish between the two moments of formalization is as follows: in the late eighteenth century, *formalization* through *calculation* organized a mode of governing that treated the population as a quantitative rational multiplicity and was future oriented. In the nineteenth century, the science of life deployed formalization through *categorization*, a modality that reconstitutes the "multiplicity of individuals" as a qualitative rational multiplicity that incorporates a specific past. An important distinction: while *calculation* is concerned with how *natural* processes affect the bodies of individuals, *categorization* focuses on the human body as a *biological* (natural) formation. Furthermore, whereas the calculations of biopower ushered in a different notion of freedom concerned with circulation, the determinative tools of the science of life map a new domain for the laws of nature (the tools of necessity).

So, then, how did both modes of formalization—calculation with its future orientation and categorization with its past implications—unfold in the political context of the pandemic? To answer this question, one must attend to how categorization operates in biopolitical decisions when there is a perceived threat to the existence of the population, that is, when it activates juridical power. This requires an approach to the biopolitical moment of decision that tracks the operation of raciality in a situation in which biopower and juridical power appear indistinguishable.

Racial Milieu

How does raciality become apparent in the decisions regarding COVID mitigation strategies? It maps the milieu and comprehends the population via categorization, as a rational qualitative multiplicity,[13] such that the distribution of harm and the accumulation of decisions that enable exclusion facilitate the unfolding of the logic of obliteration. Raciality operates as a political symbolic tool that awards necessity to the social situations and events it explains; to be more precise, it has to do with how necessity figures in the grid of racial difference. In the modern episteme, necessity as it is presumed in the sovereign will (which is that of the divine law) also unfolds in time (as that of the transcendental spirit, which establishes its actualization in history and its expressions in the determination of nature, in the nineteenth century). For that reason, determinative (scientific) tools of knowledge, which have the force of necessity, have not only a descriptive and an operative but also a prescriptive function. This is an important distinction because it accounts for how difficult it is to grasp how raciality operates in the liberal political architecture.

In the nineteenth century, the analytics of raciality emerged in projects of knowledge that re-presented the population as a complex rational formation, that is, as Marx's concentration of many determinations, or a "unity of the diverse." The scientists of man deployed the tools of the classical (disciplinary) order, which were recuperated in the qualitative scientific program assembled by Cuvier and converted into descriptive tables that inscribe necessity into the categories (of the table) that correspond to the "essences" of what they named. Cuvier's table of difference could now be deployed safely, alongside of the Hegelian recuperation of time as the site of actualization of the Human's self-producing capacity but also as the fundamental force at work in Nature. In other words, what Cuvier inscribes in the forms of the living body is time, in the realm of necessity—a move that yields the organic, which refigures the formal efficacy of time in the arrangement of its parts and movements; while Darwin's refiguring of the science of life transforms living nature into a graph of effects of temporal differentiation. Borrowing from both, the science of man's grid of racial categories produces different temporal moments of human mental (moral and intellectual) development (evolution) in the organic forms and functions of existing human bodies, which is explained by the determinative strategies of the science of life. It is crucial, however, to recall that these determinative tools reproduced the liberal subject of politics as the transparent I, that is, as a *natural* entity that is mentally superior to the rational determinative forces of nature, that is, the laws and forms that govern reality.

The first and crucial analytical move is then to foreground the symbolic moment of the liberal political architecture by attending to how it is organized by the two notions that would rule the post-Enlightenment episteme, namely, the organic (parts and functions) and the historic (living whole) in which temporal linearity is added to the spatial linearity of rules and laws organizing Foucault's thinking. While the historic (living whole) refigures the principle of liberty as it is formulated in Hegel's world history's rendering of finality, the organic refigures the principle of necessity, the ground for scientific knowledge, as it is rendered in the formal efficacy of Cuvier's science of life and the efficient finality of Darwin's thesis on evolution. When assembling the analytics of raciality, the scientists of man gathered this latter (temporal) articulation of necessity in an arsenal that attaches necessity to the connections it establishes between the forms of the bodies and corresponding degree of development of their mental (moral and intellectual) capacities.

Let me expand on this argument in connection with Foucault's analysis of power. According to Foucault, what characterizes security is that it regulates by targeting the (efficient) causes of the concerning events (pandemic, scarcity, etc.), which it assumedly cannot prohibit or control, much less eliminate. Regulation relies on the *determinative graph*, which gives "the distribution of cases in a population circumscribed in time or space" (quantification). When considering the pandemic, I am interested in how raciality plays out in regards to the *determinative graph*: how it produces the graph, what positions it establishes (its form), and how a racial category (not a person) will then be positioned in it (efficacy) both in terms of what is said about it (Indigenous, white, Black, Latinx) and tells about her as a person (her social position and predispositions) and how it plays in the logic underlying the calculation that produces the graph. Another way of saying this is that, beyond registering the differences in performance (spatially on the table) or numbers between Indigenous and Latinx (to track its descriptive function) persons, for instance, it is also necessary to follow how each racial category appears temporally in the graph (to track its operative function) and how it performs against or in the universal (abstract) measurements (to track its prescriptive function). For the latter shows how political decisions (sovereign, disciplinary, security) affect a "population," as raciality (as an index of objective causality) functions in support of the political authority, when it justifies decisions that, in the case of COVID-19, had life-and-death implications for tens of millions of persons.

Models and Maps

In order to account for how raciality maps the milieu from which the data used in the models are generated, the biopolitical analysis should not treat the "population" as a mathematical (scientific) abstraction but as (to paraphrase Marx) the concentration of many political (juridic) decisions. The crucial analytical move here, then, is to focus on how raciality operates as an index of objective causality—that is, how it signals that what happens is not the effect of liberty (of free only-future-oriented decision making) but of necessity (of the accumulation of the past effects of decisions made based on racial difference). To do so, one attends to how, as in the global pandemic, the specific temporality inscribed in racial difference organizes the context of public health decisions. In the maps of the distribution of the disease, it does so as a grid produced from previous (accumulated, or past) calculations of exclusionary decisions. And in the curves of the possible evolution of the spread of the disease, it does so as a line and as a projection of future infections and deaths. The abstract presentations that guide the calculation, such as models or curves, also involve determinations, which in themselves also gather—as they are building with—many other determinations. These data express the effects of previous decisions and form the basis for further decisions. It would be uncanny if many critics of modern thought had not mapped how abstract, transparent images of events and existents correspond to the metaphysic of linearity supporting modern representations of space and time: on the one hand, the models produced by algorithms graph temporally (spatial events), and, on the other hand, maps and tables, both of which can also reflect distribution frequencies, gather spatially (temporal events). Basically, when looking at a *model* one can gather a sense of the temporal unfolding of something that is happening somewhere (country, city, neighborhood) while when looking at a *map* or a *table* one has a snapshot of the spatial distribution resulting from the accumulation or lack of something that happened sometime (unemployment rate, educational level).

Racial data mattered at both ends of the evolution of the pandemic—as the basis for and outcome of the biopolitical decisions that shaped the chosen mitigation strategy. The emergency decisions about how to deal with the pandemic occurred in a context already racially mapped by previous decisions. Instead of taking one or a series of policy decisions as one-off and circumstantial, when attending to the racial map –what I am calling the social index—one finds an accumulation of effects of decisions. The past-oriented social index (geographic locations corresponding to education, health, income, the criminal justice system, occupation), which is a mathematical gathering of the effects of previous biopolitical decisions, now immediately and necessarily establishes how the decisions regarding the COVID-19 pandemic, which are based on models—that is, on future-oriented projections—will differentially affect the population.

My goal here, it should be noted, is not to advance an empirically based argument; instead, I am calling attention to how the evidence (examples, cases, etc.) used in such kinds of argument is already politically inscribed, that is, constructed by the symbolic (social scientific) tools of raciality. Every chief of state who decided not to intervene, like Donald Trump, Jair Bolsonaro, and even Boris Johnson (initially)—and who decided to ease mitigation measures even given the evidence of risk—who did so in light of what was known, also included a racial calculus, guided by a social index (as a proxy for lethality). They used a social index in the same way as the science of man used a facial index (as a proxy for mental attributes) in its mapping of humanity.[14] The facial index gave the necessity (natural/universal) element to the science of man's claims of moral difference being an effect of natural processes, which are not alterable. The social (health, housing, occupation, income) index, which informs today's biopolitical decisions, similarly, because it is a referent of necessity, allows for indifference (the not taking of preventive action and allowing death to occur) by the state and also by those who are supposed to act, such as teachers, physicians, and nurses.[15] Put differently, biopolitical decisions reflect two orders of abstraction: racial data (which hides previous decisions, which become datum and operators of necessity) and pandemic data (which will guide current and future decisions and the basis of liberty—that is, the future determination of the course of events—which are based, respectively, on categorization and calculation).

Every major and minor decision about the management of the COVID-19 pandemic has been informed by these two figurations of death—the curve and the map—both of which are also recalled in the moment of justification of the decisions, which have as their primary objective the health of the economy, as Boris Johnson said to the *Guardian* in the piece cited earlier. What the theoretical and analytical moves proposed here do is show how the decisions around the pandemic—both the first major decision (between mitigation and eradication) and the subsequent ones (on lockdowns, etc.)—were not responses to a threat to human life, a sort of lethal equalizer. Instead, by attending to how raciality maps the milieu from which the data used in the building of models were generated, it becomes evident how these latest global biopolitical decisions were in fact instances of a calculus made in the face of mathematical (abstract) representations of individual cases, which were in fact formalizations of a racially mapped population.

IV

Declaring COVID-19 a pandemic was a political decision based on a calculation of economic costs versus human costs. Economic costs, you may recall, were the reason that most countries—with notable exceptions such as New Zealand and Australia (initially)—chose mitigation over suppression. Most governments, that is, chose to manage a public health crisis that some had anticipated but most did not believe would ever actualize. They treated it as a crisis that was "deadlier" to (repeating Boris Johnson's words quoted earlier) those "more vulnerable than others," to those with a certain social profile, that is, workers in health care and other named essential services, those with serious health conditions (diabetes, cancer, and other diseases), and those living in small and overcrowded housing.

The COVID-19 pandemic is a major global event, and, as such, its morbidity indexes exemplify how raciality operates in tandem with humanity in the post-Enlightenment political architecture. That the COVID-19 global pandemic was going to have the greatest impact upon Indigenous, Black, and other nonwhite populations was something studies of the 1918 pandemic had already anticipated. How the emergency decisions regarding how to manage the spread of the disease would have that effect has to do with the context for these decisions: how racial difference, as given by the social index, already mapped the milieu.

In the United States, for instance, Indigenous, Black, and Latinx persons are a part of data and maps; more so, they are part of the maps hidden in the pandemic data. As we all learned very quickly, variations on the average number of infections, hospitalizations, and deaths became the basis for the implementation of policies designed to limit the spread, such as quarantine, shelter in place, mandatory masks, etc. That same attention to the average, which, according to Foucault, characterizes an apparatus of security, became ubiquitous even in daily conversations among those who could shelter in place. At the same time, those who could not isolate learned also very quickly, with disastrous consequences, that decisions made on the basis of mathematical images of the pandemic, such as models and curves, do not take place in a homogeneous space; that is, these formalizations do not plot on a transparent set.

Another abstract image, one Foucault did not take into account, is racial categorization: it compounds the average number of infections, and, as such, it is an implicit (in the social index) basis for the calculations that underlie the decision regarding how to manage the COVID-19 pandemic. Here again raciality checked humanity (the ethical concept) in the very movement of including everyone in the calculations. The ethical notion of humanity addresses the single individual person, the one that matters or counts, the one whose dignity is at stake or life is at risk, the one whose trajectory temporally unfolds like that of a sole straight line. How does it happen? How is the logic of obliteration, the tool of necessity, activated in governmental decision making triggered by the COVID-19 pandemic? The logic of obliteration addresses the person, each person and every person, as a part of an aggregate, of a social group or of a population (data or maps) that (the social index shows) will *necessarily* perish and as such is counted but does not count.

Notes

1. There seems to be no definitive definition of public policy, and most are as vague as, for example, "the set of activities that governments engage in for the purpose of changing their economy and society." B. Guy Peters, *Advanced Introduction to Public Policy* (Cheltenham: Edward Elgar, 2021), 1.
2. I use "nonwhite" (and "nonhuman") because there is no value attached to the undeterminate prefix "non." It has the advantage of not being something that signals difference without indexing identity, without forcing what it names to signify the opposite, such as that of the white or of the human. The term that does that work is "not."
3. For the date of the official announcement of the COVID-19 lockdown I use March 16, 2020. On that day, Imperial College–University of London released a report titled "Report 9—Impact of Non-pharmaceutical Interventions (NPIs) to Reduce COVID-19 Mortality and Healthcare Demand," https://www.imperial.ac .uk/mrc-global-infectious-disease-analysis/covid-19/report-9-impact-of-npis -on-covid-19. This report laid out the two options before public health officials across the planet: mitigation (measures that would control and minimize the spread of the new disease) or suppression (measures that would stop the spread and eventually eliminate the virus). A few days after the release of this report, the majority of the countries started implementing mitigation measures, which included shelter-in-place orders and closing their borders.

4. Molly Blackhall, "Boris Johnson 'Optimistic' About Easing Some England Lock-down Measures," *Guardian*, February 13, 2021, https://www.theguardian.com/world/2021/feb/13/boris-johnson-optimistic-easing-some-england-lockdown-measures.

5. I find an advantage in Foucault's concept of the biopolitical—vis-à-vis Mbembe's notion of the necropolitical and Agamben's formulation of biopower. See, generally, Achille Mbembe, *Necropolitics* (Durham, NC: Duke University Press, 2019); and Giorgio Agamben, *Homo Sacer* (Stanford, CA: Stanford University Press, 1998). His tripartite analytics of power support my view that the decision (to act or not to act) would remain the crucial event in the post-Enlightenment political architecture. His analytic of (juridico-politico, disciplinary, and bio) power allows us to take into account how, even in this managerial (biopolitical) presentation, the modern authority is also defined by the right to decide on life and death (as the old monarchical sovereign).

6. Michel Foucault, *Security, Territory, Population* (New York: Palgrave Macmillan, 2009), 20–21.

7. Precisely this centrality that movement and time would acquire in the following decades, I find, also exposes the limit of the Foucault's tripartite analysis of power, a limit that reflects more of how his framework is a bona fide modern epistemological product than any limitation of his knowledge of the periods; better put, it is not that Foucault would have a better framework if he had more information but that his conception of modern power missed a crucial concept.

8. The notion of the modern episteme is elaborated in Michel Foucault, *The Order of Things* (London: Routledge, 2001). The argument presented here is based on a distinct mapping of post-Enlightenment thought provided in Denise Ferreira da Silva, *Unpayable Debt* (Berlin: Sternberg, 2022).

9. For his argument regarding the notion of race, see Michel Foucault, *Society Must Be Defended* (London: Penguin, 2004).

10. For a development of this argument, see, generally, Denise Ferreira da Silva, *Toward a Global Idea of Race* (Minneapolis: University of Minnesota Press, 2007); and Ferreira da Silva, *Unpayable Debt*.

11. For a discussion of these two logics (exclusion and obliteration) of operation of racial power, see Ferreira da Silva, *Toward a Global Idea of Race*, part 2.

12. Karl Marx, *Grundrisse: Foundations of the Critique of Political Economy (Rough Draft)*, trans. Martin Nicolaus (New York: Penguin, 1973).

13. Foucault indicates this when he says that "the problem of the town"—as the home of disease—is "at the heart of" the "mechanisms of security." What Foucault's account of the problem of circulation suggests is the image of a multiplicity of individuals moving about the town, from home to work, and the need to control and regulate this circulation in a way that nullifies not only the possibilities of diseases and death but also of revolt.

14. The head index and the facial index were the main tools of the science of man, which were used in racial classification, as they served as a proxy for mental capacity. See Ferreira da Silva, *Toward a Global Idea of Race*, chap 4.

15. By now there is a large number of news reports and academic texts that comment on how exposure to and mortality from COVID-19 infection is racially

mapped. For this reason, I am not including any such references. What I will highlight, however, is the fact that this racial map was available to Trump, Bolsonaro, Trudeau, and Johnson at the time they decided for mitigation over suppression of the new virus. The most important measure highlighted by the Imperial College report (see note 3), social distancing, was precisely that measure those who were the worst affected by the pandemic—Black, Indigenous, Latinx, and other working-class nonwhite persons—could not follow because of their overcrowded housing conditions or employment in the so-called essential services.

PART TWO

Purity and Mixture

SIX

Biometric Hybridity

Anglo-Indians, Race, and National Science in India, 1916–1969

PROJIT BIHARI MUKHARJI

"THAT IS WHAT we are—Mongrels!" declared the famous Indian anthropologist Irawati Karve in 1963.[1] Writing in 2008, the eminent feminist sociologist and activist Nandini Sundar cited Karve's declaration as evidence that, notwithstanding her lifelong interest in the biological study of human difference, Karve's findings were "anti-racist and anti-caste."[2] It is this positive valence that is given, in both scholarly and progressive activist circles, to biological espousals of racial mixture that I want to interrogate in this article.

Sundar is not alone in seeing "hybridity" in a positive light. From the recondite postcolonial theory of Homi K. Bhabha to the recent assertion by the bestselling author and journalist Tony Joseph that "we are all migrants, and we are all mixed," hybridity has often been the trope that has grounded visions of progressive political futures in South Asia.[3] As Anthony Easthope points out, with regard to Bhabha's framing of hybridity, it is predominantly an "adversarial discourse" that operates by "playing off ambivalence of various kinds against fixity."[4] Its redemptive political charge accrues from rejecting fixed, pure identities.

The problem with such theorizations is that they tend to dehistoricize "hybridity" and present it, misleadingly, as a monolithic concept. Does hybridity *always* map clearly onto progressive politics? Are all articulations of hybridity the same? Can hybridity really interrupt the politics of racialization?

My answers to these questions are more pessimistic than Sundar's or Bhabha's. I would argue that the faith in hybridity being always already opposed to racialization is founded in fact upon a fundamental misunderstanding of "race," in its historically shifting "scientific" senses. After the horrors of the Holocaust, race science—especially in its geneticized forms—reinvented itself by moving away from the strictly "typological race concept" and gradually embracing a "populational race concept." As Lisa Gannett points out, there were two key differences in these two concepts. First, whereas the typological concept held the total number of races to be fixed, the populational one understood the total number to be a shifting, dynamic number. Second, whereas the former concept was defined by essential and unchanging characteristics, the latter was marked by the frequency with which a characteristic appeared in it.[5] According to Nadia Abu El Haj, this populational concept "molecularized" race, thereby reconstructing it as a "fundamentally statistical" entity where quanta of admixture rather than the simple fact of it is at issue.[6]

The simple fact remains that "hybridity" is as much a biological concept as it is a cultural or literary one. Moreover, the biological concept has continued to evolve along fairly precise technical lines. This technical evolution of hybridity in biology was linked to the histories of race science and genetics. South Asian scholars and activists have largely ignored this scientific and technical history and continued to see hybridity as a surrogate for ambivalence, indeterminacy, and plurality, even as biologists, physical anthropologists, and geneticists have increasingly fixed hybridity within highly quantified, statistical grids that have frequently intensified racialization.

In speaking of "racialization," I draw upon the work of Suman Seth on race and race science. Seth argues that we need to recognize that "race science" covers a plurality of race concepts and that these do not simply conform to the nineteenth-century typological ideas of race or indeed the Nazi ones derived from it. What justifies analytical classification of various ideas about human difference as "race science" is that biologically inherited differences between groups of people are seen as a *cause* rather than an *effect* of other forms of social difference.[7] In my case, "racialization" entailed a repeated insistence that a range of social differences between groups were essentially grounded in inheritable biological differences. In the specific case of the Anglo-Indians, on whom I focus in this article, this meant insisting that their distinctiveness within the subcontinental peoples was grounded in an inherited, biological heritage rather than simply social, cultural, or indeed educational or religious differences. My deployment of the word "intensification" draws on Jasbir Puar's suggestion that "intensification" helps us understand the violence of race not as a singular event that ruptures the continuum of social existence but as an "ongoing, non-linear process."[8] Intensified racialization, therefore, constantly reaffirms biologically inheritable differences as underwriting social differences. Even when individual studies seem open-ended and willing to acknowledge the possibility that the biological differences are not that great between any two specific groups, the sheer fact of repeating such studies reaffirms the basic idea that the inheritable biological difference is stable, traceable, and the ultimate arbiter of whether certain socially legible differences are meaningful or not. Labeling particular groups as "hybrids," in other words, must be seen as a way of amplifying the claim that the socially legible characteristics that mark them out as a group, at least potentially, has some biologically inheritable basis to them.

I agree wholeheartedly with Anil and Vinay Lal that "hybridity itself needs to be hybridized, made heteronomous to itself," but rather than do so purely in relation to a reconceptualized "economic space," I would suggest that, especially in talking of race, we need to attend to the historically specific ways that scientists themselves have articulated notions of "hybridity."[9] In what follows, I will describe a very specific type of scientific investment in racial hybridity. I call this "biometric hybridity." This description will unfold over three sections. In the first of these, I will give a potted account of the global interest in racial admixture. In the next section, I will describe the technopolitical visions of biometric nationalism that took root in mid-twentieth-century South Asia. Finally, in the third section I will describe the way biometric hybridity was dually articulated within both the global tradition of racial hybridity studies and the Indian frameworks of biometric nationalism. In this last and most substantive section, I will dwell principally upon studies done on the Anglo-Indian community of India.

The period in which this story of biometric hybridity is set is a particularly important if, until recently, historically under-researched period for South Asia. The end of the Great War had witnessed significant changes in both the political and scientific institutions in British India. Politically, elected ministries took power at provincial levels from the 1920s and thereby, gradually, ushered in electoral practices and mass politics. Scientifically, the end of the Great War witnessed the rapid Indianization of the main scientific institutions. As a result, an increasing number of Indians entered various kinds of scientific research during the period. The numerous social challenges, opportunities, and conflicts generated by the combination of these two factors meant that this was a time when much of what the postcolonial South Asian nation-states and their sciences would look like was hammered out. This was particularly true for the ways that science and technical expertise were increasingly mobilized to grapple with complex issues of national belonging, especially as competing nationalisms—usually either religiously and linguistically defined ones—came to compete with other forms of political ideologies like communism and fascism, caste assertion, etc. The complex negotiations between the competing forms of the political ascertainment of community boundaries and national belongings naturally did not end immediately upon the political end of empire. In some ways they are still with us. But the debates that started in the post–World War I era mostly ended by the end of the 1960s as the political and scientific generation of the interwar years began to be replaced by a new generation of postcolonial figures. The death of Prime Minister Nehru in 1964 and his successor Prime Minister Lal Bahadur Shastri soon after in 1966 signaled the change of guard in conspicuous and dramatic fashion. It therefore makes good analytical sense to trace the history of biometric hybridity in this period of political, scientific, and national ferment, roughly between the Great War and the death of Shastri. The exact dates I have chosen here are marked by particularly important scientific studies of hybridity, but the general frame of reference that animated these discussions arose from the intersecting shifts in scientific and political balance.[10]

Hybrid Studies

Writing in 1968, Dibyendu Kanti Bhattacharya, then a young physical anthropologist at the Delhi University, stated that "hybrid studies represents an important field in physical anthropology."[11] Bhattacharya dated "hybrid studies" as commencing with Eugen Fischer's 1913 study of the "Rehoboth Bastards" of Namibia, who claim to be descended from Boer fathers and Khoekhoe mothers. To be sure, there had been studies of racial mixture even before Fischer's study of the Basters. Franz Boas's 1894 study titled *The*

Half-Blood Indian, for instance, remains one of the pioneering biological studies of racial mixing. But as Fischer himself pointed out, where his study was utterly novel was in taking up for investigation an entire "hybrid race."[12] In other words, hybridization had here gone beyond individuals and produced a new "race." Later, others would study groups such as the Indos of Indonesia, the Burghers of Sri Lanka, the Mestizos of Kisar, the Anglo-Indians of India, the Pitcairn Islanders, and others as "hybrid races."

The influence of Fischer's Namibian study derived in part from his later leadership of the Kaiser Wilhelm Institute of Anthropology, Human Heredity, and Eugenics, in Berlin. During the interwar years, the institute was "Europe's main center of research on race and physical anthropology." His students and those affiliated with the institute would go on to study other "hybrids." One of his students, Wolfgang Abel, for instance, would study and recommend the sterilization of the children of French African soldiers and German women during the Nazi era.[13] In fact, Bhattacharya's citing of Fischer's study as the *fons et origo* of "hybrid studies" was possibly connected to these pedagogical networks. In the early 1960s, when Bhattacharya conducted his own researches into hybridity as a student of the Delhi University's anthropology department, the department was still headed by its founder, P. C. Biswas, a man trained at the Kaiser Wilhelm Institute in Berlin.[14]

The popularity and success of Fischer's study did not, however, eclipse the slightly earlier tradition, evinced in Boas's study, of researching cases of individual hybridity. In fact, the two traditions continued to develop in tandem and on occasion intersected within the field of interwar hybrid studies.[15] A review article published in 1926 by Earnest Hooton and focusing mainly on studies done at Hooton's home institution, Harvard University, listed a wide number of past, current, and proposed studies of hybridity. Among these were studies of Hawai'ians with mixed Hawai'ian, Chinese, Japanese, and European ancestries; Pitcairn and Norfolk Islanders; U.S. "Mulattoes" with mixed white and Black parentage; those with mixed Finnish, Lapp, and Swedish ancestries; those descended jointly from Arabs and Berbers; Mexicans and Central Americans deriving from mixed white, Black, and Indigenous ancestors; and groups with shared Arab and non-Arab ancestries in Egypt and Libya.[16] Some of these groups, such as the Pitcairn Islanders, who were descended famously from the liaisons between Tahitian women and the mutineers of the HMS *Bounty*, had already become a fairly homogenous group, that is, in Fischer's terminology, a "hybrid race." Others, such as those of mixed Arab and Berber descent, were still seen as individualized "hybrids."

At this time, race had not yet completely transitioned from its typological to its transitional concept. Hooton distinguished between "primary" and "secondary" race. The former were the unchanging, typological ones, while the latter were the dynamic groups produced through the historical mixture of the primary races. "Man," Hooton observed, was a "migratory, promiscuous animal," and this made primary races "well nigh anthropological abstraction[s]." Yet he still held on to the belief that the number of such abstract primary races was finite and could be determined with certainty. Accordingly, abstracting from the actually existing races of the world on the basis of "physical affinities" Hooton identified four "classes" or primary groupings: Whites, Negroids, Mongoloids, and Intermediates.[17]

Each of these abstract primary races was in fact constituted by a number of actual, secondary races. The White race, for instance, was made up of Alpine, Armenoid, Mediterranean, Nordic, and "possibly others." The Negroids were made up of African Negroes, Melanesian Papuans, and Negritoes, while the Mongoloids were composed of the "Mongoloid proper," Indo-Malayans, and the American Indians. The fourth, or Intermediate, race, by contrast to the others, was constituted of secondary races that were themselves "probably the stabilized result of racial hybridizations."[18] Such stabilized hybrid races, according to Hooton, included examples such as the Ainos, the Dravidians, and the Australians.

Building on this basic schema, Hooton further pointed out that "racial mixtures" could be divided into two types. First were the "radical mixtures between races physically far removed from each other," and second were the "mixtures between allied races." Thus, hybridization between Alpine and Nordic Whites would be considered in the second category, whereas mixtures between Alpine Whites and African Negroes would be viewed as "radical mixtures." The mixtures between allied races being allegedly socially and physically unremarkable has, Hooton imagined, continued since prehistoric times, with the progeny being absorbed into the parental stocks. As a consequence, the original distinguishing characteristics of each race have been so far obscured that the very possibility of finding any primary race in pure form is doubtful. Thus, to Hooton it seemed "no longer possible" to study such intermixtures. Such hybridity was simply too prolonged and too pervasive to be amenable to scientific investigation.[19]

In stark contrast, Hooton argued that radical race mixtures invited more vigorous scientific investigation. Such mixtures had "long been viewed with alarm by statesmen, sociologists and anthropologists." Since radically different races had, "as a rule," developed "dissimilar cultures and social traditions," their hybridization was viewed as a "social anomaly." Yet the "pious hope that human hybrids would prove to be like mules, 'ornery but sterile,'" had become unsustainable in the face of extensive census data. Consequently, Hooton thought, scientifically exploring the fertility, vitality, and mental capacities of such hybrids produced through radical race mixture was an urgent necessity.[20]

One key aspect of these studies of racial hybridity in the interwar period was the connection that was often drawn between racial hybridity and cultural progress. As Forrest Clements pithily pointed out, there were essentially two broad schools of thought on this question. One held racial purity to be necessary for cultural advancement, whereas the other thought that it was actually hybridity and mixture that fostered cultural achievements.[21] Politically and socially, therefore, hybridity studies often had mutually orthogonal valences. Clements pointed out that scholars belonging to both schools of thought carefully chose their case studies to advance opposite positions.

Writing nearly a decade later, Melville J. Herskovits still held on to an almost identical classification of racial admixtures. He asserted that while "all human groups living today are of more or less mixed ancestry" and absolutely no "pure" races can be said to actually exist, "we must recognize the fact" that in some cases racial admixture has occurred between two peoples of "more dissimilar physical characteristics" and that it is mainly to these latter groups that the scientific concept of "race crossing" should be limited. Herskovits also suggested that instead of individual cases of race crossing, scientific energy should focus on racially hybrid populations that had emerged around the world in stable form. Among these he identified the African Americans, the Spanish-Mayan people of the Yucatan, the South African Basters, the Mestizos of the island of Kisar, and the descendants of the *Bounty* mutineers on Pitcairn and Norfolk islands. These populations, Herskovits insisted, were immensely valuable to any student of human heredity, especially to those interested in genetics. He pointed out that unlike fruit flies, guinea pigs, and white rats, human beings cannot be experimented upon in laboratories. Hence, these historically produced populations were the only ways that scientists could test theories of human heredity and particularly Mendelian laws.[22]

Where Herskovits and others writing in the mid-1930s had begun to differ from the earlier students of racial hybridization was not so much in the way they conceptualized primary and secondary races but rather in redefining the reasons for studying such hybridity. Whereas Hooton had insisted on the social and political utility of such studies in determining their desirability or otherwise, Herskovits emphasized the value of such studies in working out intellectual problems in human biology, especially questions around Mendelian genetics.

As we move from the 1930s into the 1940s, the kind of distinction between primary and secondary races that sought to reconcile earlier typological conceptions of race with emerging populational concepts becomes progressively redundant. A new generation of geneticists, such as Theodosius Dobzhansky, L. C. Dunn, and others, increasingly refashioned race as a "Mendelian population." That is, it was now viewed as a reproductively isolated gene pool rather than an eternal type. A mere few hundred years of reproductive isolation through geographic or social barriers were now thought to be enough to engender a race.[23]

Alongside the eclipse of the older typological conception came other changes. Two of these are particularly significant. First, cultural, social, and even intellectual rationales for studying race crossing gave way to increasingly medical arguments. Reviewing the state of the field in 1959, the Nobel Prize–winning Australian virologist Sir Macfarlane Burnet referred to Rh incompatibility, sickle-cell anemia, and even allegedly teeth that were too large for their jaws, which mixed-race people occasionally had as a consequence of hybridity.[24] Second, rather than linking these questions of race crossing to philosophical questions about cultural achievements or even arcane intellectual questions about Mendelian law, they now came to be rationalized by arguments about practical utility for the state. The highlighting of medical dimensions of the problem was part and parcel of this emphasis on applied benefits. In fact, the comments by Burnet were made at the Australian Citizenship Convention organized by the Department of Immigration.

It is also worth remembering that "hybrid studies" were not the exclusive preoccupation of European, American, or Australian scholars. A number of colonized, decolonizing, and postcolonial national intellectuals became interested in "hybrid studies." In 1940, for instance, Tao Yunkui, a

Chinese anthropologist who had trained with Fischer, published a study on children of a mixed Chinese-European couple.[25] Similarly, Jaehwan Hyun has recently described how Japanese scientists like Suda Akiyoshi and Hoshi Hiroshi studied *konketsuji*, or "hybrid children," well into the 1970s. Hyun also draws attention to what he calls the "double play" of such research, whereby global interest in seemingly antiracist studies of human adaptation were aligned with local Japanese investments in pure-blood, racialized nationalism.[26] Likewise, in 1960s Brazil, ideas about *mestiçagem* remained "central" to the thinking of Brazilian geneticists who conceptualized their nation as an "immense mosaic." The Brazilian researchers Francisco M. Salzano and Newton Freire-Maia profiled multiple different "hybrid" populations, such as "trihybrid (White/Indian/Negro)," "mulatto," "dark mulatto," "light mulatto," and "mestizos."[27] The Indian studies of "race mixture" we will discuss later in this chapter were part of this larger global uptake of "hybrid studies." As Soraya de Chadarevian has recently observed, while twentieth-century racial sciences were "per definition transnational and comparative in their outlook," their "findings [always] carried special meaning at the national level."[28]

A detailed history of the global evolution of "hybrid studies" through the middle decades of the twentieth century would be out of place here. Instead, what I have provided are some signposts sketching out the larger contours of these studies through roughly the first five decades of the twentieth century. There are two general issues that I want to underline within these contours.

First are a set of issues that emerge from the general fact of the existence of hybrid studies as a subfield of research. Foremost among such issues is the simple fact that "hybridity" here is no mere cipher for rampant biological fluidity. Rather, it is a highly technical discourse with precise boundaries and methods that sought to render the fluidity of human reproduction measurable and fixed. Indeed, one of the explicit concerns of these studies was to understand how "hybridity" became "stabilized" and thus gave rise to new, fixed groups with reasonably stable characteristics. Likewise, it is also worth pointing out that the simple existence of these studies undermines any claim that recognition of hybridity interrupted the racialization of identities. In truth, we see that each of these scholars called for further studies of this nature, and projects studying hybridity proliferated. Hybrid studies, in a very literal sense, intensified research into race.

The second issue is the historical differences between different eras of hybrid studies. We have seen, for instance, that in the earliest studies scholars like Boas had mainly been concerned about individuals of mixed racial ancestry. By the time of Fischer, there emerged more of an interest in stabilized racial hybridity that could give rise to new racial groupings. By the 1920s, we also notice an attempt to reconcile older typological conceptions of race to newer populational ones by distinguishing between primary and secondary races. Such distinctions persisted into the 1930s, though the rationale for being interested in them began to alter. From Hooton writing in the mid-1920s, to Herskovits writing in the mid-1930s, and to Burnet writing in the late 1950s, we notice a rapid and significant shift in the rationale for pursuing hybridity studies. The importance of large philosophical questions in the 1920s gave way to the more specific technical questions in the 1930s, and by the end of the 1950s, practical, medical, and statist rationales had come to predominate. Moreover, by the 1960s and 1970s a number of Asian and Latin American states had incorporated these studies into their postwar nation-building projects.[29] These shifts demonstrate that "hybridity" was not only a precise technical discourse but was in fact a heteronomous and historical notion. Rather than a single precise, technical discourse on hybridity, there existed a number of historically specific, if overlapping, forms of hybridity in genetics.

Biometric Nationalism

While the global evolution of hybrid studies offers one kind of context for the emergence of hybrid studies in India, another more immediate context was provided by biometric nationalism. Biometric nationalism was a "heterogeneous body of techniques assembled around the idea that political questions about national belonging were susceptible to technocratic clarification through the use of biometric measurements."[30] Unlike the biometric state described by Keith Breckenridge, biometric nationalism was not putatively grounded in state-held repositories and databases.[31] It was a scientific, technocratic, and ideological apparatus seeking to derive the biocultural boundaries of the nation, rather than the borders of the state as such.

Biometric nationalism was grounded in the ideas of the Bengali and Indian polymath Sir Brojendra Nath Seal. He had articulated these ideas most succinctly in a speech given at the Universal Races Congress in 1911.[32] There, Seal had elaborated on how nations evolved, biologically and culturally, through stages and out of earlier forms of sociality as a unified, homogenous population. Though in his speech he had not metricized this process, he did suggest that at some future point it might be possible to derive a mathematical expression of the process. This in turn would help settle the difficult political questions of various nationalisms.

Two men, both in their own different ways personally inspired by Seal's ideas, would go on to try and give shape to his ideas.[33] The first of these was Biraja Shankar Guha, who became the first Indian to earn a doctorate in anthropology in 1924, while the second was Prasanta Chandra Mahalanobis, who today is remembered mainly as a pioneering statistician. Guha's Harvard PhD dissertation was titled "The Racial Basis of the Caste System in India," and Earnest Hooton was a member of his dissertation committee. Upon his return to India, he worked for the census department before eventually convincing the dying colonial state to establish an Anthropological Survey of India, which would finally establish a formal link between academic anthropology and formal statecraft. This institution, which, by the 1980s, was the world's largest single employer of anthropologists, would go on to play a huge role in sponsoring and sustaining biological researches into human variation in early postcolonial India.[34] Guha helmed the survey in its initial years and thereafter retained a significant amount of power in it until his untimely death in a railway accident in 1961.

Mahalanobis enjoyed even greater political power in postcolonial India. Not only did he establish a flagship institution known as the Indian Statistical Institute to promote the study and application of statistics, but he was also one of the earliest appointees to the powerful Planning Commission of India after the country became independent. Though race was eventually only a small part of his overall research interests, it was the very first topic upon which he published, and he continued to publish on the subject throughout the 1930s and 1940s. Even more importantly, like in Guha's case, he mentored an entire generation of younger researchers to take up the questions of race, racial mixture, and nationalism.[35]

By the time biometric nationalism begun to emerge, race science itself was rapidly changing. "In the 1920s, blood group researchers promised to put racial science on a more scientific footing."[36] Thus, by the 1930s new serological techniques increasingly, though not completely, overshadowed earlier anthropometric measurements. This was the period when biometric nationalism begun to emerge. As a result, it came to be largely instantiated through seroanthropological studies.[37] Older elements of colonial race science, such as a notion that Indian castes were actually the products of racial intermixture between more primordial racial groups, such as Aryans and Dravidians, came now to be reframed in seroanthropological terms.[38] Over the next three decades, discussion of fixed, primordial types would recede, but the use of seroanthropology to determine the racial distinctiveness of individual Indian caste, religious, or linguistic communities would only intensify.

With decolonization and especially because of the political clout of some of the founding figures such as Guha and Mahalanobis, biometric nationalism moved from the fringes of the state to institutions directly linked with it. Yet it somehow failed to produce at the time a biometric state. Instead, it continued to be a largely technoscientific discourse that produced technoideological insights into national belonging, racial identity, etc. without acquiring direct applications in the form of large-scale databases. Though recent initiatives for producing a biometric basis for citizenship through massively centralized databases certainly draw on this earlier technoideological tradition, the fact that similar database projects had not taken off earlier is also important.[39]

In any case, the popularity of biometric nationalism meant that even without the emergence of a biometric state, the number of institutions and individual researchers devoted to biometric explorations of racial identity grew steadily in postcolonial India. Aside from specialized institutions such as the Anthropological Survey of India or the Human Genetics Unit at the Indian Statistical Institute, the number of university departments conducting research in the subject also climbed rapidly. Starting with the first department of anthropology in the early 1920s, by 1965 there were fifty-eight university departments across the country that were granting PhDs in anthropology. Race and racial history were taught and researched in most of these programs.[40]

The growing number of individual departments, research institutions, and researchers, without any clear centralized biometric state to serve, meant that biometric nationalism continued to be a polymorphic technoideological discourse. The thematic and political orientation of the research varied widely. What remained constant were a set of technical tools and a belief that these tools were producing valuable, objective, and measurable insights into racial identities.

As we shall see in the next section, hybrid studies formed a distinct part of biometric nationalism from the beginning and continued well into the 1950s and 1960s. Indeed, a larger number of studies on the subject were produced in these two initial postcolonial decades. The absence of a centralized biometric state might have helped the persistence and indeed intensification of such research into hybridity, but irrespective of the reasons, it clearly demonstrates that contrary to what some scholars have said, there was no uniformly global move away from biological studies of race mixture in the 1950s.[41]

Biometric Hybridity

Hybridity, in one sense, was central to British Indian raciology since the late nineteenth century. Many colonial raciologists held that the various Indian castes were the product of different primordial or ancestral races hybridizing to different extents. This theory remained popular with several later Indian researchers as well. Guha's Harvard thesis, for instance, was a rigorous and detailed elaboration of essentially this theory locating the birth of caste in the admixture of ancestral races.

Yet racial hybridity per se had seldom in itself been an object of systematic study in British India before the Great War. Though there had been occasional stray descriptions of racial hybridity in the works of colonial ethnologists, there was little systematic study of such subjects. Moreover, these references were usually to individuals and families rather than entire groups or communities. E. H. Man, writing in 1889, for example, had described a family on Nicobar Island that was of mixed Afro-Nicobarese descent. Later, Edgar Thurston, writing in 1909, noted the existence of a few families in the Nilgiris who were descended from the marriage of freed Malay Chinese convict laborers and local Pariah women. These stray references were first collated by Usha Deka in 1954 in the process of reformatting racial hybridity as an independent research topic.[42] Following Deka, others would follow up some of these earlier references and do some fairly limited studies on these earlier noted individuals and families.[43]

The earliest systematic attempt to study racial hybridity at a group level was undertaken around 1916 by Nelson Annandale, a talented young Scottish zoologist working at the Zoological Survey of India. Annandale had systematically gathered a large set of biometric data on the Anglo-Indian community of Calcutta. In 1921, during a meeting of the Indian Science Congress Annandale met the young and then virtually unknown Mahalanobis. The latter convinced Annandale that a statistical analysis of the data would be most useful. Annandale agreed.[44] This resulted in Mahalanobis's very first scientific publication.[45]

In this lengthy article of nearly a hundred pages, Mahalanobis demonstrated, using statistical methods, that the Anglo-Indians were in fact a homogenous racial group. This was important precisely because the transition from the typological to the populational concept of race had not yet happened, and there was skepticism about whether a group that was at best only a couple of centuries old could be considered a race properly so-called. Annandale commented on this aspect of Mahalanobis's findings in a foreword he penned for the publication.[46]

Another interesting claim Mahalanobis advanced in this early study was that while every race has an internal range of variations, "the more civilized races have greater variabilities than average." He surmised that this greater variability might result in "greater adaptability." Yet, this greater variability within the "civilized races" did not "occupy the extreme ends of the table," and thus "a higher state of civilization" is not associated with "extreme variability."[47]

Interesting as these observations were, the true value of Mahalanobis's publication was to provide a clear statistical test for when hybridity has been sufficiently stabilized to create a new homogenous racial group. It also allowed Mahalanobis to develop new statistical tools for comparing the distance between different racial groups. Thus, despite Annandale's sudden and untimely death at forty-seven in 1924, Mahalanobis continued to publish on the data Annandale had collected over the next two decades. He even published the raw data in the hope that others would work on it. In some sense, the Anglo-Indians became a kind of model organism to study racial variation using statistical tools.[48]

By the 1930s, new researchers began to collect new forms of data that had not formed part of Annandale's set. By then, seroanthropology was ascendant among raciologists, and therefore the blood group frequencies, rather than the length of the head or stature, were becoming the basis of new claims of race. Lt. Col. S. D. S. Greval and Capt. S. N. Chandra, both members of the Indian Medical Service and based at the Imperial Serologist's Laboratory in Calcutta, published the first set of blood group data from over three hundred Anglo-Indians alongside similar data from a number of other communities residing in the city of Calcutta in 1939–1940.[49] Greval and Chandra did not, however, single out the Anglo-Indian data or seek to draw conclusions about racial hybridity from it. That was done, soon after, by S. S. Sarkar.

Sarkar was the first scholar to receive a PhD in anthropology at an Indian university. He would later go on to work at the Anthropological Survey of India before returning to his alma mater, Calcutta University, and becoming an influential professor. Sarkar had also been a member of a small, elite Eugenics Society in Calcutta in the early 1930s and later, just before the Second World War, held a fellowship at the Kaiser Wilhelm Institute in Berlin. Numerous genetic anthropologists and seroanthropologists were trained by Sarkar at Calcutta University, and his role in shaping seroanthropological research in India is second to none.[50] In one of his earliest single-authored publications in 1942–1943, he discussed the problem of "race crossing" in view of the available serological and anthropometric data on Anglo-Indians. His main interest, at the time, was in determining which element "predominated" in hybrid races. Based upon Greval and Chandra's data, Sarkar pointed out that serologically the Anglo-Indians were much closer to Europeans than other North Indians.[51]

A few years later, Sarkar would follow up these initial comments and publish a much more comprehensive study of the Anglo-Indians of Calcutta. Sarkar was avowedly inspired by the 1918 study of the Mestizos of Kisar by Ernest Rodenwaldt.[52] In this paper, authored jointly with two junior colleagues, Sarkar combined both morphological and serological data to argue that the British element was dominant in the Anglo-Indians, thereby making them biologically closer to Britons than to Indians. Interestingly, and contrary to popular opinion, Sarkar also suggested that the Anglo-Indians had more upper-caste than lower-caste ancestry.[53]

The largest number of studies of racial hybridity using Anglo-Indians as empirical subjects was conducted in the following decade in Delhi by D. K.

Bhattacharya. The department of anthropology at the Delhi University had commenced in 1947, the same year the country became independent. Among its early students was Dibyendu Kanti Bhattacharya, the son of a high school Sanskrit teacher who also officiated as a Hindu priest. Anthropology had not been Bhattacharya's first choice of subject. But to his father's consternation he had missed the deadlines for college admission because of his involvement with a radical land-redistribution movement called the Bhoodan Movement, which sought to convince large landowners to willingly give up land to the poor.[54] Anthropology was the only department still admitting students, so that is what he enrolled in. Once enrolled, Bhattacharya gravitated toward physical anthropology and wrote a PhD on the Anglo-Indians. This in turn resulted in slew of publications on different aspects of Anglo-Indian genetics in leading international journals.

Unlike the studies of Annandale and Greval and Sarkar, Bhattacharya did not examine Anglo-Indians from Calcutta, the erstwhile imperial capital and a city with a large and diverse Anglo-Indian population. Instead, he studied Anglo-Indians from Lucknow, Kanpur, Delhi, and the state of Maharashtra. Also, unlike the earlier studies Bhattacharya's research entirely omitted traditional anthropometric measurements such as stature, head length, etc. Instead, he mapped an entire repertoire of genetic traits. Blood groups were the oldest genetic traits used to map racial identity. Bhattacharya comprehensively mapped not only the ABO blood groups but also MN groups and the Rh factor. Besides the traditional serological data, among his Anglo-Indian subjects he also mapped a number of newer genetic traits in the form of frequencies of various patterns in finger and palm prints, the ability to taste the bitter-tasting chemical phenythiocarbamide, rates of colorblindness, and even relatively rarely studied genetic traits such as the thickness of the upper cutaneous lip.[55]

I will not go into the details of Bhattacharya's extensive researches into the Anglo-Indians here. Instead, I will only underline a few general facets of his studies. First, his studies were almost entirely genetically focused and included reasonably sophisticated statistical analysis of the data. The statistical framing of hybridity developed by Mahalanobis in the 1920s had clearly survived and flourished into the late 1960s, even as the popularity of the traditional anthropometric measurements Annandale had collected had waned. Second, in line with all the other studies since Mahalanobis,

Bhattacharya insisted that even when the samples were too varied, rendering the conclusions devoid of predictive power, the studies did establish the biological distinctiveness of the Anglo-Indians from other Indians. Finally, notwithstanding the lack of any invocation of primary races, Bhattacharya, like those before him, was never in doubt about the biological determinability of national races like "Indian," "English," etc.

It is this technological framing of hybridity as a way of stabilizing the biological basis of national categories that I call "biometric hybridity." This biometric hybridity is certainly not the open-ended, fluid identity that postcolonial scholars like Bhabha describe. Nor indeed is it necessarily the foundation of some kind of all-are-welcome national imaginary. Rather, racial hybridity in this case is given a very specific meaning that operates not only to reemphasize the fixed biological basis of national communities but also often subtly reinserts hierarchies between different racialized groups.

The hybrid group holds value in these research programs not as an independent community with its distinctive history and culture but as an experimental tool: a model organism that can simultaneously serve scientific and national interests. There is practically nothing in any of the studies conducted on Anglo-Indians to suggest that these studies would in any way even notionally benefit the community itself. Rather the emphatic declarations are always about how the study will illuminate scientific problems or national identities. This is particularly significant since the Anglo-Indian community suffered a fairly precipitous decline immediately after Indian independence given the public perception of their proximity to the erstwhile colonizers.[56] Indeed, Bhattacharya referred to the Anglo-Indians as an "experimental hybrid group"—bringing to mind Herskovits's quip that human beings cannot be "experimentally" produced in laboratories like guinea pigs, fruit flies, or white mice.[57]

Pace common understandings of hybridity, biometric hybridity was something that had to be accessed through strict technical protocols and through the mediation of a range of laboratory-based and statistical techniques. These mediations transformed the socially available group into an "experimental hybrid" from which the truths of biometric hybridity then could be derived. Both the tools and the protocols shifted over time, as did the rationales as to why such truths were sought. Yet what remained constant throughout the four decades we have studied here is that hybridity had a precise technical meaning that rendered it legible, fixed, and measurable in a way that made it useful for both science and nation defining.

Conclusion

Discussions of racial hybridity are often embraced as a progressive moment that pushes back against espousals of rigid racial purity. Postcolonial theorists all too often valorize hybridity as a cipher for fluidity. An exploration of the actual uses of hybridity by scientists studying race in the first half of the twentieth century, however, gives us much cause for caution.

Far from being a simple rejection of biologically fixed racial identities, hybrid studies emerged as a global field of knowledge production on race that, perhaps counterintuitively, stabilized and fixed the meaning of racial hybridity. By remaking the concept as a technical discourse, a range of evolving tools and techniques were deployed to define, classify, and measure hybridity. In so doing, let alone undermining racial identities or race science, it greatly intensified race research.

The emergence of a molecularized and geneticized notion of race around the 1940s made hybridity a productive experimental resource. Living communities with rich histories of social and cultural braiding were massaged into being "experimental hybrids." Such experimental hybrids were of use not only to scientists but also to the newly decolonized nation in India.

Much of the positive spin given to hybridity emerges from the experience of race in Euro-Australo-American contexts, where the lines between white and Black are often stark and where any assertion of mixture is usually a politically progressive move. In countries such as India, however, racial admixture has a very different valence. Recently, Latin American scholars have made a very similar point, arguing that in countries like Brazil, Colombia, and Mexico, "mestizo genomics" carries a very different set of political and intellectual stakes.[58] South Asian uses of the race concept, particularly in the late-colonial and postcolonial periods, are only just beginning to be studied. In order to understand them, however, we must begin by rethinking the place and politics of hybridity in the history of race thinking.

The first step toward such a rethinking will entail a more archivally grounded excavation of the notion of the hybrid, particularly in the archive of race science. It will require a clear mapping not only of what hybridity means but also of how and by what tools it is accessed and mobilized. Most importantly, however, it would need careful and critical examination of who is called a "hybrid" and what pure identities are imagined as standing behind the so-called hybrid. This is what, I hope, the delineation of biometric hybridity does for us. It presents us with a historically and archivally specific figure of hybridity.

Notes

1. Irawati Karve, "Racial Factor in Indian Social Life," in *Studies in Social History (Modern India)*, ed. O. P. Bhatnagar (Allahabad: St. Paul's Press Training School, 1964), 36.
2. Nandini Sundar, "In the Cause of Anthropology: The Life and Work of Irawati Karve," in *Anthropology in the East: Founders of Indian Sociology and Anthropology*, ed. Patricia Uberoi, Nandini Sundar, and Satish Deshpande (Ranikhet: Permanent Black, 2007), 402. For a more nuanced account of Karve's involvement with race science, see Thiago Pinto Barbosa, "Racializing a New Nation: German Coloniality and Anthropology in Maharashtra, India," *Perspectives on Science* 30, no. 1 (2022): 137–66.
3. Homi K. Bhabha, *The Location of Culture* (London: Routledge, 1994); Tony Joseph, *Early Indians: The Story of Our Ancestors and Where We Came From* (New Delhi: Juggernaut, 2018), 52.
4. Anthony Easthope, "Bhabha, Hybridity, and Identity," *Textual Practice* 12, no. 2 (1998): 341.
5. Lisa Gannett, "The Biological Reification of Race," *British Journal for the Philosophy of Science* 55, no. 2 (2004): 323–45.
6. Nadia Abu El-Haj, "The Genetic Reinscription of Race," *Annual Review of Anthropology* 36 (2007): 283–300. For a fuller discussion of the genomic ideas about race, see Dorothy E. Roberts, *Fatal Invention: How Science, Politics, and Big Business Recreate Race in the Twenty-First Century* (New York: New Press, 2011).
7. Suman Seth, *Difference and Disease: Medicine, Race, and the Eighteenth-Century British Empire* (Cambridge: Cambridge University Press, 2020), 170.
8. Jasbir K. Puar, *Terrorist Assemblages: Homonationalism in Queer Times* (Durham, NC: Duke University Press, 2007), 195.
9. Anil Lal and Vinay Lal, "The Cultural Politics of Hybridity," *Social Scientist* 25, nos. 9/10 (1997): 69.
10. For a discussion of the period, its specificities, and its investments in race, see Projit Bihari Mukharji, *Brown Skins, White Coats: Race Science in India, c. 1920–66* (Chicago: University of Chicago Press, 2022).
11. D. K. Bhattacharya, "A Somatometric Study of the Anglo-Indians of India," *Journal of the Anthropological Society of Nippon* 76, no. 2 (1968): 21.

12. Eugen Fischer, "Racial Hybridization: Studies of the Offspring of Boers and Hottentots—Mendel's Law Seen in the Inheritance of Many Characters—Racial 'Prepotency' Not Found to Exist—Mankind a Single Species," *Journal of Heredity* 5, no. 10 (1914): 465.

13. Barbosa, "Racializing a New Nation," 143–44.

14. On Biswas and the "strongly physical anthropological" bent of the department he founded see Barbosa, "Racializing a New Nation," 144.

15. Warwick Anderson, "Hybridity, Race, and Science: The Voyage of the Zaca, 1934–1935," *Isis* 103, no. 2 (2012): 229–53.

16. Earnest A. Hooton, "Progress in the Study of Race Mixtures with Special Reference to Work Carried on at Harvard University," *Proceedings of the American Philosophical Society* 65, no. 4 (1926): 312–25.

17. Hooton, "Progress in the Study of Race Mixtures," 312.

18. Hooton, "Progress in the Study of Race Mixtures," 313.

19. Hooton, "Progress in the Study of Race Mixtures," 314–15.

20. Hooton, "Progress in the Study of Race Mixtures," 315–16.

21. Forrest Clements, "Race Mixture," *American Anthropologist* 33, no. 4 (1931): 649–50.

22. Melville J. Herskovits, "Race Crossing and Human Heredity," *Scientific Monthly* 39, no. 6 (1934): 541.

23. Veronika Lipphardt, "Isolates and Crosses in Human Population Genetics; or, A Contextualization of German Race Science," *Current Anthropology* 53, no. suppl. 5 (2012): S69–82. See also Alexandra Widmer, "Making Blood 'Melanesian': Fieldwork and Isolating Techniques in Genetic Epidemiology," *Studies in History and Philosophy of Biological and Biomedical Sciences* 47, no. A (2014): 118–29.

24. Macfarlane Burnet, "Migration and Race Mixture from the Genetic Angle," *Eugenics Review* 51, no. 2 (1959): 93–97.

25. Jing Zhu, "Measuring Non-Han Bodies: Anthropometry, Colonialism, and Biopower in China's South-Western Borderland in the 1930s and 1940s," *History of the Human Sciences*, advance access (n.d.): 9.

26. Jaehwan Hyun, "In the Name of Human Adaptation: Japanese American 'Hybrid Children' and Racial Anthropology in Postwar Japan," *Perspectives on Science* 30, no. 1 (2022): 167–93.

27. Rosanna Dent and Ricardo Venturo Santos, " 'An Immense Mosaic': Race Mixing and the Creation of the Genetic Nation in 1960s Brazil," in *Luso-Tropicalism and Its Discontents: The Making and Unmaking of Racial Exceptionalism*, ed. Warwick Anderson, Ricardo Roque, and Ricardo Ventura Santos (New York: Berghahn, 2018), 136.

28. Soraya de Chadarevian, "Commentary: Nationalism and Transnationalism in Anthropological Research," *Perspectives on Science* 30, no. 1 (2022): 198.

29. This was part of a much wider move linking genetics and nationalism within late and postcolonial settings. Elise K Burton, *Genetic Crossroads: The Middle East and the Science of Human Heredity* (Stanford, CA: Stanford University Press, 2020). Thiago Pinto Barbosa, "Making Human Differences in Berlin and Maharashtra: Considerations on the Production of Physical Anthropological Knowledge by Irawati Karve," *Südasien-Chronik—South Asia Chronicle* 8 (2018): 139–62; Jaehwan Hyun, "Blood Purity and Scientific Independence: Blood Science and

Postcolonial Struggles in Korea, 1926–1975," *Science in Context* 32, no. 3 (2019): 239–60; Projit Bihari Mukharji, *Brown Skins, White Coats: Race Science in India, c. 1920–66* (Chicago: University of Chicago Press, 2022).
30. Mukharji, *Brown Skins, White Coats.*
31. Keith Breckenridge, *Biometric State: The Global Politics of Identification and Surveillance in South Africa* (Cambridge: Cambridge University Press, 2014).
32. Projit Bihari Mukharji, "Profiling the Profiloscope: Facialization of Race Technologies and the Rise of Biometric Nationalism in Inter-War British India," *History and Technology* 31, no. 4 (2016): 376–96; Projit Bihari Mukharji, "The Bengali Pharaoh: Upper-Caste Aryanism, Pan-Egyptianism, and the Contested History of Biometric Nationalism in Twentieth-Century Bengal," *Comparative Studies in Society and History* 59, no. 2 (2017): 446–76; Mukharji, *Brown Skins, White Coats.*
33. Mukharji, "The Bengali Pharaoh," 452.
34. Mukharji, *Brown Skins, White Coats.*
35. Mukharji, "Profiling the Profiloscope."
36. de Chadarevian, "Commentary: Nationalism and Transnationalism in Anthropological Research," 195.
37. Mukharji, *Brown Skins, White Coats.*
38. Nicholas B. Dirks, *Castes of Mind: Colonialism and the Making of Modern India* (Delhi: Permanent Black, 2002); Crispin Bates, "Race, Caste, and Tribe in Central India: The Early Origins of Indian Anthropometry," in *The Concept of Race in South Asia*, ed. Peter Robb (Delhi: Oxford University Press, 1995), 219–59. Sayori Ghoshal, "Race in South Asia: Colonialism, Nationalism, and Modern Science," *History Compass* 19, no. 2 (2021): 1–11.
39. On the recent schemes of biometric identification of all postcolonial Indian citizens, see Lawrence Cohen, "The 'Social' De-Duplicated: On the Aadhaar Platform and the Engineering of Service," *South Asia: Journal of South Asian Studies* 42, no. 3 (2019): 482–500. On the continuities between biometric nationalism and the recent schemes, see Mukharji, *Brown Skins, White Coats.* For biometric citizenship projects in postcolonial Pakistan, see Zehra Hashmi, "Making Reliable Persons: Managing Descent and Genealogical Computation in Pakistan," *Comparative Studies in Society and History* 63, no. 4 (2021): 948–78.
40. Mukharji, *Brown Skins, White Coats.*
41. Hans Pols and Warwick Anderson, "The Mestizos of Kisar: An Insular Racial Laboratory in the Malay Archipelago," *Journal of Southeast Asian Studies* 49, no. 3 (2018): 445–63.
42. Usha Deka, "Early Instances of Race-Crossing in India," *Man in India* 34, no. 4 (1954): 271–76.
43. S. S. Sarkar, "Tamil-Chinese Crosses in the Nilgiris, Madras," *Man in India* 39, no. 4 (1959): 309–11.
44. Mukharji, "Profiling the Profiloscope."
45. P. C. Mahalanobis, "Anthropological Observations on the Anglo-Indians of Calcutta—Part I. Analysis of Male Stature," *Records of the Indian Museum* 23 (1922): 1–96.
46. Mahalanobis, "Anthropological Observations—Part I."
47. Mahalanobis, "Anthropological Observations—Part I," 66–67.

48. Mahalanobis, "Anthropological Observations—Part I"; P. C. Mahalanobis, "Anthropological Observations on the Anglo-Indians of Calcutta—Part II. Analysis of Anglo-Indian Head Length," *Records of the Indian Museum* 23 (1931): 97–149; P. C. Mahalanobis, "Anthropological Observations on the Anglo-Indians of Calcutta—Part III. Statistical Analysis of Measurements of Seven Characters," *Records of the Indian Museum* 23 (1940): 151–87.

49. S. D. S. Greval and S. N. Chandra, "Blood Groups of Communities in Calcutta," *Indian Journal of Medical Research* 27 (1940): 1109–16.

50. Mukharji, *Brown Skins, White Coats.*

51. S. S. Sarkar, "Analysis of Indian Blood Group Data with Special Reference to the Oraons," *Transactions of the Bose Research Institute* 15 (1942): 1–15.

52. On the Rodenwaldt study, see Pols and Anderson, "The Mestizos of Kisar."

53. S. S. Sarkar, Bhuban Mohan Das, and Kamal Krishna Agarwal, "The Anglo-Indians of Calcutta," *Man in India* 33, no. 2 (1953): 93–103.

54. Bimal Kumar Mandal, "Bhoodan Movement of India and Its Impact," *Proceedings of the Indian History Congress* 76 (2015): 837–43.

55. D. K. Bhattacharya, "Tasting of P.T.C. Among the Anglo-Indians of India," *Acta Geneticae Medicae et Gemellologiae* 13, no. 2 (1964): 159–66; D. K. Bhattacharya, "Finger Dermatoglyphic Study of the Anglo-Indians of India," *Zeitschrift für Morphologie und Anthropologie* 54, no. 3 (1964): 346–54; D. K. Bhattacharya, "The Palmar Dermatoglyphics of the Anglo-Indians of India," *Zeitschrift für Morphologie und Anthropologie* 55, no. 3 (1964): 357–67; Bhattacharya, "A Somatometric Study of the Anglo-Indians of India"; D. K. Bhattacharya, "The Anglo-Indians in Bombay: An Introduction to Their Socio-Economic and Cultural Life," *Race* 10, no. 2 (1968): 163–72; D. K. Bhattacharya, "A Study of ABO, RH-HR, and MN Blood Groups of the Anglo-Indians of India," *Human Biology* 41, no. 1 (1969): 115–24.

56. Noel P. Gist and Roy Dean Wright, *Marginality and Identity: Anglo-Indians as a Racially-Mixed Minority in India* (Leiden: Brill, 1973).

57. Bhattacharya, "The Palmar Dermatoglyphics of the Anglo-Indians of India," 361.

58. Peter Wade et al., "Introduction: Genomics, Race Mixture, and Nation in Latin America," in *Mestizo Genomics: Race Mixture, Nation, and Science in Latin America*, ed. Peter Wade et al. (Durham, NC: Duke University Press, 2014), 1–32.

SEVEN

"Multicultural Genes in Our Blood"?

Genetic Governance and Biocultural Purity in South Korea

JAEHWAN HYUN

"MULTICULTURAL GENES ARE carved in the [Korean] history," says Lee Chan-Wook, director at the Institute of Multicultural Content Research of Chung-Ang University.[1] According to his historical analysis, an influx of immigrants from Asia had already happened in the ancient period, and thus, since that period, Koreans were already a multicultural nation (*Tamunhwa minjok*).[2] Lee found traces of "multicultural genes" in the Korean people by illuminating Korean family names known to have originated from Chinese, Mongol, Jurchen, Uighur, Arab, Vietnamese, and Japanese lineages. Although in the first half of the twentieth century Japanese colonialism and the recent national division after World War II provoked strong, pure-blood nationalism in South Korea, Koreans have always been multicultural both in terms of culture and biology.[3]

Lee's research tells us two things about multiculturalism in South Korea. First, the multicultural is often equated with the multiethnic. Koreans have long considered themselves a distinct ethnic group sharing a unified bloodline and a singular culture. In this sense, new cultures came to the Korean peninsula mostly as a result of the influx of new ethnic groups. In this vein, international couples with their mixed-origin children are categorized as "multicultural families" (*Tamunhwa kajŏng*).[4] Second, genetic discourses play a role in delineating national identity in terms of multiculturalism. The multiethnic past of Koreans is represented in the term "multicultural genes." This is closely linked with the fact that Korean geneticists have continuously reported the heterogeneous origins of the contemporary Korean population via their genetic history research.[5] Genes serve as the most authoritative voice with which to refute the old pure-blood nationalism and reveal South Korea's true character: the multicultural or, more correctly speaking, multiethnic nation.

In recent years, with the global rise of multiculturalism, science and technology studies (STS) scholars have found the ways multicultural norms and logics of group inclusion became incorporated into biomedical genetic research. Already during the last two decades of the twentieth century, resonating with the refashioning of the United States as a multicultural society during that period, inclusionary policies and practices considering ethnic differences were promoted in U.S. biomedical research and health care.[6] In the countries where multiculturalism was already well institutionalized, human genomic research was framed and legitimatized by multicultural logics. In Canada, census classifications including "visible minority" and "aborigine" categories, which had originally been produced in multicultural politics, were widely adopted, and such "nonwhite" groups are included in designating large-scale human genomic research to gain public legitimacy.[7] In the United Kingdom, the aim and results of one genetic history research project studying premodern settlements in Britain were interpreted and presented in multicultural terms. Pointing to the waves of migrations and mixed populations in the premodern period, scientific enterprise served as proof of the innate multicultural nature of British society.[8] The global turn toward multiculturalism has also influenced human genomic research in Latin American countries such as Columbia, Brazil, and Mexico, where race mixture was highlighted and celebrated as a core value for nation building already from the mid-twentieth century. In particular, the language of purity, sometimes applied to indigenous groups in past genetic research, has faded, giving way to greater emphasis on mestizos as a biomedical object.[9] The STS scholar Amy Hinterberger brilliantly coined the term "molecular multiculturalisms" to describe such an emerging involvement of human genetic research with multicultural politics.[10]

This chapter contributes to the previous literature by focusing on the flipside of molecular multiculturalisms—the role of genetic sciences, technologies, and thinking in governing new multicultural subjects. It examines how genetic sciences and discourses play out in immigration policy making, legal procedures, and biomedical research in contemporary South Korea, with a focus on the country's recent multicultural turn. By doing so, this chapter argues that, despite the recent discursive shifts toward the more "multicultural" or "multiethnic" nature of Koreans, a microethnic classification centered on the purity of Koreans—what I call biocultural purity—is widely shared and used in practices of DNA testing for citizenship applications, criminal investigation, and medical and forensic genetic research. It is necessary to make a distinction between the terms "biocultural" and "biosocial," as used by anthropologists of science and biomedicine, and the "biocultural" conception of purity, which I will use in this chapter. Anthropologists have coined diverse concepts such as "biosocial becomings" and "local biologies" to problematize the dichotomy of nature and culture and decentralize molecular reductionism.[11] In delineating those concepts, they highlight the "inseparably entangled" social and biological process resulting in human biology.[12] In contrast, here I use the "biocultural" in the sense that Korean officials, medical and legal authorities, and scientific experts identify the microbiological aspects of cultural differences among human groups and use them as criteria to classify "Koreans" and non-Koreans in liberalist governance and biomedical research. I specifically use the term "purity" because these officials, authorities, and experts define the biocultural properties as being unique and as only belonging to Koreans.

This paradoxical situation is closely linked with the longstanding politics of Korean national identity. Korean nationalism has been fed by the widespread belief in ethnic homogeneity (*Tanil minjok*), that is, claiming Koreans share a single bloodline and thus are biologically homogenous. For this reason, mixture was often considered a threat to national integration, and mixed-origin people were socially discriminated against.[13] In the 1990s and 2000s, when South Korea went through economic liberalization, such claims were considered the biggest drawback to the Korean government's new, "multicultural" immigrant policy, intended as a way to resolve labor shortages and increase birth rates through labor migration and intermarriages from low-income Asian countries.[14] Supporting and coinciding with this political shift, geneticists reported on the outcomes of Y chromosomal and mitochondrial DNA research indicating that Koreans were not a single-origin race but an outcome of mixed ethnic groups from northern and southern Asia.[15] While the genetic argument of multiethnic origins has been widely accepted and used to support the multicultural policy by the Korean government and related experts, governmental authorities and scientists have continued to use the categories of "Korean" and "non-Korean" in practices of public health, legal governances, and biomedical research. As a result, despite ethnic mixture being now regarded more positively in general, the idea of purity still underlies this ostensible celebration of the multicultural and multiethnic past and present of Korean people.

Recasting the Idea of Mixture and Inventing Multicultural Families

Until the early 2000s, South Korea was considered one of the least ethnically diverse countries, along with its northern counterpart and Japan.[16] The statistics, however, do not paint the whole picture, failing to reflect the fact that South Korea's ethnic homogeneity was a post–World War II outcome of the South Korean government's systematic exile of nonethnic Korean minorities and the deprivation of their political rights.[17] Irrespective of the recent origin of ethnic homogenization, such a demographic "fact" contributed to a narrow conceptualization of Korean national identity that defines true Koreans such that they "not only must have Korean blood, but must also embody the values, the mores, and the mind-set of Korean society."[18]

This cultural and biological essentialism of Koreanness owes to the long-term usage of the concept *minjok*, which was originally derived from the German ethnonational concept of *Volk* imported via its Chinese transliteration *minzu*.[19] The trope of *tanil minjok*, a single ethnonational people originated from a single mythic ancestor, Grandfather Tan'gun, invented by early nationalist historians to combat imperialism a century ago, has survived and continued to interact with the political mobilizations of pure-blood nationalism in the independent movement during the colonial period and the authoritarian government in the postcolonial period.[20] Pure-blood nationalism within the country proceeded in tandem with hatred directed at "mixed-blood" people.[21]

The concept of blood mixture (*honhyol*) was recast as a desirable and positive value for the first time in South Korean history in the mid-2000s. The South Korean government decided to redirect immigration policy to widen the scope of citizenship and residence rights to include Korean diasporas and foreign migrants in the early 2000s. This was partly inspired by the Roh Moo-hyun administration's (2003–2008) progressive human rights policy orientation. The main motivation was economic, or, more specifically, the aim was to utilize migrants to boost national economic growth.[22] In April 2006, President Roh Moo-hyun announced that "the transition of South Korea into a multicultural and multiracial society is unchallenge-able," so Koreans "need to integrate foreign migrants into the Korean society by promoting the multicultural policy."[23] Defining beliefs about blood purity as the biggest problem to tackle for the social transition, the government ordered the deletion of pure-blood nationalistic expressions, including Korean ethnic homogeneity, from school textbooks.[24] In the same month, a population economics research group speculated that "Kosians," mixed-origin Koreans from Asian-migrant mothers and Korean fathers, would make up over one-third of the total infant population and 3 percent of the total population by 2020. The Korean media reported this speculation as the scientific prospect of the rise of mixed-origin Koreans as a political force playing a similar role to the one played by Hispanic groups in U.S. politics.[25] Social welfare theorists and other social scientists inspired by multicultural studies in the United States, Canada, and Australia, supported this government-led multiculturalism while claiming South Korea was transitioning from a pure-blood nationalist society into a mixture-based multicultural one. Celebrating mixed-origin Korean children as core members of South Korea, the government and social welfare theorists introduced blood mixture into the discourse of Korean national identity.

The success of the transition should not be exaggerated. Anti-immigration and antimulticultural sentiment continued to appear, in particular in online forums dominated by far-right male users.[26] The xenophobic fear of multiethnic coexistence renewed old hatred for ethnic mixture.[27] Nevertheless, at least in mainstream newspapers and government policy documents, blood mixture became seen as an unprecedented keyword for South Korea's multicultural future.

The new immigrant policy produced "new" Koreans and foreign residents from the Korean diaspora and former foreign workers.[28] Since 2008, when it comes to social welfare policy, the government has governed these new national members under the category of the "multicultural family." The 2008 Support for Multicultural Families Act confines its subject to the families of marriage migrants and naturalized citizens. In national census surveys, however, the category includes the families of international marriage couples, ethnic Koreans moving to South Korea, foreign workers, and North

Korean refugee families. In 2019, the number of members of "multicultural families" passed 1.06 million and 2 percent of the total population of South Korea (51 million).[29] It was a significant rise, given that in 2005, the total number of foreign residents was only 0.5 million.

Multicultural families emerged as a new population for the government's concern. The South Korean government had to take care of this new population's political rights and health and deal with their crimes. In this context, granting citizenship to the population's members was the first hurdle to overcome in this new immigration governance. As we shall see, in contrast to the explicit, public objection to the language of purity, in citizenship-granting procedures, the government continued to use group classifications that were centered on ethnic purity. It was this use of purity-centered classification that would bring genetic technologies and ways of thinking to the legal process.

Governing Via Microtaxonomy

A distinct feature of South Korea's recent migration pattern is the importance of intra-Asian migration. The majority of labor and marriage migrants and members of the Korean diaspora come from China, Vietnam, and other Southeast Asian countries given their geographical proximity.[30] For instance, the majority of the Korean diaspora who "restored" their South Korean citizenship were Chinese Korean migrants—often called *Chosŏnjok*—whose ancestors migrated from colonial Korea to Manchuria and other northern Chinese areas during the first half of the twentieth century. In rural areas, "international brides" mostly originate from China, Vietnam, Cambodia, and the Philippines. This all contributes to the construction of the microtaxonomy of Koreans within "an Asian race." This does not, however, mean that there is no racial recognition in South Korea. Indeed, skin-color-based racial discrimination is very prevalent and has been seriously criticized by both domestic and international observers.[31] Nonetheless, the majority of new migrants are Asians. As a result, in bureaucratic and everyday practices, the government and laypeople still distinguish between "pure" Koreans, ethnic Koreans from abroad, and other Asia-origin naturalized Koreans, despite the fact that the national census puts all non-"pure" Koreans into the "multicultural family" category.

Irrespective of its new gesture toward celebrating mixture, the South Korean government has continued to use ethnic purity as one of the most important criteria in setting the legal boundaries of national belonging. While the requirements for nonethnic Koreans wishing to obtain permanent residency visas have been lowered, it is still difficult for them to acquire citizenship. This is related to the fact that South Korean nationality law adopts a *jus sanguinis* (right of blood) principle that confers citizenship based on a person's ancestry. A person is deemed to be a South Korean national only if one or both parents are South Koreans.

The recent diaspora-management policy, as a way to embrace the overseas Korean capital and labor as a new source for the South Korean economy, has been extended to include descendants of Korean overseas migrants before the establishment of South Korea in 1948.[32] The Overseas Korean Act, passed by the Korean National Assembly in 1999 and revised in 2004, defines overseas Koreans as "a person prescribed by Presidential Decree from among those who, having held the nationality of the Republic of Korea (including those who had emigrated abroad before the Government of the Republic of Korea was established) or as their lineal descendants, have acquired the nationality of a foreign country."[33] Members of the Korean diaspora can acquire work and residential permits much more easily than nonethnic Koreans and are eligible for citizenship after only three years' residence in South Korea. This simplified naturalization process is only open to members of the Korean diaspora since they are supposed to share the bloodline. In the same rationale, North Korean refugees, who are considered full-blooded Korean people, are automatically conferred as South Korean citizens in the legal sense. Within this ethnicity-based citizenship policy, the most plausible way for nonethnic Koreans to acquire South Korean citizenship is to establish kinship via marriage to a Korean national.

In this context, genetics has entered into the bureaucratic process. In the U.S. immigration legal process for permanent residency acquisition and family reunification, biomedical criteria play a crucial role in conferring citizenship, such that they form a basis of "biopolitical citizenship."[34] In the same way, DNA testing has become a governing tool for nationality decision making in South Korea. In 2005, DNA testing entered into nationality application procedures.[35] The government allowed Korean Chinese migrants to restore their South Korean citizenship if they could offer DNA test results that proved their relationship with native family relatives residing in South

Korea. Coincidently, not only ethnic Koreans outside of the Korean peninsula but also mixed-origin children fathered by South Koreans living in Vietnam and the Philippines started using DNA testing to claim Korean citizenship.[36] Genetic testing became the most cost-efficient evidence that applicants for nationality could use to prove their ethnic membership.

Even though members of the Korean diaspora were able to obtain South Korean citizenship, they are still not considered "authentic" South Koreans, specifically in the case of those coming from China or North Korea. In South Korea, the hierarchy of Korean nationhood works as an underlying principle for social inclusion and exclusion.[37] Some diasporic communities, mainly those from the United States and other wealthy European countries, are "accorded privileged status," while other communities are "seen and treated as inferiors or easy-to-exploit, fully expendable labor" despite their "shared blood" with South Koreans.[38] This mirrors the South Korean government's economic motivations for opening their citizenship to Korean diaspora populations in the first place, such that its diaspora engagement policy specifically targeted diasporic elites living in the wealthiest countries. On the flip side, this same policy excluded other diasporas, mainly those populations in China who lacked higher education and capital. North Korean refugees are in the same situation as Korean Chinese migrants. Although, in contrast to the Korean Chinese, North Korean refugees are automatically conferred South Korean citizenship, they are still considered inferior not only in terms of culture but also in terms of "biology" because of what is considered their malnutrition and poor public health practices. For instance, South Korean news reports in 2017 on a North Korean defector highlighted his miserable health condition, including his much lower height and weight than the average South Korean male and the enormous number of parasitic worms found in his stomach.[39] It is unsurprising that this hierarchy is also applied to nonethnic Koreans who have Korean citizenship and their mixed-origin children.[40]

The visible and invisible stratification among Korean nationals relied on the idea of purity in terms of biology and culture: North Korean refugees, Chinese Korean diasporas, Asian-origin marriage migrants, and "Kosians" living in and outside Korea. Chinese Korean migrants and North Korean refugees are not considered authentic Koreans because of their cultural differences and different physiques, which embody the different socionutritional environments they might have lived in. Multicultural families consisting of marriage migrants and mixed-origin children are considered to be different from "normal" South Korean families because of their cultural differences and different hereditary properties. In both cases, they are classified as non-"pure" Koreans following the logic of biocultural purity.

In this sense, the biological and cultural essentialism of Koreanness is not dismissed by the new inclusive immigration regime. Instead, it remains embedded in the emerging governance of these "new" Koreans. In the latter part of this chapter, I will show how the microclassification based on biocultural purity operates within scientific research on the governing of new Koreans, specifically in the domains of public health and police enforcement.

The Microtaxonomy in Biomedical Genetic Research

The STS scholar Steven Epstein coined a concept of "categorical alignment" to describe the way in which scientists, health movement activists, and health authorities superimpose social categories of human difference from everyday life and state bureaucracies onto biomedical research in the United States.[41] They used race, ethnicity, and other grouping categories as a proxy to identify biological differences at the population level. In the United States, the categories of differences used in identity politics were deliberately introduced in human genomic research, as a way to challenge health inequalities and connect benchwork to social justice.[42] In a similar vein, Korean medical researchers have introduced the social categories of "new" Koreans into their research and have framed their work as part of multicultural welfare services.

The foremost biomedical research is the Emigrant and Immigrant Studies of the Korean Genome and Epidemiology Study (KoGES). The KoGES is the government-led large-cohort project that collects epidemiological and genomic data and biospecimens from the Korean population "to identify gene-environmental factors and their interactions in common chronic diseases" prevalent in South Korea. About 245,000 participants have undergone medical examinations and biospecimen collection through this project since 2001.[43] In 2005 and 2006, when the government announced the multicultural turn of Korean society, two prospective cohort studies—the Emigrant and Immigrant Studies—began to study gene-environment interactions in common chronic diseases. The Emigrant Study targeted the Korean diaspora and their descendants residing in Changchun, China, and Kobe, Japan, for more than fifteen years. The immigrant study recruited women subjects who were marriage immigrants from Southeast Asia and their mixed-origin children.[44]

Chinese Koreans and Asian-marriage immigrant women, as subject groups, were not randomly selected. Chinese Koreans had come to the attention of doctors in South Korea's hospital context, coinciding with the large-scale influx of those immigrants to the South Korean labor market in the early 2000s. Medical researchers became fascinated with apparently different incidence rates to those displayed in South Koreans despite the supposed genetic homogeneity between the two populations, so that already before the initiation of the KoGES Emigrant Study they had started to carry out preliminary comparative genetic association studies between Chinese Koreans and South Koreans. KoGES researchers expected that collecting the epidemiological and genomic data of the Chinese Korean population would enable them to separate environmental factors from genetic ones in chronic diseases.[45]

Immigrant women from Southeast Asia were important in two aspects. First, the new genetic history research reported that the modern Korean population had originated from northern Asian and southern Asian lineages.[46] While the northern Asian cohorts had been intensively aggregated by the Chinese government, southern Asians were relatively understudied. The KoGES scientists claimed the need for collecting southern Asian epidemiological and genomic data to understand the impact of the southern Asian lineage–derived genetic factors. In particular, the immigrant women would be more likely to modify their lifestyle, including nutritional behaviors, to align much more closely with South Koreans. It would show which environmental factors of South Koreans could be the cause of chronic diseases.[47] Second, with the rise of "multicultural" policy, the marriage-immigrant women's health became a focal point specifically in relation to their reproductive potential. These women's resettlement support program, the main part of the multicultural policy, was a policy made in response to the rapid fertility decline and drastic aging of the population in South Korea.[48] Thus, the KoGES Immigrant Study offered gynecological examinations and consultations.[49]

The cohort study was also intended to give the Korean diaspora and marriage immigrants a sense of national belonging. The cohort surveys consisted of medical checkups and health consultations that on the whole Chinese Koreans and marriage immigrants would not otherwise have had access to given the relatively high cost in comparison to their low income. These free, regular medical checkups and consultations were intended to inspire these "new" Koreans' loyalty to the state.[50]

As we have seen, the KoGES cohort project classified Chinese Koreans and Asian-marriage immigrant women differently from South Koreans. The biological and cultural differences were assumed in the microclassification. The Chinese Koreans were considered a distinct population from the South Korean population given their culture. They had different dietary habits and lifestyle patterns, closer to those in mainland Chinese than to South Koreans.[51] Moreover, the Chinese Koreans mostly lived in northern China, where there was a colder and supposedly more polluted environment with poorer public health systems than South Korea's. The result of their Chinese-like lifestyle and health care system access was revealed by their health conditions, which differed accordingly from South Koreans. Asian marriage immigrants, on the other hand, were differentiated from the South Korean population because of their biology.[52]

In the categorizing work, racial and ethnic categories were used quite arbitrarily and were dependent on social and material confinements surrounding the cohort projects. For instance, in the Immigrant Study, the mixed-origin children of the Asian immigrant women were classified as being genetically different from South Koreans because of their non-Korean maternal lineage. In the Emigrant Study, however, some Chinese Korean participants mothered or fathered by nonethnic Korean parents were counted as ethnic Koreans. A pragmatic reason for this was that researchers were unable to recruit enough Chinese Koreans with two Korean parents, thus leading to KoGES researchers' lowering their biological homogeneity criteria.[53]

The category of "Southeastern Asians" in the Immigrant Study was also quite problematic in terms of consistency. In the first pilot cohort study, KoGES researchers used the term "Southeastern Asians" to define the cohort subjects, immigrant women from Vietnam, Cambodia, Thailand, and the Philippines.[54] The following year, they replaced the term "Southeastern Asians" with "East Asians," although the latter term had mostly previously been used to indicate Chinese, Japanese, or Koreans in South Korea.[55] There is no explanation given for the change in the KoGES reports, but it is probably related to the fact that *Tongnama*, the abbreviation for "Southeastern Asians" in Korean, has been widely used as a derogatory term for nonethnic Koreans in South Korea. In this context, "East Asian immigrants" would be more politically correct terminology, even though the classification sounds socially and academically incorrect.

Genetic Testing and Identification of "Multicultural" Crime

Although forensic DNA testing had already become integrated into criminal investigation in this country from the early 1990s, genetic technologies became increasingly crucial in tandem with the multicultural turn. Korean criminologists became wary of the increase in foreign, or "multicultural," crimes that coincided with the ease of immigration rules and influx of foreigners.[56] Such initial concerns turned out to be unfounded, since the rising rate of foreign crime has always been much lower than that of South Koreans. Irrespective of the statistics, their concerns continued to rise, in part in relation to the national biometric database.[57] According to the Korean national identity program, all Korean citizens over seventeen years old must register their fingerprints at a local government office. This data is part of the national biometric database for fingerprints used in criminal investigations. Foreigners represented a threat to this. In 2003, the government suspended this requirement for foreigners entering South Korea, meaning that unlike native South Koreans, they were not required to register in the national fingerprint database. This continued to be the case until 2012, when a new biometric regulation of fingerprints and facial photo collection was enforced as part of the entry inspection process.[58]

In this context, in light of the lack of a foreign biometric database, the Korean National Forensic Service (NFS) began to consider the use of DNA testing to identify suspects' nationality, albeit to a limited extent. For instance, NFS scientists used the Y-DNA database of the South Korean population and Korean-specific haplogroups to confirm a male suspect's nationality—whether the suspect was a South Korean or not—in sexual violence and homicide cases.[59] An opportunity to expand the use of DNA testing for foreign identification came to forensic scientists soon enough. The

DNA Identification Act, which contained the rules that governed the collection of DNA samples from criminal suspects and how the information from the samples should be stored, was enacted in 2010.[60]

Korean forensic scientists had been particularly interested in introducing ancestry informative markers (AIMs) analysis in forensic practices since the late 2000s.[61] AIMs are a set of genetic markers exhibiting different frequencies between human populations. In its early infancy, short-tandem repeats on the Y chromosome (Y-STR) were widely used. Soon, single nucleotide polymorphism (SNP) markers replaced Y-STR markers. The analysis process was simple: a comparison of a sample of unknown origin with the relative frequencies of the world populations' AIMs to infer the sample's ancestry admixture ratios. This was originally developed by U.S. forensic geneticists seeking a way to identify the "ethnicity" of unknown suspects of crimes using DNA evidence.[62] Korean researchers believed that South Korea's crime investigation should also be equipped with ethnic profiling in order to respond to the country's demographic transition to a multiethnic coexistence.

The only part of the AIMs analysis that Korean forensic researchers were unsatisfied with was its classification using the concept of "biogeographical ancestry"—which, in human genomic research, was often introduced as an alternative to race.[63] Biogeographic ancestral groups consisted of "African, African-American, European, European-American, Native American, Hispanic, East Asian, and South Asian heritage."[64] This classification was an American invention that proved useful in the U.S. context of racial composition but was not suited to the Korean multiethnic scene: the majority of South Korea's immigrants are from China and southeastern Asian countries. Mongolian, Chinese, Pakistani, and Vietnamese laborers and "illegal" immigrants were recognized as the populations demonstrating a high crime rate. In particular, Chinese Koreans and Mongolians were marked groups watched by criminologists and the media.[65] For this reason, Korean scientists sought a way to develop statistical methods to identify intra-Asian differences and to discover Korean-specific genetic makers allowing for a differentiation of "pure" Koreans from neighboring Asian groups like the Chinese and Japanese.[66]

Forensic scientists in the NFS and universities have strived to develop genetic ancestry testing while explicitly claiming its relation to the possible rise in multicultural crimes. A three-year project titled "The Development of DNA Profiling for Identification of Ethnic Affiliation," supported by the NFS from 2015 to 2017, might be an example of this. The project aimed to establish the multiplex genotype system using mtDNA, Y-DNA, and the autosomal DNA database and develop ancestry inference software in order to help criminal investigators specify a perpetrator's ethnicity. In the project report, its necessity was justified by the increasing number of crimes committed by Asian immigrants.[67] Another rationale was to mitigate racial discrimination in the early phase of criminal investigation. South Korean police officers, often unable to understand Asian immigrants' language and customs, might unnecessarily discriminate in cases where immigrants were listed in a suspect pool. Ethnic profiling would prevent such situations and thus contribute to making criminal investigation more adaptable to the multiethnic future.[68]

Although the ancestry inference software was released as the project's outcome in 2017, the software is not yet widely used in forensic practice. Nevertheless, NFS researchers are still devoted to customizing the AIMs analysis to identify intra-Asian differences. In a recent paper cowritten with Australian forensic researchers, NFS scientists reported finding new AIMs for differentiating between Asian populations from "China, Japan, Indonesia, Philippines, South Korea and Thailand."[69]

An intriguing point in this customization process is the slippage between ancestry, ethnicity, and race that occurred when translating the ancestry concept into Korean. Korean forensic scientists understood the ancestry concept to be equal to *minjok* in Korean. As we saw, the term *minjok* has ethnonational and sometimes racial connotations. By translating the term "biogeographic ancestry" as *minjok*, forensic researchers opened a possibility to use the three different concepts of race, ethnicity, and ancestry (in English) as one identical entity. In this context, they translated the word Korean *minjok* as "Korean race" and sometimes "*minjok* profiling" (*minjok sikbyeol*) as "ethnic profiling."[70]

In criminology and forensic science, as in biomedical research, biocultural purity works as a criterion to differentiate "multicultural" crimes from domestic ones. The Korea National Statistics Office counts naturalized Chinese Koreans and Asian-origin marriage migrants as "naturalized foreigners," one category of foreign nationals. Crime statistics applies the same logic so that criminal cases committed by naturalized Chinese Koreans are counted as crimes committed by Chinese nationals residing in South Korea.

A similar thing happens in the ethnic-profiling work. In the criminal DNA database, Chinese Koreans who reside in South Korea using a special work permit for members of the Korean diaspora but who are not yet naturalized are categorized as Chinese nationals. As a result, their DNA information is considered non-Korean. This happens in part because of the legal status of not-yet-naturalized Korean Chinese. But beyond that, forensic scientists exclusively use samples and genetic data obtained from "authentic" South Koreans in their search for Korean-specific markers, as they have done in previous decades. By not changing the sampling and collecting practices, forensic scientists do not offer any possibility of including the "new" Koreans in the Korean DNA database. Once again, biocultural purity functions as a criterion to differentiate "new" Koreans, produced by an, in theory, inclusive multicultural policy, from authentic South Koreans.

Concluding Remarks

This paper has examined how the idea of purity operates as an underlying principle for the classification of Koreans and how this purity-based taxonomy intersects with the genetic sciences and technologies in legal, scientific, and bureaucratic forensic practices. The Korean sociologist Jinwoong Kang points out that the multicultural turn visible in education did not root out the beliefs about Korean ethnic homogeneity in South Korean school textbooks. Rather, the South Korean state "renationalized" Koreans by highlighting the differences between immigrants and South Koreans: the elementary school–level social studies textbook states the need for showing tolerance toward the cultural behaviors of North Korean refugees and other "new" Koreans while still clearly differentiating them from the "we" of South Koreans.[71] Together with this, the hierarchical stratification of Korean nationals from Asian migrants and the Korean Chinese diaspora to "authentic" South Koreans serves as a taxonomy for differentiating them in the country's social and political life.[72] In the same way, intersecting with the new technoscientific possibilities such as genomic epidemiology and AIMs, the government, scientists, and forensic practitioners reproduce this purity-based taxonomy in their bureaucratic and scientific activities.

South Korea's molecular multiculturalism is worth dwelling on from a comparative perspective. It shows how new scientific and technological possibilities' entwinement with South Korea's intra-Asian circulation-centered immigration has led to using a microclassification within the larger category of Asians. It has forced biomedical and forensic scientists to designate research that assumes or identifies microgenetic differences—most of which do not appear at a phenotypic level—within Asians. This contrasts greatly to the categorization used by Anglo and Latin American countries, where multicultural subjects are distinguished from the majority population in phenotypical terms. Despite phenotypical similarities, however, in South Korea the biological sense of purity is not removed from immigration governance and related scientific research. Although the South Korean state and major media outlets loudly celebrate blood mixture as being the genuine nature of their population and as the future direction of the state, this does not translate into abandoning the idea of ethnic purity, a core set of South Korea's nation building, and instilling it in the new immigration governance.

Another distinct feature of the South Korean case is its lack of ethical engagements in medical and genetic research projects. In Anglo and Latin American countries, the inclusion of minorities—beneficiaries of institutional multiculturalism—is used as a way to gain public legitimation and secure ethical grounds for human genomic research. In contrast, the KoGES and AIMs development projects in South Korea included immigrant populations in their research to facilitate the state's control of them. The nonethical and state-centric nature of inclusive research is closely linked with the fact that multiculturalism is not a result of minority politics from the bottom but an outcome of top-down immigration policy. The South Korean government promoted the multicultural policy as a new drive for national economic growth in the mid-2000s, so the inclusive gesture in the medical and forensic research projects can only be understood in terms of the government's immigrant management or possible commercial profits. There is thus no place here for social justice. Unfortunately, such a lack of connection to social justice is a characteristic of South Korean multicultural science.

Notes

1. Chan-Wook Lee, "Naturalized Surname and Multiculture of Korea," *Journal of Multicultural Contents Studies* 17 (2014): 253–77 (in Korean).
2. *Minjok* is a local grouping category that includes the ideas of race, ethnicity (or tribe), and nation in English usage. I will deal with the issue in more detail in

the main text. For its conceptual history, see Gi-Wook Shin, *Ethnic Nationalism in Korea: Genealogy, Politics, and Legacy* (Stanford, CA: Stanford University Press, 2006).

3. Lee, "Naturalized Surname and Multiculture of Korea," 272.

4. Nora Hui-Jung Kim, "Korea: Multiethnic or Multicultural?," in *Multiethnic Korea? Multiculturalism, Migration, and Peoplehood Diversity in Contemporary South Korea*, ed. John Lie (Berkeley: Institute of East Asian Studies, University of California, Berkeley, 2014): 79–94.

5. Jaehwan Hyun, "Tracing National Origins, Debating Ethnic Homogeneity: Population Genetics and the Politics of National Identity in South Korea," *Historical Studies in the Natural Sciences* 49 (2019): 351–83.

6. Steven Epstein, *Inclusion: The Politics of Difference in Medical Research* (Chicago: University of Chicago Press, 2008).

7. Amy Hinterberger, "Categorization, Census, and Multiculturalism: Molecular Politics and the Material of Nation," in *Genetics and the Unsettled Past: The Collision of DNA, Race, and History*, ed. Keith Wailoo, Alondra Nelson, and Catherine Lee (New Brunswick, NJ: Rutgers University Press, 2012), 204–24.

8. Catherine Nash, "Genome Geographies: Mapping National Ancestry and Diversity in Human Population Genetics," *Transactions of the Institute of British Geographers* 38 (2013): 193–206.

9. Peter Wade, *Degrees of Mixture, Degrees of Freedom: Genomics, Multiculturalism, and Race in Latin America* (Durham, NC: Duke University Press, 2017), 181.

10. Hinterberger, "Categorization, Census, and Multiculturalism," 218–20.

11. Tim Ingold and Gísli Pálsson, *Biosocial Becomings: Integrating Social and Biological Anthropology* (Cambridge: Cambridge University Press, 2013); Margaret Lock and Vinh-Kim Nguyen, *An Anthropology of Biomedicine* (Malden, MA: Wiley-Blackwell, 2010).

12. Margaret Lock, "Comprehending the Body in the Era of Epigenome," *Current Anthropology* 56 (2015): 161.

13. Nadia Y. Kim, *Imperial Citizens: Koreans and Race from Seoul to LA* (Stanford, CA: Stanford University Press, 2008).

14. Andrew Eungi Kim, "Global Migration and South Korea: Foreign Workers, Foreign Brides, and the Making of a Multicultural Society," *Ethnic and Racial Studies* 32 (2009): 70–92.

15. Han-Jun Jin, Kyoung-Don Kwak, Michael F. Hammer, Yutaka Nakahori, Toshikatsu Shinka, Ju-Won Lee, Feng Jin, Xuming Jia, Chris Tyler-Smith, and Wook Kim, "Y-Chromosomal DNA Haplogroups and Their Implications for the Dual Origins of the Koreans," *Human Genetics* 114 (2003): 27–35.

16. Alberto Alesina, Arnaud Devleeschauwer, William Easterly, Sergio Kurlat, and Romain Wacziarg, *Fractionalization*, NBER Working Paper Series 9411 (Cambridge, MA: National Bureau of Economic Research, 2002).

17. John Lie, "Introduction: Multiethnic Korea," in *Multiethnic Korea? Multiculturalism, Migration, and Peoplehood Diversity in Contemporary South Korea*, ed. John Lie (Berkeley: Institute of East Asian Studies, University of California, Berkeley, 2014), 8–15.

18. Timothy Lim, "Who Is Korean? Migration, Immigration, and the Challenge of Multiculturalism in Homogeneous Societies," *Asia-Pacific Journal* 30 (2009): e3192.

19. Gi-Wook Shin, "Nation, History, and Politics: South Korea," in *Nationalism and the Construction of Korean Identity*, ed. Hyung Il Pai and Timothy R. Tangherlini (Berkeley: Institute of East Asian Studies, University of California, Berkeley, 1998), 148–65.

20. Kim, *Imperial Citizens*, 23–25.

21. Jaehwan Hyun, "Blood Purity and Scientific Independence: Blood Science and Postcolonial Struggles in Korea, 1926–1975," *Science in Context* 32 (2019): 239–60.

22. Jack Jin Gary Lee and John D. Skrentny, "Korean Multiculturalism in Comparative Perspective," in *Multiethnic Korea? Multiculturalism, Migration, and Peoplehood Diversity in Contemporary South Korea*, ed. John Lie (Berkeley: Institute of East Asian Studies, University of California, Berkeley, 2014), 301–30.

23. "The 74th Presidential Agenda Meeting: A Report on Support Measures of Social Integration of Mixed-Blood People and Migrants," April 26, 2006, Presidential Committee on Social Inclusion, http://pcsi.pa.go.kr/publish/chp03.asp?ex=v&ex2=1&seq=3132.

24. "The 74th Presidential Agenda Meeting."

25. "A Possible Rise of Kosians as a Political Force," *Jungang Daily*, July 2, 2006, https://news.joins.com/article/2250434.

26. Jiyeon Kang, "Reconciling Progressivism and Xenophobia Through Scapegoating: Anti-Multiculturalism in South Korea's Online Forums," *Critical Asian Studies* 52 (2019): 87–108.

27. Jin-Gu Kang, "Critical Study on Discourse on Korean Society's Anti-Multiculturalism," *Journal of Multicultural Contents Studies* 17 (2014): 7–37 (in Korean).

28. EuyRyung Jun, "Tolerance, Tamunhwa, and the Creating of the New Citizens," in *Multiethnic Korea? Multiculturalism, Migration, and Peoplehood Diversity in Contemporary South Korea*, ed. John Lie (Berkeley: Institute of East Asian Studies, University of California, Berkeley, 2014): 79–94.

29. Korean Statistical Information Service (KOSIS), "Multicultural Family Statistics," Statistics Korea, August 28, 2020, https://kosis.kr/statHtml/statHtml.do?orgId=101&tblId=DT_1JD1501&conn_path=I3.

30. From 1990 to 2017, the number of migrants from East and Southeast Asia was one million and 91 percent of the total number of migrants (about 1.1 million) to South Korea. "Migration to South Korea 1990–2017," https://worldmapper.org/maps/migration-to-south-korea-1990–2017.

31. Hye-Soon Kim, "Migrant Brides and Making of a Multicultural Society: Sociological Approach to Recent Discourse on 'Multicultural Korea,'" *Korean Journal of Sociology* 42 (2008): 36–71 (in Korean).

32. Changzoo Song, "Engaging the Diaspora in an Era of Transnationalism: South Korea's Engagement with Its Diaspora Can Support the Country's Development," *Evidence-Based Policy Making in Labor Economics: The IZA World of Labor Guide* 64 (2015): 74–76.

33. "Act on the Immigration and Legal Status of Overseas Koreans," Act No. 6015, September 2, 1999.

34. Sarah Morando Lakhani and Stefan Timmermans, "Biopolitical Citizenship in the Immigration Adjudication Process," *Social Problems* 61 (2014): 360–79.

35. Jaeeun Kim, "Establishing Identity: Documents, Performance, and Biometric Information in Immigration Proceedings," *Law and Social Inquiry* 36 (2011): 760–86; Chulwoo Lee, "The Law and Administrative Practice of Classification, Boundary Delimitation, and Membership Determination of Koreans," *Korean Journal of International Migration* 1 (2010): 5–36 (in Korean).

36. Chun Ho Lee, "A Study of Vietnam Lai Dai Han," *Journal of Humanities and Social Science* 9 (2018): 496–98 (in Korean); Choong Nam Ji, "An Exploratory Study on Kopinos in the Philippines: Reality and Support," *Oughtopia* 30 (2015): 286–87 (in Korean).

37. Dong-Hoon Seol and John D. Skrentny, "Ethnic Return Migration and Hierarchical Nationhood: Korean Chinese Foreign Workers in South Korea," *Ethnicities* 9 (2009): 158–62.

38. Timothy C. Lim and Dong-Hoon Seol, "Explaining South Korea's Diaspora Engagement Policies," *Development and Society* 47 (2018): 634.

39. In English, see "Defector's Condition Indicates Serious Health Issues in North Korea," *Voice of America*, November 19, 2017, https://www.voanews.com/east-asia -pacific/defectors-condition-indicates-serious-health-issues-north-korea.

40. Kim, "Migrant Brides and Making of a Multicultural Society," 60–63.

41. Epstein, *Inclusion*, 90–93.

42. Catherine Bliss, *Race Decoded: The Genomic Fight for Social Justice* (Stanford, CA: Stanford University Press, 2012).

43. Yeonjung Kim, Bok-Ghee Han, and KoGES Group, "Cohort Profile: The Korean Genome and Epidemiology Study (KoGES) Consortium," *International Journal of Epidemiology* 46 (2017): e20–e20.

44. Korean National Institute of Health, "KOGES: Data Sharing," May 14, 2019, https:// cdc.go.kr/contents.es?mid=a50401010400.

45. Myung Hee Shin, Mi Kyung Kim, and Chung Min Lee, *A Report on Genomic Research Program: International Collaboration Cohort Study I: Korean Emigrant Cohort* (Seoul: Korea Center for Disease Control and Prevention, 2006) (in Korean).

46. For the survey of genetic research on the dual origins of the Korean population, see Hyun, "Tracing National Origins," 367–74.

47. Hye Won Jung, *A Report on Genomic Research Program: Cohort of Southeast Asian Immigrant* (Seoul: Korea Center for Disease Control and Prevention, 2006), 3–4 (in Korean).

48. This framework of immigrant women's role as reproductive mothers in multicultural policy was highly criticized by domestic feminist activism but remains dominant. See Jung-Sun Kim, "The Critical Study of 'Korean-Style' Multiculturalism as Welfare Policy Excluding Citizenship," *Economy and Society* 92 (2011): 205–46 (in Korean).

49. Jung, *Cohort of Southeast Asian Immigrant*, 3–4; Hye Won Jung, *A Report on Genomic Center Research Program: Cohort of East Asian Immigrant in Korea* (Seoul: Korea Center for Disease Control and Prevention, 2007) (in Korean); Hye Won Jung, *A Report on Genomic Research Program: Cohort of Intermarried Women in Korea* (Seoul: Korea Center for Disease Control and Prevention, 2009) (in Korean).

50. Shin et al., *Korean Emigrant Cohort*, 5; Jung, *Cohort of Southeast Asian Immigrant*, 5–6.

51. Shin et al., *Korean Emigrant Cohort*, 72–77.

52. Jung, *Cohort of Southeast Asian Immigrant*, 1–4.

53. Shin et al., *Korean Emigrant Cohort*, 9.

54. Jung, *Cohort of Southeast Asian Immigrant*, 61.

55. Jung, *Cohort of East Asian Immigrant*, 3–6, 16; Jung, *Cohort of Intermarried Women*, 4–7, 12.

56. Soon-Seok Kim, "A Study on the Current Condition of Crimes in a Multi-Cultural Society and Police Countermeasures," *Police Science Journal* 5 (2010): 101–27 (in Korean); Deok In Lee, "A Study on the Multi-Cultural Society and the Multi-Cultural Crime by Foreigners Living in Korea," *Kyungpook National University Law Journal* 38 (2012): 73–106 (in Korean).

57. Lee, "A Study on the Multi-Cultural Society and the Multi-Cultural Crime," 94; Si-Soo Park, "Foreigners Face Fingerprinting Rule," *Korea Times*, May 21, 2008, https://www.koreatimes.co.kr/www/nation/2020/08/113_24546.html.

58. Jongmin Lee, "Foreigners Must Register Fingerprints, Facial Scans," June 30, 2011, *The Korea Herald* (www.koreaherald.com/view.php?ud=20110630000668).

59. Interview with Kim Wook by author, August 13, 2018.

60. Korean National Forensic Service, "Forensic DNA Division," https://nfs.go.kr/site /eng/01/10102010000002017092005.jsp.

61. Soong Deok Lee, *A Report on the Biomedical Technology Development Program: Development of New DNA Technology* (Seoul: Ministry of Education, Science, and Technology, 2012) (in Korean).

62. Lisa Gannett, "Biogeographical Ancestry and Race," *Studies in History and Philosophy of Science Part C: Studies in History and Philosophy of Biological and Biomedical Sciences* 47 (2014): 175.

63. For the problematic nature of biogeographic ancestry in relation to the return of race concepts in genomic research, see Joan H. Fujimura and Ramya Rajagopalan, "Different Differences: The Use of 'Genetic Ancestry' Versus Race in Biomedical Human Genetic Research," *Social Studies of Science* 41 (2011): 5–30.

64. Gannett, "Biogeographical Ancestry and Race," 175.

65. It would be worth noting that the number of criminal cases committed by U.S. citizens was almost equal to the number committed by Chinese immigrants, but American immigrants were not discussed in the "multicultural" crime discourse.

66. Lee, *Development of New DNA Technology*.

67. Wook Kim, *An Annual Report on the Long-Term Forensic Technology Development Program: Development of DNA Profiling System for Identification of Ethnic Affiliation* (Wonju: National Forensic Service, 2015) (in Korean).

68. Kim, *An Annual Report on the Long-Term Forensic Technology Development Program*, 8–10.

69. Ju Yeon Jung, Pil-Won Kang, Eungsoo Kim, Diego Chacon, Dominik Beck, and Dennis McNevin, "Ancestry Informative Markers (AIMs) for Korean and Other East Asian and South East Asian Populations," *International Journal of Legal Medicine* 133 (2019): 1717.

70. Hyo Jung Lee, Sun Pyo Hong, Soong Deok Lee, Hwan Seok Rhee, Ji Hyun Lee, Su Jin Jeong, and Jae Won Lee, "Evaluation of the Classification Method Using

Ancestry SNP Markers for Ethnic Group," *Communications for Statistical Applications and Methods* 26 (2019): 1–9; Kim, *DNA Profiling System*.

71. Jin Woong Kang, "The Elementary Social Studies and Multicultural Education in South Korea: The Paradox of Denationalization and Renationalization," *Social Studies Education* 55 (2016): 1–19 (in Korean).
72. Seol and Skrentny, "Hierarchical Nationhood," 162–64.

EIGHT

The Dilemmas of Racial Classification in Brazil

Reflections on Two Contemporary Case Studies

JOÃO LUIZ BASTOS AND RICARDO VENTURA SANTOS

BRAZIL IS CURRENTLY going through a significant social and ideological reconfiguration in which the theme of *mestizaje* has increasingly been displaced from the center of the nation-state's pantheon of symbols.[1] This has been happening since the 1990s, when multicultural strands became increasingly influential in Brazilian public policies. At the same time, social movements, including the Black movement, gained prominence in the context of the redemocratization process after the end of the military dictatorship (1964–1985). Such a prominence also helped build different narratives on the importance and role of *mestizaje* for the social dynamics and public policies aimed at mitigating racial inequities in Brazil.

Though we might run the risk of an oversimplification and schematization of greatly complex historical and sociopolitical processes, the social and ideological reconfiguration to which we referred may be briefly described in the following terms.[2] Perhaps it is not completely inaccurate to contend that from the beginning of the twentieth century to the present, Brazil has experienced three moments pertaining to the role and valuation of *mestizaje*. Since the late nineteenth century until about the early 1930s, when ideologies of racial determinism impregnated by eugenic perspectives were commonplace, *mestizaje* was perceived as an evil to be cut off from the nation, with the "whitening" of the population assumed to be the route to be taken for national redemption. With the cultural turn of the 1930s, *mestizaje* became valued to the point that it became a national symbol of major importance. In the latest period, over the past three decades, *mestizaje* turned out to be again associated with a negative connotation, largely seen as supporting notions of racial democracy that prevailed in the country during most of the twentieth century. In the current Brazil of racial quotas, the emphasis is mostly on the white-Black polarity. The *mestizo*, previously a symbol of the nation, becomes increasingly an element of "noise" that "pollutes" the system and has to be dealt with by the Brazilian state in the process of public policy implementation.

In this chapter, we will describe and contextualize two contemporary case studies in which, using an expression by Hartigan, " 'race' is currently 'gaining in reality.' "[3] The first relates to the implementation of public policies aimed at increasing access to higher education among African Brazilian (*pretos* and *pardos*) students, in which defining the classification of color or race[4] has led to the creation of "race classification boards" (Comissões de Verificação da Autodeclaração Etnicorracial) to confirm the candidates' (self-)declarations of ethnoracial belonging; the second, at the interface among biomedical studies, race, and genomics, is linked to recent initiatives to investigate the relationship between genomic ancestry and self-classified color or race, under which biological-scientific criteria are activated, or at least alluded to, as potentially useful in "solving" the fluidity of the Brazilian ethnoracial classification.

We argue that, in both case studies, those who have the authority to define the classification move from the individual experiencing the classificatory process to something external to them, be it members of a race classification board or a group of scientists in their scholarly publications. This is expressed in the paradoxical circumstance that the individual's self-classification undergoes a process of "confirmation" that is foreign to him/her. In addition, we emphasize that both situations fall into a sociopolitical context in which *mestizaje*, which was culturally and politically constructed and reiterated as a positive feature of the Brazilian nationality over much of the twentieth century, is currently associated with the reinforcement of inequalities. Furthermore, it is seen as an impeditive force against efforts at social transformation, in particular the confrontation of racism and social hierarchies associated with race. In this sense, we will see that in both case studies, the *mestizo* is taken as an element of "instability," something to be organized by the state or a group of interested social actors by way of classificatory efforts.

Case Study 1: The Permanent Commission for Evaluating Racial
Self-Declarations of the Federal University of Rio Grande do Sul

On February 7, 2018, the official website of the Permanent Commission for
Evaluating Racial Self-Declarations of the Federal University of Rio Grande
do Sul (hereafter abbreviated as UFRGS, according to the name of the university in Portuguese) one of the most prestigious Brazilian higher-education
institutions, published a note on the procedures that must be undertaken
to "ascertain" the color or racial classification of those applying for racial
quotas. The note states:

> For black and brown candidates, the evaluation will be effected through the ver-
> ification that the student concerned is seen socially as a member of the racial
> group to which he/she declared that he/she belonged, based on his/her pheno-
> type. Apart from skin color, other characteristics will be taken into account, such
> as hair type, shape of nose and lips. This procedure is silent, and the candidate
> will not be interviewed . . . The racial-ethnic self-declaration will be filled in and
> signed in the presence of the Permanent Commission for evaluating Racial
> Self-Declarations.[5]

The statement issued by UFRGS also mentions that the so-called Permanent
Commission for Evaluating Racial Self-Declarations was established in 2017
and is composed of sixteen members, including faculty, staff, students, and
"members of the Black Movement who have ties to UFRGS."

The UFRGS case is just one among numerous Brazilian higher-education
institutions that, over the past few years, have implemented a quota system
that follows ethnoracial and socioeconomic criteria, according to a federal
law of 2012, the so-called Quotas Law, number 12.711/2012.[6] The "Quotas
Law" demands that part of the spots offered by federal public institutions
must be reserved for "self-declared black, mixed race, indigenous, and disabled people," using as parameters their frequencies in each of the federa-
tion units, as estimated by the most recent national demographic census,
in this case, the 2010 Brazilian national demographic census.

The "Quotas Law" is part of a legal framework that was established in Brazil throughout the last two decades, from a scenario in which the issue of
combating racism has become increasingly central in public policies. Amid
the multicultural perspective that influenced constitutional reforms in

several Latin American countries, the most recent Brazilian constitution, promulgated in 1988, gave major emphasis to the ethnoracial dimension. The period after the military dictatorship, which controlled the country between 1964 and 1985, generated a strengthening of social movements, including the Black movement. In the late 1990s and in the first decade of this century, during the terms of presidents Fernando Henrique Cardoso (the first Brazilian president to officially recognize that racism is an important issue in the country) and Luiz Inácio Lula da Silva (Lula), the issue of confronting racism gained increasing visibility. Among some of these initiatives was the creation of the Special Department for Policies That Promote Racial Equality, with the status of a federal ministry. Furthermore, we saw the implementation of initiatives and the approval of laws aimed at addressing racial inequality and racism in the fields of health, work, and education, among others.

In a scenario of legal transformation concerning ethnoracial issues in Brazil, the "Quotas Law" presents particularities with respect to the relationship between the state and its citizens. In the case of the National Policy for the Integral Health of the Black Population, for example, the emphasis is on the implementation of programs for the "black population" in general.[7] This policy includes, for example, initiatives to improve the identification and treatment of sickle cell anemia, as well as strategies to combat institutional racism, among others. The "population" is the target, with no emphasis on initiatives that depend upon the identification of individuals according to their ethnoracial belonging in order to provide them with specific benefits or differentiated attention. In the case of the "Quotas Law," on the other hand, the question of the "individual" and their classification of color or race gains prominence; more so, they become central in the process of implementing public policy.

During the first few years after the promulgation of the "Quotas Law," the selection processes that took place at dozens of the existing federal institutions were primarily based on the self-declarations of color or race. Many educational institutions, like UFRGS, began to establish instances and additional steps, however, such as the "race classification boards." Defended and supported by sectors of the social movements, such commissions have been justified on the grounds that many candidates who are applying for the spots (i.e., non-Black and non-Brown students) would be "cheating" in their statements of racial belonging so as to benefit from the quotas system. This matter gained such a degree of political importance that the federal government issued a "guidance policy" in 2016, defining the "rules for measuring the truth of self-declarations provided by black candidates [for the purposes of the Quotas Law]" and specifying that "the forms and criteria for the verification of the self-declaration's veracity should only consider the applicant's phenotypic aspects, which must mandatorily be checked with the candidate's presence."[8]

As can be imagined, the creation of these commissions has generated heated debate both in the academic realm and in society in general, ranging from positions of fierce criticism to those of extreme support. In a rather polarized political climate, some critical voices coming from social science and humanities scholars, many of them supportive of a "social quota" system aimed at the poor, regardless of color or race, have been sometimes labeled as "conservative."[9] It is claimed, for example, that they are using the emphasis on the complexity of the Brazilian racial classification system as a subterfuge to hinder the implementation of public policies. This point can be illustrated in the commentary by David Lehmann in one book chapter, "The Politics of Naming: Affirmative Action in Brazilian Higher Education," in which he states that critics of the racial quotas "had a point, but missed the point."[10] It is worth noting that even in the sectors that are most favorable toward the implementation of racial quotas, as is the case with Black movement leaders, the narrative is that the "investigatory commissions" are a "necessary evil." As the journalist Ana Paula Blower reported: "Expert in affirmative actions, the director of the NGO [nongovernmental organization] Educafro, Frei David Santos, says he was radically in favor of self-declarations as the only way to guarantee the spots for blacks and indigenous [applicants]. After encountering several cases of fraud [that is, non-Blacks declaring as Blacks], he became radically in favor of the investigatory commissions."[11] On the other hand, the lawyer Wanda Gomes Siqueira commented: "This [the race classification board] is generating a feeling of hatred between black and mestizo people, who are treated as fraudsters."[12]

Case Study 2: Facing the Continuum and Racial "Fluidity" in Science Labs

The second case study is also closely related to color or race classification procedures, but in the context of scientific research on ethnoracial classification and genomics in the health field. We will focus on a specific example that, we argue, might be seen as representative of a much broader context of increasing use of information about genomic ancestry in Brazilian epidemiologic research. We argue that classificatory efforts by a group of scientists are employed to address the subjectivity, indeterminacy, and contextual dependency of the Brazilian ethnoracial classification. We further contend that efforts to fit participants into "unambiguous" racial categories are a means to make study results more legible for an international readership. Again, *mestizaje* is taken as a source of "noise" that "pollutes" the system and negatively affects communication among scientists from around the globe.

Specifically, we will address a study entitled "Genomic Ancestry and Ethnoracial Self-Classification Based on 5,871 Community-Dwelling Brazilians (The Epigen Initiative)." This article was based on data from three of the leading ongoing cohort studies on the health of human populations in the country. According to the authors of the study:

> The composition of the Brazilian population is mixed, and its ethnoracial classification is complex. Previous studies showed conflicting results on the correlation between genome ancestry and ethnoracial classification in Brazilians . . . We used 370,539 Single Nucleotide Polymorphisms to quantify this correlation in 5,871 community-dwelling individuals in the South (Pelotas), Southeast (Bambui) and Northeast (Salvador), Brazil.[13]

This study should be briefly situated in the more general scenario of research on population genetics and public health in Brazil. In the context of Latin America, Brazil has a singular trajectory concerning the history of human population genetics, which was established and expanded from the 1950s onward, through a complex network of exchanges between Brazilian and foreign researchers (mostly from the United States).[14] Since its beginnings, in addition to research on indigenous populations, a recurring theme has been to investigate, through genetic markers, what scientists have called "the biological processes involved in the formation of the Brazilian population," focusing on the differential contribution to the genetic pool of European, African, and Amerindian populations.[15] Commenting on two influential books published in the 1960s and 1970s by the geneticists Francisco Mauro Salzano and Newton Freire-Maia, who sought to summarize decades of research on human population genetics, Dent and Santos observed:

[the books] *Populações brasileiras* and *Problems in Human Biology* constructed a new kind of national population, one legible in genetic terms. Moreover, by drawing on discourses that defined a tri-hybrid origin and valorized race-mixing as a central aspect of Brazilian identity, Salzano and Freire-Maia's monographs contributed to the consolidation of the authority of genetics as a site to discuss and comment on the biological nation.[16]

This research agenda related to "bionarratives" about the nation was maintained and updated with the expansion of research based on genomic markers that began in the 1990s.[17] Based on autosomal markers, mitochondrial DNA, and Y chromosomes, Brazilian geneticists continued to emphasize research on the population's "processes of biological formation," often highlighting the strong influence of *mestizaje*. In this scenario, a specific theme grew in the agenda of Brazilian population genetics research: studies focusing on the interrelationships between color or race classification (either self- or heteroclassification) and the genomic profiles of individuals. Specifically, issues pertaining to the associations between "physical appearance" and "genetic makeup" of individuals had been receiving attention from geneticists since the 1960s, but the technical possibilities of "individualization" in genomics made this issue even more central in the last two decades, particularly in the context of studies in pharmacogenomics.[18] In recent years, a growing incorporation of genomic technologies in the context of epidemiologic research has been observed in Brazil. Currently, some of the major cohort studies on human health in the country, such as those associated with the "EPIGEN-Brazil Initiative," have emphasized genetic aspects:

> The EPIGEN-Brazil Initiative is so far the largest Latin American initiative in population genomics and genetic epidemiology. Its main goal is to study the association between genetic variants found in the Brazilian population and complex diseases. . . . Population genetic includes genomewide genotyping of 6,487 individuals and high-resolution whole genome sequencing from 30 individuals from three population-based Brazilian cohorts: Salvador, Bambuí, and Pelotas.[19]

As in previous periods of human population genetic research, the question of *mestizaje* remains a central aspect in the justifications currently presented by geneticists. As can be read on the EPIGEN-Brazil Initiative website: "Its main goal is to study the association between genetic variants found in the Brazilian population and complex diseases, taking into account one of the most important characteristics of this population: its admixture."[20] In this scenario, we would like to suggest that the article "Genomic Ancestry and Ethnoracial Self-Classification" is illustrative of the complex exchanges between race or color classification and genomics in current epidemiologic research in Brazil. The focus remains on *mestizaje*, but other elements come into play, particularly the emphasis on what we would call the "individualization of classification processes." In this sense, although not explicitly stated, one of the goals of the researchers is to investigate the degree of association between color or race self-classification and classification schemes based on genomic ancestry. As was pointed out: "The composition of the Brazilian population is more mixed, and its ethnoracial classification is more complex and fluid than in those countries where segregation was imposed by law. This was to such a degree that it has been questioned whether—and how—ethnoracial classification in Brazil correlates with genomic ancestry."[21]

The research by Lima-Costa and colleagues involved comparing the self-classification of color or race (indicated as "White," "Mixed," and "Black") from a set of 5,871 people, using data from three cohorts that are a part of EPIGEN.[22] A point worth mentioning is that the statistical models presented by the authors do not consider socioeconomic characteristics or even the sex/gender of the study subjects. This is of particular interest, given that for many decades anthropological and sociological studies conducted in various Brazilian regions point to economic, educational, and gender factors, among others, as major drivers of individual color or race classification.[23] Therefore, drawing from analyses comparing self-reported color/race vis-à-vis genomic profiles, without consideration of important covariates, some of the main findings of the study were the following: "First, the association between the phenotype and genome ancestry was statistically significant, but the strength of the association varied largely across populations; second: the association between Black and White self-classification with ancestry was most consistent in the extremes of the high and low proportion of African ancestry."[24]

"Genomic Ancestry and Ethnoracial Self-Classification" goes beyond an exercise of comparing classification systems within a given country, in this case Brazil, contrasting the individual attributes of subjects on color or race as evidenced by a set of genomic markers. In the bigger picture, what is at stake is the search for classification criteria, considered in the eyes of scientists, as "objective" and "reproducible," aiming to overcome the "fluidity" that arises from a classification system considered to be (excessively) subjective.

*   *   *

Both situations we have described have the common background of the growing prominence of the ethnoracial dimension in public policy debates in Brazil. They also present an important parallel: the fact that the subjectivities involved in the color and racial classification process become subordinate to external instances. Be it in the implementation of quotas or in scientific practice, at stake are the continuous and complex dilemmas of, in Hacking's words, "making up people."[25] A common issue is the establishment of "infrastructures" of classification through which, as noted by Loveman, "directly and acutely . . . particular categories are attached to individual bodies and used to structure administrative allocation or other dimensions of lived experience."[26]

For the authors of this text, and perhaps for many of those who read us now, the feeling of strangeness that both situations evoke might be associated with the fact that, in a way, we are going through a moment of breaking a "social pact" established since the mid-twentieth century. In this "pact," the views of the subjects about themselves were assumed to be the main reference, almost like a "gold standard." It is not our intention to identify in detail when this was established. An undoubtedly important milestone, however, was the debate about "race" and related concepts in the period following the Second World War, which resulted in several declarations about race from UNESCO.[27] The sociologist Mara Loveman, through her reflections on the history of the official censuses in Latin America, historically situates this issue, closely relating it to the mid-twentieth century's international debates:

> Contemporary theorists insist that race is a social relationship forged through a negotiated interaction between the "self" and "others." . . . The use of self-identification in collection of racial and ethnic population data . . . became the internationally recommended and recognized standard in the 1940s; before then,

it was standard practice for census enumerators to fill in household census schedules, including classification of individuals' race.[28]

While defended by segments of social movements, the existence of race classification boards produces discomfort for many scholars. In particular, historians, anthropologists, and sociologists see similarities between their classification procedures and those used during the strongest moments of scientific racism, from the end of the nineteenth century to the mid-twentieth century, when the aim was to produce differences and hierarchies.[29] Ironically, even though there is not a deeper historical criticism in the institutional frameworks for the current classification procedures, the actions are seen as a way to implement a public policy toward producing equality. The justification of social movements in practice and in the everyday routine of contemporary policy making is pragmatically guided by other justifications: if, in everyday relationships, the bodies of Blacks, individually and collectively, are constantly racially marked and classified in Brazil (by the police, at bank entrances, etc.), why cannot the same be done for the purposes of public policies that attempt to reduce inequalities?

As one might imagine, the degree of controversy associated with the use of genomic criteria to define racial classification in biomedical research in Brazil, at least explicitly, is not even remotely close to the debates about racial quotas. Perhaps, it is safer to say that there are no controversies underway. This, despite that the subject in itself carries elements that refer to important debates between the social and human sciences and biomedicine, in particular the use of biological criteria in the definition of social identities.[30]

In the broader context of knowledge production and uses of genetic science, the fact is that the prospect of using genomic methods to make inferences about the physical characteristics of individuals (as in the case of missing and/or absent people) is widespread, as in forensic genetics routines.[31] In this sense, the Brazilian research that we refer to in this text could be seen as a kind of "racial profiling." One particularity is that, in the examples of Brazilian research that we refer to, the individuals are not missing or absent. It is his/her subjectivity that produces forms of classification that are not "reliable," "reproducible," or even "legible" within the expected canons in some fields of science (biomedicine, for example).

Would it be possible to contend that the current debates about race in Brazil, which we argue are associated with a negative connotation toward *mestizaje*, have some resonance with the research at the interfaces between epidemiology and genomics? We would argue yes, but through other means and processes. In an interview in the context of a research project on pharmacogenomics and race in Brazil, the Brazilian scientist Guilherme Suarez-Kurtz, from the National Cancer Institute, made a telling comment on the inclusion of Brazilian samples in multicentric research on warfarin, the most widely used anticoagulant in the world. Suarez-Kurtz stated:

> Only a small portion of the Brazilian samples sent were utilized in the [warfarin] multi-centric study, and most of the data from individuals [racially] classified as "Intermediate" were not included. In fact, in the article the data that are analyzed are those referred to as "White," "Asian" or "Black," whereas "Mixed or missing data" constitute a combined category not used in the analyses.[32]

Santos, Silva, and Gibbon comment, "That is, when they enter the 'political economy' of publications in pharmacogenomics that involve extensive international collaboration, often the Brazilian samples, especially those of 'mestizos,' are not included because they seem not to fit the required 'racial framework.'"[33]

*   *   *

Borrowing the subtitle of the well-known and insightful text by Amade M'Charek entitled "Beyond Fact or Fiction?," the two case studies explored in this paper make us reflect on "the materiality of race in practice" in contemporary Brazil.[34] Despite being different in many ways, they involve processes that seek the "normalization" and "stabilization" of identitary categories.[35] In both instances, those who experience the "classification rites" start from a position in which their individual subjectivity about their respective "selves," in terms of ethnoracial belonging, is questioned. In the classifying exercise (that might often be confrontational), "the eyes of the beholder" pass from a regime "of themselves, and about themselves" to one "of the other, and about the other," reconfiguring and repositioning the gaze with authority, so as to determine "who is who." In addition, be it in the daily lives of "racial classification boards" or genomic sequencings to determine biological ancestries, continuum and racial fluidity, once regarded as hallmarks of Brazilian culture and society, are questioned. Instead, the polarities of white/not white and/or Black/not Black increasingly prevail in justifications, narratives, and practices associated with the implementation of a broad range of public policies.

Postscript

It has been almost five years since we drafted the first version of this chapter. In May 2023, revising the paper once again, this time before final publication, we felt the need to add a few lines about the trajectory, or perhaps the "social life," of the classificatory processes we address in our essay. To say the least, a lot has unfolded in this half a decade.

One of our case studies relates to what we perceived, when we originally wrote our paper, as an emerging trend in Brazilian epidemiologic research toward an expansion in the use of genomic markers as a means to replace the long-established indicators of ethnoracial belonging. But what we have seen over these years is that the utilization of genomic information has not taken over from self-declared ethnoracial classification as the most common approach, with all its imbedded subjectivity, indeterminacy, and contextual dependency.[36] In fact, and perhaps as an indication of the reinforcement of the "social," we see a growing recognition of the need to be more inclusive toward ethnoracial minorities across important Brazilian research groups in epidemiology, in some cases with the adoption of perspectives such as intersectionality as relevant methodological and theoretical approaches.[37] Thus, at least so far, our prognosis about the expansion of genomics as classificatory tools in the definition of ethnoracial belonging has not turned into reality in scientific practices in Brazilian public health research.

As for the other case study we discuss in our text, race classification boards are now a widespread reality across Brazilian higher-education institutions and, as a matter of fact, much beyond. Amid the implementation of race-related affirmative action initiatives, increasingly supported by legislation, race classification boards have become an integral part of employee selection processes of government agencies and departments at the municipal, state, and national levels. The constitution and practices of these boards have become a main topic of research in the emerging field of affirmative action studies in Brazil, whose lines of inquiry have largely been supported by national and international funding bodies. Therefore, over the past five years, race classification boards have become institutionalized routines in the context of race-based affirmative action policies in the country. This has provided much context and information for analyses in the social and human sciences, leading to reflections not only on the historical and political implications of these practices, including the impacts upon reparation and social justice aiming at reducing ethnic-racial inequities in Brazil, but also about the inescapable tensions and disputes involved.[38]

Notes

1. In Brazil as well as in other Latin American countries, multiple meanings and uses have been attached to the concept of *mestizaje* over the past centuries. In general, *mestizaje* has been used in reference to racial, cultural, social, linguistic, national, and ethnic mixture. According to prevailing social and political contexts, in different historical periods it has been associated with either negative or positive perspectives about social collectivities. For discussions of the term and its historical and sociopolitical uses and implications, particularly considering the Brazilian context, see Sérgio Costa, "Da mestiçagem à diferença: nexos transnacionais da formação nacional no Brasil," in *O Brasil em dois tempos. História, pensamento social e tempo presente*, ed. Eliana F. Dutra (Belo Horizonte: UFMG, 2013), 301–20; and Antônio Sérgio Alfredo Guimarães, "Colour and Race in Brazil: From Whitening to the Search for Afro-Descent," in *Racism and Ethnic Relations in the Portuguese-Speaking World*, ed. Francisco Bethencourt and Adrian Pearce (British Academy, 2012), 16–34.

2. Ricardo Ventura Santos, Michael Kent, and Verlan Valle Gaspar Neto, "From 'Degeneration' to 'Meeting Point': Historical Views on Race, Mixture, and the Biological Diversity of the Brazilian Population," in *Mestizo Genomics: Race Mixture, Nation, and Science in Latin America*, ed. Peter Wade, Carlos López Beltrán, Eduardo Restrepo, and Ricardo Ventura Santos (Durham, NC: Duke University Press, 2014), 33–54; Guimarães, "Colour and Race in Brazil," 16–34.

3. John Hartigan, "Is Race Still Socially Constructed? The Recent Controversy Over Race and Medical Genetics," *Science as Culture* 17, no. 2 (2008): 166.

4. Use of the terms "color" and "race" as interchangeable concepts in Brazil stems from the longstanding belief that race mixing has been so intense among the population that Brazilians may not be differentiated according to race but with reference to the color of their skin. Melissa Nobles, *Shades of Citizenship: Race and the Census in Modern Politics* (Palo Alto, CA: Stanford University Press, 2000).

5. UFRGS, http://www.ufrgs.br/ufrgs/noticias/afericao-etnico-racial-de-classificados-no-vestibular-com-ingresso-em-2018-1-estende-se-ate-9-de-fevereiro.

6. Presidência da República, Casa Civil, Subchefia para Assuntos Jurídicos, *Lei No. 12.711*.

7. Marcos Chor Maio and Simone Monteiro, "Tempos de racialização: O caso da 'saúde da população negra' no Brasil," *História, Ciências, Saúde-Manguinhos* 12, no. 2 (2005): 419–46.

8. See Regulatory Guideline No. 3, issued on August 1, 2016, Ministry of Planning, Development, and Management; Secretaria de Gestão de Pessoas e Relações do Trabalho no Serviço Público, *Orientação Normativa N. 3, de 1 de Agosto de 2016* (Brasília: Imprensa Nacional, 2016).

9. Peter Fry, Yvonne Maggie, Marcos Chor Maio, Simone Monteiro, and Ricardo Ventura Santos, eds., *Divisões perigosas: políticas raciais no Brasil contemporâneo* (Rio de Janeiro: Civilização Brasileira, 2007).

10. David Lehmann, "The Politics of Naming: Affirmative Action in Brazilian Higher Education," in *The Crisis of Multiculturalism in Latin America*, ed. David Lehmann (New York: Palgrave Macmillan, 2016), 179–221.

11. Ana Paula Blower, "Ações afirmativas—cotas para quem precisa: ao menos 18 instituições federais já têm comissões de análise e candidatos para evitar fraudes," *O Globo* (Rio de Janeiro), March 11, 2018, 36.

12. Blower, "Ações afirmativas."

13. M. Fernanda Lima-Costa, Laura C. Rodrigues, Maurício L. Barreto, Mateus Gouveia, Bernardo L. Horta, Juliana Mambrini, Fernanda S. G. Kehdy, et al., "Genomic Ancestry and Ethnoracial Self-Classification Based on 5,871 Community-Dwelling Brazilians (the Epigen Initiative)," *Scientific Reports* 5 (2015): 9812.

14. Vanderlei Sebastião Souza and Ricardo Ventura Santos, "The Emergence of Human Population Genetics and Narratives About the Formation of the Brazilian Nation (1950–1960)," *Studies in History and Philosophy of Science Part C: Studies in History and Philosophy of Biological and Biomedical Sciences* 47, part A (2014): 97–107.

15. Rosanna Dent and Ricardo Ventura Santos, " 'An Immense Mosaic': Race Mixing and the Creation of the Genetic Nation in 1960s Brazil," in *Luso-Tropicalism and Its Discontents: The Making and Unmaking of Racial Exceptionalism*, ed. Warwick Anderson, Ricardo Roque, and Ricardo Ventura Santos (New York: Berghahn, 2019), 135–66.

16. Dent and Santos, " 'An Immense Mosaic.' "

17. Michael Kent, Ricardo Ventura Santos, and Peter Wade, "Negotiating Imagined Genetic Communities: Unity and Diversity in Brazilian Science and Society," *American Anthropologist* 116, no. 4 (2014): 736–48; Peter Wade, Carlos López-Beltrán, Eduardo Restrepo, and Ricardo Ventura Santos, "Genomic Research, Publics and Experts in Latin America: Nation, Race and Body," *Social Studies of Science* 45, no. 6 (2015): 775–96.

18. Flavia C. Parra, Roberto C. Amado, José R. Lambertucci, Jorge Rocha, Carlos M. Antunes, and Sérgio D. J. Pena, "Color and Genomic Ancestry in Brazilians," *Proceedings of the National Academy of Sciences* 100, no. 1 (2003): 177–82; Ricardo Ventura Santos, Gláucia Oliveira Silva, and Sahra Gibbon, "Pharmacogenomics, Human Genetic Diversity, and the Incorporation and Rejection of Color/Race in Brazil," *BioSocieties* 10, no. 1 (2015): 48–69.

19. Epigen, https://epigen.grude.ufmg.br.

20. Epigen, https://epigen.grude.ufmg.br.

21. Lima-Costa et al. "Genomic Ancestry and Ethnoracial Self-Classification," 9812.

22. Lima-Costa et al., "Genomic Ancestry and Ethnoracial Self-Classification."
23. Edward Eric Telles, *Race in Another America: The Significance of Skin Color in Brazil* (Princeton, NJ: Princeton University Press, 2004); Guimarães, "Colour and Race in Brazil."
24. Lima-Costa et al., "Genomic Ancestry and Ethnoracial Self-Classification."
25. Ian Hacking, "Making Up People," in *Reconstructing Individualism*, ed. T. L. Heller, M. Sosna, and D. E. Wellbery (Palo Alto, CA: Stanford University Press, 1985), 161–71.
26. Mara Loveman, *National Colors: Racial Classification and the State in Latin America* (Oxford: Oxford University Press, 2014).
27. UNESCO, *Four Statements on the Race Question* (Paris: Oberthur-Rennes, 1969).
28. Loveman, *National Colors*, 171.
29. In 2016, the Brazilian Anthropological Association (Associação Brasileira de Antropologia) issued a note criticizing the implementation of the "racial classification boards" (see http://www.portal.abant.org.br/images/Noticias/68_Nota_Diretoria_ABA_Igualdade_Racial.pdf). See also Elena Calvo-González and Ricardo Ventura Santos, "Problematizing Miscegenation: The Fact/Fiction of Race in Contemporary Brazil," *Journal of Anthropological Sciences* 96, no. 2018 (2018): 247–54.
30. In conversations with social epidemiologists who develop research on ethnoracial inequalities in Brazil, an issue we hear about that has been of concern is that in biomedical research there is a lack of criticism toward employing genomics ancestry classifications in health outcome analyses, its determination being predominantly sociohistorical, for example, for hypertension and diabetes.
31. Charles E. McLean and Adam Lamparello, "Forensic DNA Phenotyping in Criminal Investigations and Criminal Courts: Assessing and Mitigating the Dilemmas Inherent in the Science," *Recent Advances in DNA & Gene Sequences* 8, no. 2 (2014): 104–12.
32. Santos, Silva, and Gibbon, "Pharmacogenomics, Human Genetic Diversity, and the Incorporation and Rejection of Color/Race in Brazil."
33. Santos, Silva, and Gibbon, "Pharmacogenomics, Human Genetic Diversity and the Incorporation and Rejection of Color/Race in Brazil."
34. Amade M'Charek, "Beyond Fact or Fiction: On the Materiality of Race in Practice," *Cultural Anthropology* 28, no. 3 (2013): 420–42.
35. See Santiago José Molina, "Amerindians, Europeans, Makiritare, Mestizos, Puerto Rican, and Quechua: Categorical Heterogeneity in Latin American Human Biology," *Perspectives on Science* 25, no. 5 (2017): 655–79.
36. Jeronimo Oliveira Muniz and Joao Luiz Bastos, "Volatilidade classificatoria e a (in)consistencia da desigualdade racial [Classificatory volatility and (in)consistency of racial inequality]," *Cadernos de Saúde Pública* 33, suppl. 1 (2017): e00082816.
37. Poliana Rebouças, Emanuelle Goes, Julia Pescarini, Dandara Ramos, Maria Yury Ichihara, Samila Sena, Rafael Veiga, et al., "Ethnoracial Inequalities and Child Mortality in Brazil: A Nationwide Longitudinal Study of 19 Million Newborn Babies," *Lancet Global Health* 10, no. 10 (2022): e1453–e62; Ricardo Ventura Santos, João Luiz Bastos, Joziléia Daniza Kaingang, and Luís Eduardo Batista, "Should

There Be Recommendations on the Use of 'Race' in Health Publications? An Emphatic 'Yes,' Especially Because of the Implications for Antiracist Practices," *Cadernos de Saúde Pública* 38, no. 3 (2022): e0002; Joao Luiz Bastos, Helena Mendes Constante, Helena Schuch Schuch, Dandara Gabriela Haag, and Lisa Jamieson, "How Do State-Level Racism, Sexism, and Income Inequality Shape Edentulism-Related Racial Inequities in Contemporary United States? A Structural Intersectionality Approach to Population Oral Health," *Journal of Public Health Dentistry* 82, suppl. 1 (2022): 16–27.

38. Neusa Chaves Batista and Hodo Apolinário Coutinho Figueiredo, "Comissões de heteroidentificação racial para acesso em universidades federais," *Cadernos de Pesquisa* 50, no. 177 (2020): 865–81; Sales Augusto Santos, "Comissões de heteroidentificação étnicoracial: lócus de constrangimento ou de controle social de uma política pública?," *O Social em Questão* 24, no. 50 (2021): 11–62; Ana Claudia Cruz Silva, Douglas Guimarães Leite, Flavia Rios, and Juliana Vinuto, "Comissões de heteroidentificação e universidade pública: processos, dinâmicas e disputas na implementação das políticas de ação afirmativa," *Mana* 28, no. 3 (2022): 1–31.

PART THREE
Past and Promise

NINE

Facing the Past

Human Skulls, Facial Reconstruction, and National Identity in the Middle East

ELISE K. BURTON

IN RESPONSE TO the proliferation of scholarship on race and genetics over the past two decades, the anthropologist Amade M'charek calls for renewed critical attention to phenotypes—bodily markers generally and facial features specifically—as key to the origin and perpetuation of beliefs in biological race.[1] Science and technology studies scholars and historians have extensively critiqued geneticists' use of racial categories as proxies for geographic origins (either of DNA sampling locations or of the ancestry of sampled individuals), health disparities, and other measures of genetic variation between individuals and populations. However, these critiques need to engage more directly with the problem that race is perceived to be biologically "real" through visual cues that people are socialized into recognizing through combinations of such characteristics as skin color, eye color and shape, hair color and texture, and nose size and shape.

These visions of race are not simply the uninformed assumptions of an ignorant public but rather a deliberately cultivated medium of scientific communication. As the philosopher Abigail Nieves Delgado shows, forensic anthropologists who specialize in reconstructing faces from skulls rely on problematic racial assumptions embedded within the standard methodologies to reproduce an individual's original phenotype. Namely, these anthropologists believe that skull morphology alone indicates an individual's race, yet they also diagnose the racial features of the skull according to where it was found geographically. This racial determination is considered essential to predict the depth and distribution of flesh that would have appeared on the living face. Finally, after completing these highly quantified stages of reconstruction, the experts decide how to represent the many features not preserved in the skull's structure—skin, eye and hair color, nose and ear shape, etc.—which notably coincide with racialized combinations of phenotypic characteristics. Through these decisions, forensic sculptors tend "to reproduce racial stereotypes" to communicate their expert assessment of the individual's ancestry as well as "to meet public expectations of facial appearance."[2] In other words, forensic facial reconstructions entail a highly circular racial reasoning to reproduce the "correct" living appearance of an individual skull.

Here, I build on Nieves Delgado's work on three-dimensional facial reconstruction as well as M'charek's theoretical observations that "the work of faces consists not only of *instructing* the viewer and *fashioning* difference but also in evoking feelings and interest." Indeed, producing an affective response is precisely "what a face can do" in a way that statistical tables, charts, and even photographs of skulls cannot. By evoking emotion and "generating engagement," facial reconstructions can recapture dwindling public attention to figures from ancient and medieval history by making them appear "concrete" and "alive."[3] This is precisely why facial reconstructions often appear in museum contexts, where they "can help highlight a shared humanity between past and contemporary populations."[4]

These aesthetic and evocative properties of the human face are historically inseparable from scientific understandings of human variation. The art historian David Bindman argues that aesthetic sensibilities of eighteenth-century European philosophers played an important role in the historical development of racial science. Analyzing the visual representations of humans that philosophers used in their published works, Bindman shows how European interpretations of physiognomy and aesthetic judgments of the relative beauty or ugliness of individual faces were eventually extrapolated to entire peoples. Many European thinkers assumed the correlation of appealing or repulsive facial traits with superior or inferior intellectual and moral character, an assumption that lumped together appearance and behavior in cultural "categorical distinctions" that ultimately lent themselves to the creation of racial hierarchies.[5] Simultaneously, Immanuel

Kant questioned the possibility of universal aesthetic judgments of the human form: he posited an "aesthetic normal idea" of beauty specific to different countries and races, corresponding to the "average form of all . . . members of that race."[6] These philosophical debates contain the kernel of recent psychological hypotheses about racialized capacities for "in-group" versus "cross-racial" facial recognition.[7] The materialist basis of modern psychology revives earlier philosophical tensions between culturally specific and psychologically universal bases for understanding how the feelings aroused by viewing facial reconstructions are filtered through racialized patterns of self-recognition. Regardless, facial reconstructions make it possible for the living viewer to make aesthetic judgments of long-deceased human remains: judgments of racial sameness or otherness, which in turn contribute toward nationalist ideologies of communal identity, autochthony, and cultural genealogy.

I apply these insights to cases from the Middle East by contextualizing regionally prominent cases of forensic facial reconstruction within the longer twentieth-century history of anthropometric research on living people and craniometric analysis of skulls excavated from archaeological sites in Iran and Lebanon. This research was performed by a combination of foreign scholars and local scientists and intellectuals, all of whom were motivated in their work by specific racial and nationalist ideologies. Beginning in the 1930s, Iranian intellectuals exhumed the remains of newly appointed national heroes like Ibn Sina (known in Latin as Avicenna) as part of a scientific and artistic process to prove the so-called Aryan racial identity of these individuals and reconstruct their physiognomies for sculptural monuments and commemorative portraits. I focus on the case of Ibn Sina, showing how nationalist contestations between Iran and Turkey over Ibn Sina's ethnoracial identity have staked their competing claims partially upon "scientific" facial reconstructions, with paradoxical results. Meanwhile, in Lebanon, anatomists, archaeologists, and historians argued over the racial classification of different Christian and Muslim communities in the context of a highly politicized debate about Phoenician versus Arab ancestry, a debate that partially fueled Lebanon's violent civil war. In present-day Lebanon, this powerful racial-national discourse continues to shape scientific and public interpretations of Phoenician facial reconstructions, which are not only displayed in museums but also increasingly juxtaposed alongside the results of genetic research in press releases and media coverage.

A Race for Ibn Sina: Iranian and Turkish Reconstructions

Beginning in the late nineteenth century, Iranian intellectuals took an interest in European race science. In particular, they participated in debates about the identity and origins of the so-called Aryan race. Linguistically, Persian is an Indo-Aryan language, and the word "Aryan" itself is derived from Old Persian. Many Iranian literati embraced the hypothesis that Central Asia (or even Iran itself) was the cradle of an Aryan race that conquered Iran and northern India.[8] In the 1920s and 1930s, the emerging Pahlavi dynasty began promoting an Aryan racial identity for strategic reasons. In particular, the Pahlavis and their supporters wanted to assert Iranians as equal to Europeans in terms of racial and civilizational status. Iranian nationalists therefore supported archaeological and anthropological investigations of Iran's pre-Islamic past, seeking to revive local and international interest in ancient Persian civilization.

One of the best-documented of these investigations was conducted by the Anglo-American anthropologist Henry Field in 1934. Field conducted a six-week expedition to determine "the physical characters of the modern inhabitants of Iran," which involved collecting anthropometric measurements from about three hundred people belonging to different communities across the country. Based on his findings, Field proposed a new racial classification: "a new, fundamental division of the White race . . . called the Iranian Plateau race. This new racial type, now for the first time, takes its place beside the long-accepted Nordic, Mediterranean, and Alpine races." However, Field emphasized the need for archaeological work to recover more ancient Iranian skulls for comparison, arguing: "Any effort to distinguish separate racial strains in modern population [sic] is seriously handicapped unless some evidence of their earliest ancestry is available. In general, there are lamentably few skeletal remains of the ancient dwellers on the Iranian Plateau." At the end of Field's research in September 1934, he met with the Iranian minister of education, Ali Asghar Hekmat, to discuss the possibility of "a detailed anthropometric survey of Iran." Field explained that only a comprehensive nationwide survey could confirm the "true racial position of the ancient and modern peoples of Iran in relation to Europe, Africa, and Asia."[9] I am not currently aware of any evidence that such a survey ever took place.

Ali Asghar Hekmat belonged to a group of Iranian politicians, scholars, and intellectuals called the Society for National Heritage (SNH; in Persian, *anjuman-i ās̱ār-i millī*). The SNH considered Iran's Aryan racial identity to be a foregone conclusion, so they had quite different priorities for craniometric research. Rather than cataloging the anthropometric traits of Iran's entire population, they focused their attention on measuring the skulls of a small number of famous individuals: poets like Firdawsi and Hafiz, statesmen like Nader Shah, and scholars like Ibn Sina. In the course of constructing new mausoleums for these esteemed figures, their remains were exhumed and studied.[10] For the SNH, a key goal of the exhumations was to transform these prominent historical characters of Islamic civilization into specifically Iranian national heroes.

I focus here on the case of Ibn Sina, the great physician and philosopher who lived from 980 to 1037 CE. During his lifetime, Ibn Sina wrote hundreds of influential texts in Arabic and Persian on topics ranging from logic, metaphysics, and theology to astronomy and medicine. He is best known for writing the *Canon of Medicine*, an encyclopedic work that became widely used as a teaching text across Europe and the Islamic world for over five hundred years. Given his worldwide fame, Ibn Sina's ethnicity remains fiercely contested by multiple Middle Eastern and Central Asian nations to this day. These contestations flared in response to nationalist discourses in the 1930s, beginning when both Iran and the Turkish Republic held major public events to commemorate nine hundred years since Ibn Sina's death. These events aimed not only to raise awareness of and encourage research about Ibn Sina's historical achievements but also to stake claims that Ibn Sina's "nationality" or ethnic ancestry was Turkish or Iranian. The anachronistic dispute over his nationality is underscored by the fact that Ibn Sina's birthplace and those of his known ancestors lie well outside the modern borders of either Turkey or Iran, in the territories of modern Uzbekistan and Afghanistan. Nevertheless, the Ibn Sina question made its way into the academic historiography of science, in the West as well as the Middle East. In 1939, George Sarton, the editor of the premier history of science journal *Isis*, published an article by his Turkish student Aydın Sayılı titled "Was Ibni Sina an Iranian or a Turk?" Sayılı, who went on to become a prominent historian of science at Ankara University, argued that Ibn Sina must have been Turkish based on a philological analysis of the name "Sina" as well as a speculative historical demography of Ibn Sina's birth region.[11]

To refute the Turkish claims, Iran's Society for National Heritage capitalized on its access to Ibn Sina's physical remains, which were buried in the Iranian city of Hamadan. In 1948, Ibn Sina's exhumed skull was sent to the anatomy department at Tehran University, whose staff photographed and measured the skull to confirm that Ibn Sina's biological traits matched those of the "Aryan" Iranian race.[12] However, the craniometric analysis alone was not sufficient for the SNH's goals to educate the Iranian public about their illustrious racial heritage. The SNH complained that the existing images of Ibn Sina in European books were contradictory and not based on any textual or physical evidence. Accordingly, the SNH decided it was "necessary to prepare a statue of Ibn Sina, which would be the first image of this great philosopher assembled with due attention to the available sources and evidence."[13] As a starting point, the SNH commissioned the artist Abu al-Hassan Sadighi to draw a portrait of Ibn Sina (figure 9.1). This portrait was based on the input of scholars familiar with historical texts about Ibn Sina. The SNH officially approved this image in 1945 (before the craniometric analysis of Ibn Sina's skull was completed) and began to widely disseminate this version of Ibn Sina's face through the production of postage stamps and commemorative medals, as well as monumental sculptures in prominent locations of Hamadan and Tehran.[14]

The reproduction of Ibn Sina's portrait in these diverse formats highlights the SNH's simultaneous commitments to science and spectacle. The architecture historian Talinn Grigor contextualizes the razing and reconstruction of Ibn Sina's Hamadan mausoleum amid the broader efforts of the Pahlavi dynasty to transform Iran's public and commemorative spaces to reflect an idealized Aryan past. The use of racial craniometry to claim Ibn Sina's remains and the historically informed reconstruction of his physiognomy were key steps toward the creation of larger-than-life statues. However, Grigor notes, the erection of human sculptures were only one component of creating a new nationalist visual culture that embraced themes of "Aryan Revival" alongside a modernist utopian "Great Civilization." Houshang Seyhoun, the European-trained architect of Ibn Sina's new mausoleum, dismissed the existence of an "Islamic architecture," arguing that the "architectural forms and elements" used during Iranian Muslim empires

FIGURE 9.1 Portrait of Ibn Sina by Abu al-Hassan Sadighi drawn in 1945 based on historical texts.
Source: reprinted from Muḥammad Taqī Muṣṭafavī, *Hagmatānah: āsār-i tārīkhī-i Hamadān va faṣlī dar bārah-i Abū ʿAlī Sīnā* (Tehran: [unidentified], 1332 [1953]).

were "mere variations" on pre-Islamic Sassanian architecture.[15] His design for Ibn Sina's new mausoleum therefore drew on a Beaux Arts eclecticism: the focal point of the site became a monumental tower, a modified "Franco-Persian" replica of an austere eleventh-century tomb in northeastern Iran. Other buildings at the site—museum, library, and enclosed tomb chamber—incorporated ancient Egyptian, Babylonian, and classical Greek elements, a pastiche of pre-Islamic civilizational genealogies claimed by white Europeans and now by Aryanist Iranian nationalists. Seyhoun claimed that his architectural choices were historically justified by Ibn Sina's "genius in fusing various sciences and arts" as well as "the fact that Persian philosophy was based on Greek philosophy."[16] The racial identification of Ibn Sina's remains and reconstructed features as Aryan-Iranian was therefore embedded in a broader aesthetic and monumental display of Iranian claims to European whiteness, as well as a nationalist refusal to acknowledge the ethnic diversity of either contemporary Iran or of the medieval Islamic civilization to which Ibn Sina actually belonged.

"Scientific" Methods in Facial Reconstruction

The SNH-approved reconstruction of Ibn Sina's likeness was ultimately an impressionistic rendering. The SNH records of the artistic process emphasize Sadighi's consultation with historians who collated written descriptions of Ibn Sina's appearance from medieval Islamic texts. However, they prioritized the artistic aesthetics that made Ibn Sina "look" Iranian, with fair skin, dark eyes and hair with a well-groomed beard, and a prominent yet straight nose—all features that could not have been determined through analysis of the skull alone. Based on the chronology of the portrait commission and the exhumation, it seems unlikely that Sadighi ever had the chance to examine Ibn Sina's skull either directly or through photographs. He and other members of the SNH did not seem to consider direct access to Ibn Sina's remains necessary for the creation of an accurate rendering of Ibn Sina's face—interpretations of text-based historical evidence were taken as sufficient.

However, around the same time that Sadighi was working on his portrait and sculpture of Ibn Sina, new methods of facial reconstruction began indexing measurements of "soft tissue" anatomy to rebuild faces directly onto skulls. Mikhail M. Gerasimov, a Russian anthropologist and archaeologist, was an important developer of such reconstruction methods in the 1930s and 1940s. His process began with the creation of casts of original skulls. Upon these casts, he applied special modeling clay to reconstruct layers of muscle and other soft tissues (e.g., cartilage and fat). To predict the shape of the nose, he placed the skull cast in profile against a sheet of paper, upon which he traced around the skull at a distance deemed average for the deceased individual's sex, age, and race, with two lines extrapolated to form the triangle of the nose. He then used this shape to rebuild the nose with clay.[17] Using this process, Gerasimov became well known for his work reconstructing faces from human remains from Stone Age archaeological sites in Siberia, as well as the faces of famous personages buried in Soviet territories, such as Ivan the Terrible and Timur (Tamerlane), the founder of the vast Timurid Empire in the fourteenth century. In 1954, the Oriental Institute at the Uzbekistan Academy of Sciences engaged Gerasimov to create a portrait of Ibn Sina based on photographs obtained from Iran of Ibn Sina's skull.[18]

Gerasimov, like other physical anthropologists, believed that he could already detect the race of a specimen based solely on the features of the skull and that the purpose of his facial reconstructions lay in capturing the unique face of an individual. Regardless, his portrait reconstruction of Ibn Sina in 1956 further fueled competing claims about Ibn Sina's racial ancestry. When Gerasimov received a copy of the photographs of Ibn Sina's skull, he characterized the skull as resembling a "Ferghano-Pamir" Indo-European racial type. He described certain features, like the cheekbones, eye sockets, brow ridge, and bridge of the nose, as having a "Mongoloid" character, and noted that "such skulls are common among Tajiks and Uzbeks."[19] Without access to Ibn Sina's skull to create a cast (the skull had already been reinterred in a newly built mausoleum in Hamadan), Gerasimov instead drew a reconstructed portrait on paper, at the same angle as Sadighi's version (see figure 9.2).

Gerasimov's analysis of Ibn Sina's skull, made credible by the former's extensive archaeological and anthropological work in Central Asia, thus laid the groundwork for the Soviet republics of Uzbekistan (with a Turkic-speaking majority population) and Tajikistan (with a Persian-speaking majority population) to join the battle to claim Ibn Sina as their own.[20] In 1965, scientists at the Andijan State Medical Institute with interests in the history of medicine built upon Gerasimov's earlier work. Yusub A. Atabekov,

FIGURE 9.2 Portrait of Ibn Sina by Mikhail Gerasimov drawn in 1956, based on a photo of the skull in profile.
Source: Reprinted from Atabekov and Khamidullin, *Abu Ali Ibn Sinoning ilmiĭ asoslangan khaĭkal.*

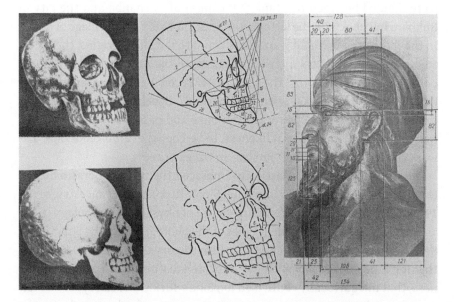

FIGURE 9.3 Three-quarter and profile photographs of Ibn Sina's skull (left), used by scientists in Soviet Uzbekistan to create craniometric diagrams (center) and ultimately to create a "scientific reconstruction" of Ibn Sina's likeness, sculpted by Ye. S. Sokolova in 1965 (right).
Source: Reprinted from Atabekov and Khamidullin, Abu Ali Ibn Sinoning ilmiĭ asoslangan khaĭkal.

the director of social hygiene, and Shavkat Kh. Khamidullin, an anatomist, aimed to transform the two-dimensional profile photographs of Ibn Sina's skull into a three-dimensional sculpture. After converting the photographs into line drawings, they took extensive measurements of the skull's features and projected these craniometric ratios into three dimensions, commissioning the Andijan institute's art instructor Ye. S. Sokolova to sculpt a bust that reconstructed Ibn Sina's likeness based on these proportions (figure 9.3). They subsequently published a description of the process of "scientific reconstruction" to create Ibn Sina's face, which began with an aluminum wire frame followed by layers of plaster to build soft-tissue features.[21]

The Andijan scholars' claims to scientific accuracy caught the attention of high-ranking authorities in the Uzbek SSR medical establishment. By the 1970s, several countries began preparing public events for 1980 to celebrate the millennium of Ibn Sina's birth. Within the Soviet Union, these events were divided between the Tajik SSR capital of Dushanbe and the city of Bukhara in Uzbek SSR (the latter being adjacent to Ibn Sina's village of birth). In the interest of upholding an Uzbek claim to Ibn Sina's origins, the Uzbek SSR Council of Ministers distributed photos of the Andijan bust to anthropologists and artists across the USSR. After receiving external assessments that the plaster bust was "scientifically valid," Uzbek government authorities ordered the bust to be recast in bronze and passed a resolution designating it as Ibn Sina's official likeness, to be copied in all promotional artwork related to the Ibn Sina "jubilee" in Uzbekistan.[22]

Turkish scholars enthusiastically embraced the work of Gerasimov and the Uzbek scientists as confirming Ibn Sina's Turkish racial ancestry. In the 1970s, Şevket Aziz Kansu, a founding figure of physical anthropology in the Turkish Republic, also obtained copies of the photographs of Ibn Sina's skull. Kansu's presentation at the Eighth Turkish History Congress in 1976 argued that Ibn Sina's "head morphology" indicated features, such as an aquiline nose, that unequivocally confirmed Ibn Sina's Turkish "racial type."[23] A few years later, during a special commemoration session for Ibn Sina during the 1980 International Congress of the History of Medicine in Barcelona, Yusub Atabekov offered the Turkish historian Arslan Terzioğlu a copy of the book on the Uzbek "scientific reconstruction" of Ibn Sina's face. Terzioğlu eagerly noted that these renderings, both the Uzbeks' artistic bust and the earlier portrait by Gerasimov, matched Kansu's predictions based on the skull. He also likened the aquiline nose of the reconstructed face to medieval portraits of Mehmet the Conqueror, insisting that the artistic representations considered together "prove that Avicenna belongs to the Turkish race."[24]

Yet Gerasimov's reconstruction of Ibn Sina has also been marshaled to support Iranian portrayals of Ibn Sina as a Persian and racially Aryan figure, as documented by the architecture historian Talinn Grigor. When Grigor visited the museum building at Ibn Sina's Hamadan mausoleum in the year 2000, she noted that a prominent display of a photograph of Ibn Sina's skull in profile is flanked by a portrait drawing captioned in English: "The face of the Ave-sina [sic] that is desig[n]ed to skull."[25] The portrait displayed is not the aesthetic drawing by Sadighi (figure 9.1) but instead the "scientific" drawing by Gerasimov (figure 9.2)—the same drawing whose aquiline nose allegedly made Ibn Sina identifiable as "racially Turkish." The plaque adjacent to the display makes no mention of the drawing's provenance. The fact that the very same facial reconstruction can be made to serve two different and competing racial narratives recalls the arbitrariness of racial typolo-gies. Gerasimov's reconstruction was accepted by both Turks and Iranians because his methods were framed as "scientific" and therefore accurate determinations of Ibn Sina's racialized features. Yet apparently Gerasimov's version of Ibn Sina's face made the latter, as an individual, recognizable to both national populations as "one of their own."

This raises important questions about differing historical trajectories and methodological preferences of the discipline of anthropology in Iran versus Turkey. In both countries, the emergence of anthropology research was driven by scholars with strong nationalist commitments who aimed to have Turks and Iranians classified as constituent branches of a white European race. However, while Turkish anthropology was originally dominated by physical anthropologists who staked their scientific claims through racial craniometry, Iranian anthropologists largely favored folklore and linguis-tic studies.[26] This history goes some way toward explaining why the Iranian SNH commissioned Ibn Sina's reconstructed portrait on the basis of textual historical evidence, without waiting for craniometric analyses of the skull. Yet while Turkish scholars seemingly fetishized the skull and Soviet recon-structions based upon craniometric studies, they too turned to alternative evidence—in Terzioğlu's case, art history—to legitimate their readings of Ibn Sina's features as racially Turkish. Finally, while postrevolution generations of Iranian scholars may have regarded the Uzbek reconstruction as suspect for its pan-Turkic implications, the reconstruction by Gerasimov—a Russian and representative of Soviet scientific achievements with less obvious polit-ical stake in his racial interpretations—could be perceived as relatively neutral and objective. To this day, Iranian and Turkish debates over Ibn Sina draw on carefully curated international scientific networks of citation and collaboration to legitimate their own racialized interpretations and down-play their nationalist interests.

Despite the lack of access to Ibn Sina's skull, new reconstructions of his face based on simulated tissue-layering methods continue to appear. In 2012, a group of Iranian neuroscientists based at the Tabriz University of Medical Sciences worked together with Caroline Erolin, a medical artist based at the Centre for Anatomy and Human Identification at the University of Dundee in Scotland, to revise the reconstruction of Ibn Sina's face. This col-laborative team described their work as the "first scholarly attempt" to produce Ibn Sina's likeness, either unaware or dismissive of the Soviet

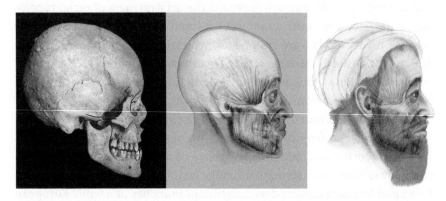

FIGURE 9.4 Reconstruction of Ibn Sina's face published by Erolin et al. (2013).
Source: Reprinted from Caroline Erolin et al., "What Did Avicenna (Ibn Sina, 980–1037 A.D.) Look Like?," *International Journal of Cardiology* 167 (2013): 1660–63, with permission from Elsevier.

reconstructions. Regardless, like Gerasimov in the 1950s, the Iranians provided Erolin only a single photograph of Ibn Sina's skull in profile, and thus she could not follow her normal three-dimensional modeling techniques. She worked instead in two dimensions, using layers of tracing paper (see figure 9.4).[27]

After the resulting portrait was published in the *International Journal of Cardiology*, several Turkish anatomists and forensic anthropologists critiqued Erolin's reconstruction on a number of points, disagreeing in particular with the subjective choices made in depicting facial hair, eyebrows, and nose shape. To reconstruct these features accurately, they insisted that an individual's ethnicity or race must be known in order to apply the correct standards for average soft-tissue thickness, hair color and texture, and eye color. However, "it would not be right to decide whether Avicenna was Persian or Turk based on his skull." Instead, the Turkish experts argued, "DNA analysis" would be necessary to resolve these most heavily racialized features.[28] Genetic technologies are now invoked, albeit implicitly, as the panacea for racial identity disputes—not simply through comparison of biogeographical markers through which tissue thickness is estimated but also through gene sequences directly relevant to phenotype. Such phenotypic data could enable the reconstruction of Ibn Sina's "true face" and therefore make possible his correct ethnic categorization. However, as long as Ibn Sina's remains stay buried, so does any potential examination of his DNA.

Although the debate over Ibn Sina's ethnoracial identity remains bizarrely persistent (notably in the pages of international scientific journals), ultimately the stakes of the Ibn Sina reconstructions are relatively low.[29] It is a dispute over political and cultural prestige between the majoritarian populations of established nation-states; no existential threat to any group or territory hangs in the balance. Turkish, Uzbek, Tajik, and Afghan claims to "ownership" over Ibn Sina are purely rhetorical, as there is no serious demand (yet) to "repatriate" Ibn Sina's remains from Iran to another country. More troubling cases of racial and facial politics arise not from the identity of esteemed individuals like Ibn Sina but rather from the uncertain ancestral relationships linking entire ancient civilizations to the living populations and territories of modern nation-states. For such an example, I turn to the discourse of Phoenicianism in Lebanon.

Lebanon and the "Phoenician Race"

Lebanon has been a particularly important site for the training of Arab physicians and anthropologists and also for the dissemination of racial discourses. In the late Ottoman period, French and American missionary universities in Beirut engaged in archaeological studies of the ancient Phoenicians, a Mediterranean civilization based in the Levant circa 2500–500 BC. These studies helped their students conceptualize a secular Syrian regional identity distinct from the rest of the Arabic-speaking world. After the First World War, the question of Phoenician ethnicity increasingly exacerbated communal tensions between the many different religious groups living in Beirut and its hinterland, an area called Mount Lebanon within the French Mandate for Syria. Many people living within the mandate hoped that Syria would become a unified and independent nation-state. Instead, the French divided the area along confessional lines to maintain a set of weak client states. Their local supporters in this endeavor were the Maronite Catholics, who had long been French protégés in the region. As the majority religious group in Mount Lebanon, they worked with the French to cultivate an independent Lebanese state separate from Syria.

To justify the creation of this "Greater Lebanon," the French and their allies turned to scientific racial classification and archaeology. Some European anthropologists used the cephalic index, a measurement of skull shape, to argue that the Maronites represented a distinct race from the Arabs. Certain Maronite elites and French educators claimed that this signified the Maronites' unique descent from the Phoenician race. The French, through the Jesuit Université Saint-Joseph in Beirut, promoted a non-Arab Phoenician genealogy for Maronites in order to justify dividing "Greater Lebanon" from the rest of the Syrian mandate. French-educated Maronite archaeologists and historians actively popularized ideas of Phoenician racialism among their coreligionists. The French-established Lebanese National Museum, founded in 1922 explicitly to house collections of Phoenician archaeology (including human remains) at the expense of Islamic and Arab material culture, came under the leadership of the USJ-trained Maronite archaeologist Maurice Chéhab in 1928. Chéhab became an extremely important figure for the dissemination of Phoenicianism through his museum curatorship, excavation work, and publications.[30]

Unsurprisingly, the many non-Maronite religious communities in Lebanon became increasingly alienated by this discourse, which devalued both their historical connections to Lebanese territory and their political aspirations to inclusion in a larger secular Arab state. The separatist Phoenician narrative was challenged by a number of faculty at USJ's rival institution, the American University of Beirut, where a major program of anthropometric research emerged at the medical school during the 1920s and 1930s. The histologist William Shanklin and the pathologist Harald Krischner measured thousands of people across the Levant, Mesopotamia, and Persia. In 1929, the Dutch neuroanatomist Cornelius Kappers visited Beirut to teach at AUB's medical school and conduct his own research in craniometry. These men recruited AUB students of different religious and ethnic backgrounds to learn anthropometric methods and become liaisons to their communities to collect additional data.[31] Shanklin, Krischner, and Kappers agreed that the Arabic-speaking communities of the Levant originated from several racial stocks. Specifically, nomadic Bedouins were considered true Arabs. In contrast, "the so-called Arabs forming the settled population" were a distinct racial group.[32] In fact, Kappers decided that the Maronites were not distinct from other Christians, whom he lumped together in a category labeled "real Lebanese," in contrast to "real Arabs."[33] These AUB-affiliated scientists never directly referenced Maronite Phoenicianism in their scientific publications, but it clearly influenced the questions they asked of their data, including: "Who are the present representatives of [the Phoenician] race?"[34]

To settle this question, Kappers secured the permission of Maurice Chéhab to photograph and measure eighteen Phoenician skulls kept in the Lebanese National Museum. He compared the cranial indices of these skulls to the average cephalic index of the many Lebanese religious groups measured by Shanklin, Krischner, and their students. Based on these comparisons, Kappers argued that the head measurements of the Bedouin tribes of the Syrian desert "strongly suggest their relation with the Phoenicians."[35] In other words, AUB's anthropometrists claimed that Lebanese Christians, including Maronites, were indeed racially distinct from Arabs, but not because the former had Phoenician ancestry. Instead, starkly rejecting the prevalent political discourse, they identified the Arab Bedouins as the living descendants of the Phoenicians.

The craniometric research fed back into the efforts of pro-Arab Lebanese intellectuals to debunk Phoenicianist politics. In his 1939 work *National Consciousness*, the AUB history professor Constantin Zurayq insisted that scientific methods of racial classification, including skull measurements (*ṭūl al-r'as wa istidāratihi*), showed that Phoenicians and Arabs belonged to the same race.[36] Ancient skulls therefore lent scientific validity to his argument that Lebanon should not be separated from the rest of Syria on the basis of racial difference. Unfortunately, French political interests ruled the day, and ultimately "Arabists" like Zurayq became reconciled to the idea of a separate Lebanon wherein they could simultaneously embrace an Arab identity and a Phoenician heritage. As a result, the discourse of Phoenicianist nationalism did not fade away after Lebanon's independence in 1943, nor did its associations with Maronite Christian supremacy and anti-Arab sentiment. Instead, these associations were cemented by the eruption of the Lebanese Civil War (1975–1990), when Maronite militias, particularly the right-wing Kata'eb (Phalanges), mobilized Phoenicianist rhetoric to disparage Arabs and Muslims and even to justify their massacres of Palestinian refugees. The idea of Phoenician ancestry thus became irrevocably tainted by this violence, and the peace accords that ended the war reasserted Lebanon's Arab political identity, without attempting to erase Phoenician heritage from Lebanon's distinct cultural identity.[37]

Facial Reconstruction in Light of Ancient DNA Technologies

Over the past twenty years, DNA sequencing projects coupled with the application of forensic reconstruction techniques to Phoenician remains have contributed to a revival of Phoenicianist national discourse in Lebanon. Interestingly enough, the Phoenician remains used for reconstruction were excavated not from Lebanon but from Tunisia, where the remains of the Phoenician city of Carthage are located. In 1994, a Punic tomb was uncovered at Carthage's citadel of Byrsa. The skeleton of the so-called Young Man of Byrsa, also called "Arish," based on the name appearing in inscriptions beside his tomb, was sent to France for reassembly and examination. Based on a craniometric analysis, anthropologists determined Arish's "racial type" to be "Mediterranean, European Caucasoid."[38] The French dermoplasty artist Élisabeth Daynès was hired to produce a full-body reconstruction of Arish's remains. Daynès, who is known for her reconstructions of prehistoric hominids, is not a trained anthropologist but nonetheless relies on methods akin to those of Mikhail Gerasimov, that is, meticulously using modeling clay to rebuild muscles and soft tissue onto casts of skulls. Artistically, however, her work is much further developed: she casts her clay models into life-size resin and silicone figures, allowing her to paint highly lifelike skin texture and complexion and attach natural-fiber hair, eyebrows, and eyelashes. The resulting sculpture (figure 9.5) appears compellingly alive compared to either the monumental busts and statues or the scientifically derived portraits of Ibn Sina. Daynès's reconstruction of Arish was placed on display at the National Museum of Carthage in 2010.

When Leila Badre, the director of the archaeological museum at the American University of Beirut, heard about the reconstruction, she immediately requested that it be loaned to Lebanon, insisting that because "Lebanon is believed to be the land of Phoenicians, it is only normal that [Arish] be brought home to the land of his ancestors before being exhibited anywhere else."[39] In her role as museum director, Badre considers the representation of Phoenician material particularly important, with the aim of correcting "misinformation/misconception about Phoenicia among the general public

FIGURE 9.5 Close-up of the life-size sculptural reconstruction of "Arish," the Young Man of Byrsa, with hazel eyes, light skin, and brown hair.
Source: Photo by M. Rais, 2010 (available via Wikimedia Commons), reprinted under Creative Commons Public Domain Dedication.

and Lebanese visitors in particular. After all, isn't the basic goal of a university museum to educate the audience with precise information often neglected to be taught, even in schools?"[40]

Although it was not possible to bring Arish's actual skeleton to Lebanon, Daynès's reconstructed model and some of the real funerary objects discovered in the Byrsa tomb arrived in Beirut for a small one-month exhibition in January 2014. The press release for the exhibition trumpeted the scientific accuracy of the reconstruction, describing it as "95 percent accurate [although] the color of skin, eyes, and hair remains uncertain and subjective."[41] These uncertainties were downplayed in the exhibit display, which stood Arish's model in a separate and nearly empty room, set apart from most of the other artifacts and information plaques. Badre's educational plan for Arish therefore involved bringing viewers, alone or in small groups, into confrontation with the reconstructed model as though it were a living person capable of speaking for itself. The only contextualizing information in the room was a looping video showing some scenes of Daynès building the reconstruction in her Paris studio, assuring viewers that the model was the product of a scientific process based on the original skull. In a brief interview with the Associated Press, Badre explained, "I thought it is very important for the Lebanese to see a real shape [sic] of their ancestors and this is why it is attracting so many [visitors] because everyone, every Lebanese wants to be identified, or wants to identify himself with this person who is his ancestor."[42] In this way Badre acknowledged the model's power to manipulate underlying Lebanese nationalist fascinations with the Phoenicians to serve the museum's educational mission and simultaneously to compel viewers to compare their own appearance to the scientifically derived "real shape" of a Phoenician face.

The exhibition opened in Beirut in the aftermath of DNA research that purported to track down the living descendants of the Phoenicians. During the ten years before 2014, the Lebanese geneticist Pierre Zalloua, sponsored by the Genographic Project, had already been working to find "Phoenician DNA" among the peoples living in Lebanon and around the Mediterranean.[43] In 2008, Zalloua's announcement that he had identified a "Phoenician" Y-DNA haplotype attracted much public attention in Lebanon, with the result that many Lebanese men asked to be tested for this haplotype.[44] Leila Badre, the AUB museum director, invited Mikhail Jouni, one of these alleged Phoenician haplotype carriers, to the opening of the exhibition of the Young Man of Byrsa. Badre told Jouni, a fisherman from the city of Tyre, that the model represented his "cousin." Jouni responded with great emotion, saying that it was "enchanting" to see the model, because "he had my eyes, my nose, and my pose." Jouni added, "I was happy when I discovered I was of Phoenician origins, but I was much more overwhelmed when I discovered that I also looked Phoenician."[45]

The features of the model that Jouni most related to, however, are speculative: nose shape and size and eye color, and, of course, any habitual pose cannot be detected from the physical form of the skull or other bones from Arish's grave. Rather, they were predicted based on archaeologists' knowledge of Phoenician art and artifacts, which include sculptural representations of people with curly hair and prominent noses. Meanwhile, the model's pale skin and hazel eyes reflect racial assumptions derived from the original analysis of the skull as possessing "European Caucasoid" traits. As Abigail Nieves Delgado shows, experts in forensic reconstruction consider this a priori categorization foundational to the process of choosing the "correct" phenotypic traits missing from the bony evidence.[46] In the case of these Phoenician remains, the light skin and eyes also reflect a racial sensibility enshrined in Orientalist scholarship that depicts Phoenician civilization as an ancestral component of Western European, not West Asian or African, cultures.[47]

About five years after Daynès built her sculpture of Arish, an international team of geneticists subjected Arish's skeleton to the delicate process of ancient DNA sequencing. Two small pieces of Arish's ribs were shipped to the state-of-the-art ancient DNA laboratory at the University of Otago in New Zealand. There, Arish's mitochondrial DNA (mtDNA) was successfully sequenced and classified as a specimen of the haplotype U5b2c1. This was a surprise for the Lebanese geneticists on the team, since this haplotype did not appear in any of the Lebanese people they had tested. In fact, this very rare haplotype is only currently known from a few individuals living in Western Europe, with one person in Portugal having the most closely related mtDNA sequence to that found in Arish.[48] The geneticists' publication and subsequent media coverage of these results advertised that this ancient Phoenician man possessed a "European" mtDNA haplotype. The implications of this association with Europe are readily visualized: press releases by the University of Otago and Lebanese American University, as well as reports by newspapers like the *Independent*, *Daily Mail*, and *Christian Science Monitor*, were illustrated by photographs of the Daynès reconstruction, even though her reconstruction of Arish did not rely on any information related to the newly sequenced DNA.[49] In fact, mtDNA contains no genes that could confirm or refute the speculative reconstruction of Arish's skin, eye, or hair color. Yet through the putative geographic origins of his mtDNA haplotype in "Europe," these reconstruction choices were retroactively justified—readers could accept the DNA findings by looking at Arish's white-racialized features.

The legibility of the Arish reconstruction as simultaneously Phoenician, Lebanese, and European reveals the insidious nature of nationalist racism in the Middle East, where most ethnoreligious communities have historically staked claims to European racial whiteness while also articulating national and religious differences through the language of biological race. Scientific approaches to forensic reconstruction have been embraced throughout the region with the belief that learning what ethnic or intellectual ancestors *truly looked like* can resolve longstanding disputes over the national belonging of great individuals like Ibn Sina or entire cultures like the Phoenicians. The fact that Gerasimov's reconstruction of Ibn Sina's face is promoted by both Iranians and Turks as evidence that Ibn Sina was a member of the Persian-Aryan "race" or Turkish "race" exposes not only the arbitrary nature of human racial classification but also the uselessness of biological phenotypes for mediating arguments that, at their core, are about cultural genealogies. It represents an unconscious admission that (for example) Iranians and Turks cannot be reliably distinguished on the basis of physiological differences allegedly detected by quantitative science. Instead, as Ania Loomba argues, cultural markers like language, religion, and behavioral stereotypes have always been intrinsic criteria for racial diagnosis, not only in postcolonial but also in premodern and non-Western contexts.[50]

It is precisely these slippages of biology and culture that undergirded the emergence of racial science, which in turn created a key ideological foundation for modern nationalism. As the scholar and artist Denise Ferreira da Silva writes, a symbolic arsenal derived from scientific racial knowledge dictates how a nation's members can "participate in the nation's present and how they will perform in its future without ever accounting for their being placed in its past."[51] Facial reconstructions created from the skulls of historical figures deemed representative of a nation's past are now part of this symbolic arsenal: Mikhail Jouni's affective response to the reconstruction of Arish highlights the political and scientific stakes of putting a face on the past. In this case, the reconstructed model has supplanted the actual human remains in its physical and photographic circulation. The speculative features of the model are ignored when the overall reconstruction meets the public's expectations of self-recognition and self-discovery. Even as new technologies of facial reconstruction emphasize their enhanced accuracy through the use of digital modeling and DNA sequencing, these technologies refocus scientific and public attention on biological phenotypes—the racialized features of old skulls and their predicted physiognomies—as sites of both national and personal identity. Through this focus, facial reconstructions can ultimately promote destructive beliefs in the reality and visibility of biological race, even when the evidence of racial ambiguity is literally staring back at us.

Notes

1. Amade M'charek, "Tentacular Faces: Race and the Return of the Phenotype in Forensic Identification," *American Anthropologist* 122, no. 2 (June 2020): 369–80.
2. Abigail Nieves Delgado, "The Problematic Use of Race in Facial Reconstruction," *Science as Culture* 29, no. 4 (October 1, 2020): 588.
3. M'charek, "Tentacular Faces," 372, 377.
4. Nieves Delgado, "The Problematic Use of Race in Facial Reconstruction," 586.
5. David Bindman, *Ape to Apollo: Aesthetics and the Idea of Race in the 18th Century* (Ithaca, NY: Cornell University Press, 2002), 12. The literary scholar David Lloyd, through close readings of texts by Immanuel Kant and Edmund Burke on aesthetics and the sublime, elaborates the implications of European aesthetic theory for the political representation of peoples marked by racial and cultural difference. David Lloyd, *Under Representation: The Racial Regime of Aesthetics* (New York: Fordham University Press, 2019).
6. Bindman, *Ape to Apollo*, 188.
7. For a critical perspective on "psychological essentialism" in hypotheses about own-race bias in facial recognition, see Jessica S. Leffers and John D. Coley, "Do I Know You? The Role of Culture in Racial Essentialism and Facial Recognition Memory," *Journal of Applied Research in Memory and Cognition* 10, no. 1 (March 2021): 5–12.
8. Reza Zia-Ebrahimi, *The Emergence of Iranian Nationalism: Race and the Politics of Dislocation* (New York: Columbia University Press, 2016).
9. Henry Field, *Contributions to the Anthropology of Iran*, Publications of the Field Museum of Natural History Anthropological Series 29.1 (Chicago: Field Museum of Natural History, 1939), 13, 507, 257, 9, 15.
10. Talinn Grigor, "Recultivating 'Good Taste': The Early Pahlavi Modernists and Their Society for National Heritage," *Iranian Studies* 37, no. 1 (March 2004): 17–45.
11. Aydin M. Sayili, "Was Ibni Sina an Iranian or a Turk?," *Isis* 31, no. 1 (November 1939): 8–24.
12. Talinn Grigor, *Building Iran: Modernism, Architecture, and National Heritage Under the Pahlavi Monarchs* (New York: Periscope, 2009), 132.

13. Ḥusayn Baḥr al-ʿUlūmī, *Kārnāmah-i Anjuman-i Āsār-i Millī az āghāz tā 2535 Shāhanshāhī/1301-1355 Hijrī-i Khvurshīdī* (Tehran: Chāpkhānah-i Mīhan, 2535 [1976]), 107.

14. Grigor, "Recultivating 'Good Taste,'" 19; Baḥr al-ʿUlūmī, *Kārnāmah-i Anjuman-i Āsār-i Millī*, 107–9.

15. Talinn Grigor, "Use/Mis-Use of Pahlavi Public Monuments and Their Iranian Reclaim," *Thresholds* 24 (January 2002): 49.

16. Talinn Grigor, "Cultivat(ing) Modernities: The Society for National Heritage, Political Propaganda, and Public Architecture in Twentieth-Century Iran," PhD diss., Massachusetts Institute of Technology, 2005, 363–64.

17. Herbert Ullrich and Carl N. Stephan, "Mikhail Mikhaylovich Gerasimov's Authentic Approach to Plastic Facial Reconstruction," *Anthropologie* 54, no. 2 (2016): 97–108.

18. U. I. Karimov, ed., *Portret Ibn Siny*, UzSSR F. A. Sharqshunoslik Instituti Asarlari 5 (Toshkent: Uzbekiston SSR Fanlar Akademiyasi Nashriyoti, 1957), 17. I thank Iman Darwish for helping me access this source.

19. Karimov, *Portret Ibn Siny*, 19, 20.

20. On the Tajik and Uzbek claims to Ibn Sina, see Mohira Suyarkulova, "Statehood as Dialogue: Conflicting Historical Narratives of Tajikistan and Uzbekistan," in *The Transformation of Tajikistan: The Sources of Statehood*, ed. John Heathershaw and Edmund Herzig (London: Routledge, 2013), 165; Shahram Akbarzadeh, "Nation-Building in Uzbekistan," *Central Asian Survey* 15, no. 1 (1996): 23–32.

21. Yusub Atabekovich Atabekov and Shavkat Khusainovich Khamidullin, *Abu Ali Ibn Sinoning Ilmiĭ Asoslangan Khaĭkal Obrazini Iaratish = Sozdanie Nauchno Obosnovannogo Skul'pturnogo Obraza Abu Ali Ibn Siny = A Bust of Abu Ali Ibn Sina: A Scientific Reconstruction of the Great Scholar's Image* (Tashkent: Izdatel'stvo "Meditsina" UzSSR, 1980), 79–90.

22. Atabekov and Khamidullin, *Abu Ali Ibn Sinoning ilmiĭ asoslangan khaĭkal*, 92–95.

23. Şevket Aziz Kansu, "İbni Sina'nın Başının Morfolojisi Üzerine Bir Gözlem," in *VIII. Türk Tarih Kongresi, 11-15 Ekim 1976, Ankara* (Ankara: Türk Tarih Kurumu Basımevi, 1983), 1:28.

24. Arslan Terzioğlu, "Ibn Sînâ (Avicenna) in the Light of Recent Researches," *Erdem* 9, no. 25 (1996): 430.

25. Grigor, *Building Iran*, 133.

26. See Elise K. Burton, *Genetic Crossroads: The Middle East and the Science of Human Heredity* (Stanford, CA: Stanford University Press, 2021), chap. 1.

27. Caroline Erolin et al., "What Did Avicenna (Ibn Sina, 980–1037 A.D.) Look Like?," *International Journal of Cardiology* 167, no. 5 (September 2013): 1660–63.

28. Çağrı Zeybek Ünsal and Nüket Örnek Büken, "Reconstruction Study of Avicenna's Face and Ongoing Discussions About His Ethnicity," *Yeni Tıp Tarihi Araştırmaları* 23 (2017): 53–54.

29. As one recent example, see the exchange between Iranian and Turkish scientists that was allowed to take place over a year's worth of issues of the *Archives of Gynecology and Obstetrics* and that cited Aydın Sayılı's nationalist work from the 1930s: Masumeh Mobli et al., "Scientific Evaluation of Medicinal Plants Used for the Treatment of Abnormal Uterine Bleeding by Avicenna," *Archives of Gynecology and Obstetrics* 292, no. 1 (July 2015): 21–35; Mumtaz M. Mazicioglu, "Ibni Sina

(Avicenna), the Most Known and Greatest Turkish Medical Doctor in Late Ancient World," *Archives of Gynecology and Obstetrics* 292, no. 3 (September 2015): 473–74; Arman Zargaran and Roja Rahimi, "Response to: Avicenna, a Persian Scientist," *Archives of Gynecology and Obstetrics* 292, no. 3 (September 2015): 475–76; M. Mumtaz Mazicioglu, "The Nationality of Ibni Sina (Avicenna)," *Archives of Gynecology and Obstetrics* 293, no. 1 (January 2016): 219–20; Arman Zargaran and Roja Rahimi, "Some Clarifications on the Avicenna's Nationality," *Archives of Gynecology and Obstetrics* 293, no. 1 (January 2016): 221–22; Mumtaz M. Mazicioglu, "Final Comment on the Nationality of Ibni Sina," *Archives of Gynecology and Obstetrics* 293, no. 4 (April 2016): 925–27. Also notable is the 2020 exchange between two Iranian research groups, based in Tehran and Tabriz respectively, in *World Neurosurgery*; what began as an attempt to correct an American publication's characterization of Ibn Sina as "Arabian" dragged out into competing explanations about why Ibn Sina, as a Persian, wrote his most famous works in Arabic. Danika Paulo et al., "History of Hemostasis in Neurosurgery," *World Neurosurgery* 124 (April 2019): 237–50; Zahid Hussain Khan et al., "A Note About the Ancestral Origin of Abu Al Husain Ibn Abdullah Ibn Sina, Avicenna (980–1037 CE)," *World Neurosurgery* 135 (March 2020): 173–75; Reza Mohammadinasab, Javad Ghazi Sha'rbaf, and Somaiyeh Taheri-Targhi, "Letter to the Editor Regarding 'A Note About the Ancestral Origin of Abu Al Husain Ibn Abdullah Ibn Sina, Avicenna (980–1037 CE),'" *World Neurosurgery* 136 (April 2020): 431; Zahid Hussain Khan et al., "In Reply to the Letter to the Editor Regarding 'A Note About the Ancestral Origin of Abu Al Husain Ibn Abdullah Ibn Sina, Avicenna (980–1037 CE),'" *World Neurosurgery* 136 (April 2020): 432.
30. Asher Kaufman, *Reviving Phoenicia: In Search of Identity in Lebanon* (London: I. B. Tauris, 2014), 123–26.
31. For more information, see Elise K. Burton, "Comparative Globalizations: Building and Dismantling Genetic Laboratories in Lebanon," *British Journal for the History of Science*, January 19, 2022, 1–19.
32. William M. Shanklin and Nejla Izzeddin, "Anthropology of the Near East Female," *American Journal of Physical Anthropology* 22, no. 3 (April 1937): 413; Harald Krischner and M. Krischner, "The Anthropology of Mesopotamia and Persia A. Armenians, Khaldeans, Suriani (or Aissori) and Christian 'Arabs' from Iraq," *Proceedings of the Royal Academy of Sciences at Amsterdam* 35 (1932): 205–17; C. U. Ariens Kappers, *An Introduction to the Anthropology of the Near East in Ancient and Recent Times* (Amsterdam: Noord-Hollandsche Uitgeversmaatschappij, 1934).
33. Kappers, *An Introduction to the Anthropology of the Near East*, 21–25.
34. Kappers, *An Introduction to the Anthropology of the Near East*, 55; William M. Shanklin, "Anthropometry of Syrian Males," *Journal of the Royal Anthropological Institute of Great Britain and Ireland* 68 (July 1938): 404. For more details on the anthropometric research conducted at AUB, see Burton, *Genetic Crossroads*, chap. 1.
35. Kappers, *An Introduction to the Anthropology of the Near East*, 48, 57.
36. Qusṭanṭīn Zurayq, *al-Aʿmāl al-fikriyya al-ʿāmma li-l-Duktūr Qusṭanṭīn Zurayq* (Bayrūt: Markaz Dirāsāt al-Waḥdat al-ʿArabīya, 1994), 1:59.
37. Kaufman, *Reviving Phoenicia*, ix; Josephine Crawley Quinn, *In Search of the Phoenicians* (Princeton, NJ: Princeton University Press, 2018), 14–15.

38. Institute INdPNH, "Exposition (octobre 2010—mars 2011): 'Le jeune homme de Byrsa' au Musée de Carthage," January 2, 2013.

39. Nadine Elali, "A Phoenician Homecoming," *NOW Media*, February 4, 2014.

40. Leila Badre, "The Archaeological Museum of the American University of Beirut and Its Educational Role: A Case Study," in *Museums and the Ancient Middle East: Curatorial Practice and Audiences*, ed. Geoff Emberling and Lucas P. Petit (Abingdon: Routledge, 2019), 203.

41. "The Young Phoenician Man of Byrsa—Carthage," press release, American University of Beirut Archeological Museum, https://www.aub.edu.lb/museum _archeo/Documents/Introduction%20to%20the%20Young%20Phoenician%20 Man%20of%20Byrsa%20-%20Carthage.pdf.

42. "Ancient Phoenician recreated and on show among his Lebanese descendents," film footage produced in Lebanon by Associated Press Television on February 25, 2014, AP Archive, Story Number 934360, http://www.aparchive.com/metadata /youtube/a85eefd36d5ffb3ac2b75042e9da43b0.

43. Pierre A. Zalloua et al., "Identifying Genetic Traces of Historical Expansions: Phoenician Footprints in the Mediterranean," *American Journal of Human Genetics* 83, no. 5 (November 2008): 633–42.

44. Marc Haber, interviewed by the author on October 8, 2018, at the Wellcome Sanger Institute in Hinxton, UK.

45. Elali, "A Phoenician Homecoming."

46. Nieves Delgado, "The Problematic Use of Race in Facial Reconstruction."

47. See Quinn, *In Search of the Phoenicians.*

48. Elizabeth A. Matisoo-Smith et al., "A European Mitochondrial Haplotype Identified in Ancient Phoenician Remains from Carthage, North Africa," ed. Luísa Maria Sousa Mesquita Pereira, *PLoS One* 11, no. 5 (May 25, 2016): e0155046.

49. University of Otago, "Ancient DNA Study Finds Phoenician from Carthage Had European Ancestry," *EurekAlert*, May 25, 2016, https://www.eurekalert.org/pub _releases/2016–05/uoo-ads051916.php; "LAU Scientist Authors Study Revealing European DNA Found in Ancient Phoenician Man," LAU News Archive, May 30, 2016, https://www.lau.edu.lb/news-events/news/archive/lau_scientist_authors _study_re/; Will Worley, "DNA of Ancient Phonecian Could Make Us Reconsider History of Human Migration," *Independent*, June 1, 2016, https://www.independent .co.uk/news/science/archaeology/dna-sequencing-ancient-phoenician -remains-could-make-us-reconsider-history-human-movement-a7048046 .html; Stacy Liberatore, "First DNA from Ancient Phoenician Shows European Ancestry: 2,500-Year-Old Skeleton Could Rewrite History of Human Migration," *Daily Mail*, May 26, 2016, https://www.dailymail.co.uk/sciencetech/article-3611457 /Ancient-DNA-study-finds-Phoenician-Carthage-European-ancestry.html; Ben Rosen, "Ancient Phoenician DNA Suggests a New Model of Human Migration," *Christian Science Monitor*, May 29, 2016, https://www.csmonitor.com/Science/2016 /0529/Ancient-Phoenician-DNA-suggests-a-new-model-of-human-migration.

50. Ania Loomba, "Race and the Possibilities of Comparative Critique," *New Literary History* 40, no. 3 (2009): 501–22.

51. Denise Ferreira da Silva, *Toward a Global Idea of Race* (Minneapolis: University of Minnesota Press, 2007), xiv, 196.

TEN

Racism and Weightism in the Māori Community

From Weight-Focused Health to Indigenous Solutions

ISAAC WARBRICK

THE NAME OF this section of the book, "Past and Promise," is appropriate for this chapter as both weightism and racism have a closely connected though not chronologically aligned past—and both are founded on a promise. Racism stepped ashore in Aotearoa (New Zealand) in the late eighteenth century when European "explorers" and "settlers" took it upon themselves to civilize unruly natives. This was not the first time European colonists had come to "darker" shores with the promise to civilize and "save" browner folk, nor would it be the last, but the effects of colonization and the subordinary views toward "those of color" continue today. Fast forward a few hundred years—to a time where colonialist ideals have firmly taken hold throughout "developed" countries and Indigenous peoples have been separated from the lands which defined their Indigeneity, shaped their worldview, and sustained their survival—and the gradual increase in waistlines, particularly among Indigenous folk and others of darker complexion, has become a major concern to health experts and systems. Though a lot of important things have happened since the arrival of European colonists so many generations ago vital to understanding the racist and weightist narratives that exist today, this chapter will not go into detail about the historical reasons for and links between racism and weightism. Instead, I hope briefly to highlight the relationship between racism and weightism and then outline possible alternative solutions to a weight-obsessed approach to health—an approach that just doesn't seem to work for "our people."

Racism and Weightism

Just as racism is underpinned by a belief and "promise" that a noble few could save the seemingly less moral masses of brown folk, weightism is an expression of a similar belief and promise—those who aren't fat providing the solutions to save those who are. Inequities between white and others somewhere along the "darkness spectrum" are significant, and this is particularly apparent when it comes to health—brown people die earlier than white people, they have a greater prevalence of most illnesses, they have reduced access to and trust in services, and—in context of this chapter specifically—they are more overweight and obese.[1] Whether it is more politically correct to use the term "obese" or refer to the issue in terms of greater BMI or body fat percentage, darker people in developed countries tend to be "visually different" in more ways than color from their "normal" white counterparts. The body mass index (BMI) has become a popular means of measuring the degree of overweight/obesity and is commonly used to make judgments about one's risk of weight-related illnesses. With the BMI comes classifications of underweight, "normal," overweight, and obese. Although some acknowledge the need for ethnic-specific adjustments to the BMI to account for ethnic differences in body composition,[2] these do not account for inter-individual variations in bone density, muscle mass, and body-fat distribution. Yet clinicians continue to use BMI as a surrogate measure for risk of lifestyle illness in assessing individuals. Thus, individuals (or ethnic groupings, in some cases) tend to be seen as "unhealthy" or "abnormal" as a result of not fitting into the "normal" or "healthy" BMI range.

While the definition of overweight and obesity is a topic unto itself, "overweight" and "obese" being defined most commonly as a BMI over 25 and 30 respectively,[3] this chapter is more concerned with how those considered obese, regardless of the definition used, are treated and managed. Just like skin color, body size, more specifically the fat part of that body size, allows people (usually the thinner, and most often whiter) to visually categorize individuals on a continuum that includes laziness, mental and emotional strength, and even morality. Well-known phrases and social media memes proclaiming a "War on Obesity" and "No Excuses" suggest we're in a battle with obese people—those who tend to make excuses for their gluttony and idleness. In fact, governments explicitly waged a "war on obesity," which has fueled fears of the overweight enemy, justified unhelpful health promotion messages, and swayed public opinion toward the evils of being overweight.[4] Lupton highlights this type of narrative in her paper about public health campaigns designed to disgust and shock "offenders" into behavioral change. She comments that "disgust can be employed as a means of distinguishing Self from Other, reinforcing prejudice and bigotry, marginalising out-groups."[5] Likewise, weight stigma has been shown to have negative effects on the mental health and social engagement of those who are targeted, so this approach is not only ineffective in many cases but harmful.[6]

Being called abnormal or idle and having wars waged against us is strangely common for brown people, though—we've already experienced these attitudes, perspectives, and comments long before we were the "fatter race." Addy criticized the fact that white people get to be "normal" and that other groups are defined against this monocultural center.[7] The same could be said of those who align with social "norms" associated with body size. Another common feature of racism and weightism is that the whiter and/or slimmer rarely recognize that the structures and perspectives they adhere to are the very cause for the thing that disgusts them most (poor Māori, sick fat people, etc.). In the case of Māori, other Indigenous peoples, and colored minorities in most places around the world, concern for weight or the health issues associated with it didn't exist before European coloni- zation. Societal racism is a process where negative attributes are ascribed to a culture, ethnicity, or race and then enacted on those prejudices through an imbalance of power. One example of this is the way Māori and other cul- tures globally are negatively represented within the media. McCreanor and Nairn also highlighted how non-Māori general practitioners (physi- cians) talk about Māori health, revealing constructions that blame Māori for their poor health, which is caused by ignorance, poverty, and/or self- destructiveness. It's as though Indigenous and minority people are seen as unable to achieve "health" without the leadership of slim, healthy, and "nor- mal" health professionals who work within a normal (and noble?) health system.[8]

In contrast with the Euro-Western focus on the biophysical aspect of health, Indigenous models of "health" are holistic in nature, emphasizing a balance and interaction between physical, emotional, psychological, social, and spiritual well-being. This holistic view also includes one's sense of cultural identity, connection to ancestral lands and waterways, and the abil- ity to access and utilize traditional knowledge and practices. What's more, practices and ideologies that blame Indigenous people and disregard their input directly conflict with Indigenous philosophies of health promotion. For example, Te Pae Mahutonga, a model of Māori health promotion proposed by Professor Sir Mason Durie, outlines (among other tasks for health promoters) the importance of "Te Mana Whakahaere" (autonomy) and "Te Oranga" (participating in society) in achieving health—both of which are underpinned by the inclusion and privileging of Māori voices and leadership over their own health.[9] McPhail-Bell and colleagues and Donatuto and colleagues also speak of the importance of autonomy for Indigenous Australians and Indigenous Americans (respectively) but also highlight the "ethical tension" of measuring "health" or defining health "risk" in ways that don't necessarily align with the perspectives of Indigenous populations and that disregard Indigenous input or approval. McPhail-Bell and colleagues also note that the "failure" of Indigenous populations to respond to (non-Indigenous) health promotion messages reinforces stigma of Indigenous peoples and causes further harm.[10]

So, if health initiatives that disregard Indigenous perspectives and Indigenous input are less effective—and may do more harm—why do we still use "weight speech" as a driver for behavioral change among Indigenous peoples? As mentioned earlier, racism is usually underscored by the notion that one group is superior to another, without the recognition of the roles that advantage and privilege play. Racism and privilege is often intermeshed with other systems of oppression, such as classism, sexism, or, in this case, "weightism," all the while compounding one's advantage and the other's disadvantage.[11] In addition to the privilege of certain parts of society, there are also the commercial benefits of weight stigma to consider. A booming weight-loss market has thrived, while gyms and food and pharmaceutical companies have been the beneficiaries of this war and our anxiety not to become victims of this "hefty" foe. Blaming individuals or one or more ethnic groups diverts attention from policies that favor commercial interest over health, with weak restrictions on the sale and marketing of certain types of food, and is an easy sell when it is the ethnic "others" who are the worst offenders. A commercially focused society maintains the profitable cycle with claims that "the overweight could succeed if they just worked harder . . . made better choices . . . were more like their thin counterparts, etc." This sounds all too similar to discourse from beneficiaries of colonization, who maintain positions of privilege with claims that "Māori [or substitute for your chosen brown population here] could succeed if they just worked harder . . . made better choices . . . were more like their white counterparts, etc." Same narrative, different angle. The problem with this perspective is that it ignores the many social determinants of health highlighted in the literature. For example, household income, which has a well-established link to health, raises concern for those with an eye on racial health inequities. Not only do white people have a greater average income and wealth compared to Indigenous and minority ethnicities, but even at the same income level, those "of color" are likely to have significantly poorer health.[12]

In summary, health inequalities among Indigenous and non-Indigenous minorities are not reduced through the current weight-focused health approach; in fact, weight-centered health may actually perpetuate discrimination and poorer health. The following section moves away from the problems associated with racism and weightism and explores instead solutions and potential health directions driven by Indigenous knowledge, values, and beliefs. Although the following discussion largely derives from an Indigenous and more particularly New Zealand Māori perspective, similar themes of racism and weightism are woven through the experiences of other Indigenous groups (Native Americans, Aboriginal Australians, etc.) and non-Indigenous "minorities" (Hispanic and African Americans in the United States and Pacific Islanders in New Zealand). It is hoped that our experiences and stories in Aotearoa, New Zealand, can help or support already established efforts by other ethnic groups, connected to us through the themes of racism and weightism, to shift the narrative and create solutions and outcomes that are meaningful within their own cultural frameworks.

Toward Indigenous Solutions

Indigenous peoples globally have expressed their thoughts and perspectives of "health" in a myriad of Indigenous health models that, though unique to each group, share a common theme of holism. Psychological, physical, spiritual, social, and emotional well-being (or a combination of similar concepts) are valued equally and are not expressed in isolation from one another. For example, Te Whare Tapa Whā is a well-known Māori health model that likens health to an ancestral house with four load-bearing walls, representing the equal importance of psychological, social, spiritual, and physical health.[13] In contrast, Robison observes, "As with other aspects of Western health care, traditional approaches to weight management are rooted in a biomedical, reductionist paradigm that separates mind from body and feelings from physiology."[14] Others have highlighted the crucial role of cultural identity and spiritual connection with land and ancestors in achieving well-being. McLennan and Khavarpour, through their research with Aboriginal Australians, suggest that "encouraging connections with ancestors and the land could be an important instrument in preventive programs and health promotion initiatives."[15] These authors also encourage health promoters to increase opportunities for and integrate into health initiatives, where possible, spiritual expression through rituals, painting, storytelling, community gatherings, dance, and visits to sacred sites. Cultural identity and connection are multifaceted and not simple to measure objectively. Durie noted more than twenty years ago that most measures of cultural identity stress "links with traditional knowledge and skills but fail to capture the range of activities, lifestyles and multiple affiliations which characterise Māori people in modern society."[16] Accordingly, Te Hoe Nuku Roa, a cultural identity measure developed by Durie and others, aimed at recognizing the diverse realities of Indigenous peoples without making a judgment on how "Indigenous" someone is or isn't. Many since have discussed cultural identity and the difficulty of "measuring" one's Indigeneity.[17] Cultural identity often includes genealogy and heritage, as well as language, practices, relationships, and beliefs. Others have highlighted a connection with the environment and link to land, water, mountains, etc. as a defining characteristic of Indigenous identity. Interestingly, non-Indigenous researchers have highlighted the close association between health and engagement with outdoor environments as well. For Indigenous peoples, this "association" between lands/environment and health includes an interconnected link with genealogical, cultural, and spiritual identity, so a shift toward "environment-based" approaches to health may positively affect Indigenous well-being at many levels.[18]

While moving away from weight-based approaches and toward Indigenous approaches to health promotion and practice has been slow, many in New Zealand (and around the world) are calling for (and developing) solutions that don't align with dominant discourse, cultural "norms," and commercial interests—in fact, the possible solutions outlined here are about decolonizing health and well-being and adopting and designing approaches that are led by Indigenous peoples and based on Indigenous values and knowledge. In essence, the "alternative" approaches proposed here aim to achieve good health without health being the focus at all. In other words, health becomes a secondary outcome in the pursuit of cultural reconnection and revitalization of cultural knowledge. Instead of drawing from "formal" research on Indigenous initiatives, which are relatively limited, the remainder of this chapter will draw upon vignettes, examples in practice, and anecdotal evidence to portray a case for Indigenous health solutions. It is hoped that the examples given will also prove helpful for other non-Indigenous minority groups, whose cultural practices and connections could provide motivation for achieving well-being and health but often go overlooked as being less valid or useful when compared to Western norms. While knowledge and practice that differ from the dominant norm (the Euro-Western biomedical norm) are often seen as having value for only those cultures they relate directly to (if they have value at all), we Indigenous peoples would argue that Indigenous knowledge and practices have value for all people.

Example 1: Dr. Ihirangi Heke and Connection with Ancestral Lands and Waterways

An example of this is a health-promotion initiative that took place in a semirural, mostly Māori community on the eastern coast of New Zealand's North Island. Tasked with improving the health of those in the area, where the prevalence of obesity was high, the health promoter Dr. Ihirangi Heke challenged residents not to exercise, go on a diet, or lose weight but instead to "get to know" their *maunga* (ancestral mountain), which rose above the community, and their *awa* (ancestral river), which flowed through it. Heke would go for walks up the mountain and invite anyone else who wanted to come, to connect with the landmark that was well known as a part of the community's cultural identity but usually just observed from a distance rather than engaged with and known intimately. Over time, this shift in focus led many in the community to "visit" their mountain or river daily—climbing, walking, and swimming not for exercise's sake but to strengthen their cultural connection to these sites and the history of their ancestors, who were constantly interacting with these places. In his words, "exercise, weight loss, and improved health became an incidental outcome of connecting with cultural knowledge."[19]

Ironically, but not so surprising, the promises of success and prosperity that came with colonization and capitalism also resulted in the separation of Indigenous people from the lands and environment with which we had always measured our health and prosperity—the lands that provided food and resources to survive and thrive and enabled us as Māori to *manaaki* (care for, reciprocate kindness toward) other peoples. Such drivers of wellness are not considered or accounted for in Western measures of health and particularly in a weight-centered model of health promotion. The change in access to traditional foods and food security directly related to the loss of land means that in many cases Indigenous peoples must rely on Western systems of food production and commercially driven food sources to sustain themselves and their families. Where there were once abundant food sources of birds, eels, fish, shellfish, and other foods available in the surrounding environment, now legal restrictions, poor management of the resource, destruction of ecosystems through environmental pollutants or "development," and a loss of transportation access to or knowledge about the resource have greatly limited the availability of traditional food sources. This has led to movements around the world reclaiming traditional food sources and Indigenous-based food processes. As Bodirsky and Johnson put it, "Fostering the resurgence of traditional Indigenous knowledge about food is necessary in healing the trauma emerging from colonialism."[20]

*Example 2: Paora Te Hurihanganui and Returning to
an Ancestral Diet*

Paora Te Hurihanganui is a Māori health researcher who applies cultural knowledge in the promotion of physical activity and health in New Zealand, as the CEO of Te Arawa Whānau Ora. His story highlights, once again, the powerful role of culturally relevant drivers in changing lifestyle and enhancing health and wellness. In short, Paora came to the realization that one particular adjustment to his lifestyle would wield significant improvements in his overall "well-being"—that realization was that he needed to eat as his *tūpuna* (ancestors) ate and reject the "food" that had been introduced as a result of colonization—"food" that has contributed to a growing and disproportionate rate of noncommunicable diseases among Māori. Despite the obvious challenges of making such a commitment and shift in modern New Zealand, where many traditional sources of food have become limited and difficult to access, Paora has found many positive outcomes as a result of adopting the diet of his ancestors. He acknowledges that he has lost a significant amount of weight, but he maintains that the weight loss is not important when compared to the opportunity to reconnect with aspects of culture and ancestral knowledge and practices that the simple act of eating food now affords. Instead of keeping track of calories and micronutrients, the adoption of (or return to) a precolonized diet affords a feeling that food—from growing and hunting to preparing and eating—can be a way of drawing closer to our ancestors and their ways of life. Paora explains,

> Tūpuna Kai (Ancestral Foods/Eating) is not only a model of behavioural change but is an entrenchment in philosophical and metaphorical understanding of Kai (food), ancestral knowledge, and alignment to the systems of the natural environment. To understand genealogical links of Kai, its individual and collective effects on people, its relationship with environment, and the development of food-related knowledge through intergenerational experiences is far more significant than simply what you are eating. Tūpuna Kai is simply one part of knowing your natural environment intimately and developing modes of practice to reach optimum wellbeing.[21]

The reality of our modern food systems is that many industrialized "foods" aren't directly linked to soil, sunlight, and other aspects of the environment, so it's of little value to food producers (and perhaps detrimental to their reputation) to highlight the process taken to get that food to your plate. While decolonizing perspectives toward and processes concerned with food seem to be positive steps in reclaiming the health and well-being of those most stigmatized by weight and race bias, often times the shift that is required is less dramatic, drawing upon parts of both Western and Indigenous paradigms at an interface between the two. The third example I'll highlight here is from an exercise study with Māori men that I oversaw personally. Though aspects of the study have been published elsewhere,[22] the following account happened after the research had formally concluded, as we continued to meet regularly with participants to support participants to continue to be active.

Example 3: The BROs and Exercise Sessions Based on
Cultural Narratives

My research team and I were conducting a gym-based exercise study with Māori men, exploring the effects of different exercise modalities on physiological and holistic markers of health. Feedback had been positive, despite initial discussions suggesting that the gym was typically an uncomfortable and somewhat foreign space for many of our Indigenous men. It became apparent to us that the nature of the modern gym space and gym culture did not "fit" most of our men, so we sought a way to make it a space where Indigenous principles could be expressed and where Indigenous people could feel comfortable and participate in physical activity in uniquely Indigenous ways. We started experimenting with an exercise structure that instead of being guided by the number of sets and repetitions was instead guided by a culturally relevant story or narrative. For example, a well-known Māori story is that of the Polynesian voyager Kupe, an ancient ancestor of many Māori in New Zealand, who, upon his sea vessel, chased a *wheke* (octopus or cuttlefish) from his home in the Pacific Islands all the way down to New Zealand. Catching up with the *wheke* at times, a battle between the two ensued, until Kupe was eventually able to overpower and kill it. This story provided the structure of the exercise session—a session based on training in pairs, where one person "chases" the other in the workout (either chasing in a shuttle run or chasing across a number of repetitions laid down by the other) and then battling together in a variety of physical challenges. All the while, the story is being told by the "trainer," who acts as both a storyteller and coach. Other stories were integrated into subsequent exercise sessions, and the participants (and the "trainer") became interested to see what the next story would be and how it would be integrated into an exercise session. The feedback from the participants was positive, a few commenting that they had known these stories as children but only now saw that they had an application to something in their lives.

While this approach within our team is still in an experimental phase—perhaps the ideal would be to have our people out of an indoor gym entirely and engaging with the natural environment during physical activity—others appear to have had success with similar approaches. Enterprises such as Patu Aotearoa (patunz.com) and Māori Movement (maorimovement.co.nz) use cultural knowledge, beliefs, and values in a gym-based setting, to create a health experience infused with Indigenous knowledge. These have become popular among Māori.

For Indigenous peoples, storytelling was and is an essential component of transferring and disseminating knowledge—particularly for cultures like Māori, who didn't use written language. While many stories have been labeled mythical and/or mystical by colonizing peoples, these stories are not seen as such by Indigenous people and were usually encoded with lessons and robust scientific principles stored within a narrative style that could be easily remembered, understood at different levels by those of various ages, and passed on in story form down the generations.[23]

An example of a story often deemed myth or legend by Western audiences that is well known among many tribes in New Zealand and that highlights the role that the pursuit of cultural knowledge plays in motivation is that of Tāne's pursuit of the baskets of knowledge. In this story, Tāne, an *atua* (a term that has been used synonymously with Euro-Western concepts of deity or god but that also relates to environmental personification and/or ancestors) is known to have ascended through twelve "heavens" to obtain three baskets of knowledge. These contained knowledge essential to survival, knowledge about rituals and invocations, and knowledge that was deemed harmful to people. In order to obtain these baskets, Tāne had to overcome adversaries and challenges that his brother Whiro sent to slow his ascent and subsequent descent through these twelve heavens. With the help of another brother, Tāwhirimatea, Tāne was able to obtain the baskets and ultimately return them to be shared with mankind and passed down through the generations. Though greatly simplified for the sake of this chapter and acknowledging the differences in worldview and cultural perspective of readers, this well-known account highlights a key motivating factor that applies to the link between cultural knowledge and lifestyle change. First, it wasn't weight loss, aesthetic beauty, fame, or health that drove Tāne in his ascent through twelve heavens; it was to obtain knowledge and share that knowledge with his community and their descendants. Second, in order to obtain this goal, support from others was a necessary part of overcoming the challenges and barriers in Tāne's way.

Likewise, those working to restore Indigenous knowledge within health research and practice have found that a connection with traditional knowledge and culture has resulted in a shift of thinking about what it means to be and become "healthy." Like other cultural stories that provide directly applicable lessons and metaphor, the story of Tāne highlights the difficulty of the pursuit of health, of restoring indigeneity, and of ascension to higher standards and understanding; the story also acknowledges the existence of unforeseen challenges, of the importance of peers, family, and societal support in overcoming those challenges, and of the values associated with applying and sharing collective knowledge and understanding with others and for future posterity. Although this chapter has offered just a few examples of the shift from the Western, weight-focused promotion of health and the management of health behaviors (exercise, diet etc.) closely tied to weight and weight loss, there are a growing number of initiatives throughout New Zealand and throughout the world where people are shifting this focus and drawing upon cultural knowledge and worldviews to do so.

Conclusion

This chapter started with claims that modern weight loss–based approaches to health perpetuate racial stigma and produce discourses and attitudes similar to racism. The fact that Indigenous people, and generally anyone of color, have a disproportionately higher prevalence of overweight and obesity than their white counterparts makes it easy to point the finger of blame at "fatter" and "darker" groups. The problem with this perspective is that it disregards the many sociocultural determinants of health, many of which derive from colonization, that influence an individual or family's health-related behavior. While the most common approach to improving health follows a Western paradigm, Māori in New Zealand and Indigenous people globally are reconnecting with cultural knowledge, values, and beliefs as a driver and source of health and well-being. Such initiatives are decolonizing in nature and acknowledge the value of Indigenous knowledge and approaches in achieving health and well-being for all people.

Notes

1. D. Williams, "Miles to Go Before We Sleep: Racial Inequities in Health," *Journal of Health and Social Behavior* 53 (2012): 279–95; C. Hamilton, *Beyond Racism: Race and Inequality in Brazil, South Africa, and the United States* (Boulder, CO: Lynne Rienner, 2001); D. Bramley, P. Hebert, R. Jackson, and M. Chassin, "Indigenous Disparities

in Disease-Specific Mortality, a Cross-Country Comparison: New Zealand, Australia, Canada, and the United States," *New Zealand Medical Journal* 117, no. 1207 (2004): U1215; J. Nazroo and D. Williams, "The Social Determination of Ethnic/ Racial Inequalities in Health," in *Social Determinants of Health*, ed. M. G. Marmot and R. G. Wilkinson (New York: Oxford University Press, 2006), 238–66; S. Crengle, R. Lay-Yee, P. Davis, and J. Pearson, *A Comparison of Māori and Non-Māori Patient Visits to Doctors: The National Primary Medical Care Survey (NatMedCa): 2001/02*, Report 6 (Wellington: Ministry of Health, 2005); L. Boulware, L. Cooper, L. Ratner, T. LaVeist, and N. Powe, "Race and Trust in the Health Care System," *Public Health Reports* 118 (2016): 358–65; A. Zilanawala, P. Davis-Kean, J. Nazroo, A. Sacker, S. Simonton, and Y. Kelly, "Race/Ethnic Disparities in Early Childhood BMI, Obesity, and Overweight in the United Kingdom and United States," *International Journal of Obesity* 39, no. 3 (2015): 520; and V. Chiavaroli, J. Gibbins, W. S. Cutfield, and J. Derraik, "Childhood Obesity in New Zealand," *World Journal of Pediatrics* (2019): 1–10.

2. J. Duncan, E. Duncan, and G. Schofield, "Ethnic-Specific Body Mass Index Cut-Off Points for Overweight and Obesity in Girls," *New Zealand Medical Journal* 123, no. 1311 (2010): 22–29.

3. World Health Organization, "BMI classification," 2017, https://apps.who.int/bmi/index.jsp?introPage1/4intro_3.

4. J. Boswell, *The Real War On Obesity: Contesting Knowledge and Meaning in a Public Health Crisis* (London: Palgrave Macmillan, 2016).

5. D. Lupton, "The Pedagogy of Disgust: The Ethical, Moral, and Political Implications of Using Disgust in Public Health Campaigns," *Critical Public Health* 25, no 1, (2015): 8.

6. L. Bacon and L. Aphramor, "Weight Science: Evaluating the Evidence for a Paradigm Shift," *Nutrition Journal* 10, no. 1 (2011): 9.

7. N. Addy, "White Privilege and Cultural Racism: Effects on the Counselling Process," *New Zealand Journal of Counselling* 28, no. 1 (2008): 10–23.

8. T. McCreanor and R. Nairn, "Tauiwi General Practitioners Explanations of Maori Health: Colonial Relations in Primary Healthcare in Aotearoa/New Zealand," *Journal of Health and Psychology* 7, no. 5 (2002): 509–18; P. Wilson and M. Stewart, "Indigeneity and Indigenous Media on the Global Stage," *Global Indigenous Media* (2008): 1–35.

9. M. Durie, "Te Pae Mahutonga: A Model for Māori Health Promotion," *Health Promotion Forum Newsletter* 49 (1999): 2–5; D. Fijal and B. Beagan, "Indigenous Perspectives on Health: Integration with a Canadian Model of Practice," *Canadian Journal of Occupational Therapy* 86, no. 3 (2019): 220–31.

10. K. McPhail-Bell, C. Bond, M. Brough, B. Fredericks, and A. Bond, "'We Don't Tell People What to Do': Ethical Practice and Indigenous Health Promotion," *Health Promotion Journal of Australia* 26, no. 3 (2015): 195–99; J. Donatuto, L. Campbell, and R. Gregory, "Developing Responsive Indicators of Indigenous Community Health," *International Journal of Environmental Research and Public Health* 13, no. 9 (2016): 899.

11. K. Crenshaw, "Race, Gender, and Sexual Harassment," *Southern California Law Review* 65 (1991): 1467.

12. J. Lynch, G. Smith, G. Kaplan, and J. House, "Income Inequality and Mortality: Importance to Health of Individual Income, Psychosocial Environment, or

Material Conditions," *BMJ* 320, no. 7243 (2000): 1200–4; K. Pickett and R. Wilkinson, "Income Inequality and Health: A Causal Review," *Social Science and Medicine* 128 (2015): 316–326. S. McKernan, C. Ratcliffe, C. Steuerle, and S. Zhang, *Less Than Equal: Racial Disparities in Wealth Accumulation* (Washington, DC: Urban Institute, 2013); L. Marriott and D. Sim, "Indicators of Inequality for Maori and Pacific People," *Journal of New Zealand Studies* 20 (2015): 24; A. Geronimus, M. Hicken, D. Keene, and J. Bound, "Weathering and Age Patterns of Allostatic Load Scores Among Blacks and Whites in the United States," *American Journal of Public Health* 96, no. 5 (2006): 826–33; D. Williams and S. Mohammed, "Discrimination and Racial Disparities in Health: Evidence and Needed Research," *Journal of Behavioral Medicine* 32, no. 1 (2009): 20–47.

13. M. Durie, "Te Whare Tapa Whā: A Māori Perspective of Health," Māori Womens Welfare League Annual Conference, Gisborne, New Zealand, 1982.

14. J. Robison, "Weight, Health, and Culture: Shifting the Paradigm for Alternative Health Care," *Alternative Health Practitioner* 5, no. 1 (1999): 45–69.

15. V. McLennan and F. Khavarpour, "Culturally Appropriate Health Promotion: Its Meaning and Application in Aboriginal Communities," *Health Promotion Journal of Australia* 15, no. 3 (2004): 238.

16. M. Durie, "Te Hoe Nuku Roa Framework: A Maori Identity Measure," *Journal of the Polynesian Society* 104, no. 4 (1995): 469.

17. H. Weaver, "Indigenous Identity: What Is It, and Who Really Has It?," *American Indian Quarterly* 25, no. 2 (2001): 240–55; L. McCubbin and T. Dang, "Native Hawaiian Identity and Measurement: An Ecological Perspective of Indigenous Identity Development," in *Handbook of Multicultural Counseling*, ed. J. G. Ponterotto, J. M. Casas, L. A. Suzuki, and C. M. Alexander (London: Sage, 2010), 269–82; B. Jacobs, "Indigenous Identity: Summary and Future Directions," *Statistical Journal of the IAOS* 35, no. 1 (2019): 147–57.

18. C. Houkamau and C. Sibley, "The Multidimensional Model of Māori Identity and Cultural Engagement," *New Zealand Journal of Psychology* 39, no. 1 (2010): 8–28; M. Durie, "Understanding Health and Illness: Research at the Interface Between Science and Indigenous Knowledge," *International Journal of Epidemiology* 33, no. 5 (2004): 1138–43; M. Greenwood and S. Leeuw, "Teachings from the Land: Indigenous People, Our Health," *Canadian Journal of Native Education* 30, no. 1 (2007); T. Pasanen, L. Tyrväinen, and K. Korpela, "The Relationship Between Perceived Health and Physical Activity Indoors, Outdoors in Built Environments, and Outdoors in Nature," *Applied Psychology: Health and Well-Being* 6, no. 3 (2014): 324–46; D. Pearson and T. Craig, "The Great Outdoors? Exploring the Mental Health Benefits of Natural Environments," *Frontiers in Psychology* 5 (2014): 1178.

19. I. Heke, personal communication, February 2, 2018.

20. M. Bodirsky and J. Johnson, "Decolonizing Diet: Healing by Reclaiming Traditional Indigenous Foodways," *Cuizine: The Journal of Canadian Food Cultures/ Cuizine: Revue des cultures culinaires au Canada* 1, no. 1 (2008); S. Grey and R. Patel, "Food Sovereignty as Decolonization: Some Contributions from Indigenous Movements to Food System and Development Politics," *Agriculture and Human Values* 32, no. 3 (2015): 431–44.

21. P. Te Hurihanganui, personal communication, June 6, 2019.
22. I. Warbrick, D. Wilson, and A. Boulton, "Provider, Father, and Bro—Sedentary Māori Men and Their Thoughts on Physical Activity," *International Journal for Equity in Health* 15, no. 1 (2016): 22; I. Warbrick, D. Wilson, and D. Griffith, "Becoming Active: More to Exercise Than Weight Loss for Indigenous Men," *Ethnicity and Health* 25, no. 6 (2018): 796–811.
23. Bodirsky and Johnson, "Decolonizing Diet"; J. Corntassel, "Indigenous Storytelling, Truth-Telling, and Community Approaches to Reconciliation," *ESC: English Studies in Canada* 35, no. 1 (2009): 137–59; J. Iseke, "Indigenous Storytelling as Research," *International Review of Qualitative Research* 6, no. 4 (2013): 559–77.

ELEVEN

After Race Classification

Grappling with South African Indigenous DNA in Practice

NOAH TAMARKIN

SINCE THE END of apartheid, genetics has become increasingly prominent as a method of knowing human diversity and history in South Africa, aided both by an intensified global appetite for indigenous DNA and by post-apartheid South African geneticists' and their collaborators' framing of genomics as a way to move beyond apartheid race classification.[1] Regardless of how they frame their work—and this varies from claims that genomics projects are thoroughly neutral to claims that they are antiracist—hierarchical race classification is the backdrop for South African genetics. In this context, South African indigenous DNA has emerged as an especially potent category.[2] The prominence of genetic ancestry in general and in relation to indigeneity in particular means that with all of its fraught baggage, just as geneticists cannot avoid histories of race classification, those who produce, design, and curate heritage in South Africa likewise cannot avoid indigenous DNA, even as they, too, seek to self-consciously break from colonial and apartheid pasts. In light of this genetic inevitability, this chapter examines how South African museums and other heritage sites engage with genetics. It asks: if race classification haunts genetic ancestry and genetic ancestry cannot be ignored, how are both reworked in practice?

This chapter argues that indigenous DNA is unavoidable in South Africa. Especially for those who seek to represent precolonial pasts or for those who have systematically and routinely been positioned as themselves embodying or representing those pasts, it isn't an option to fully disengage from genetic forms of knowing. However, I argue that these reworked genetic engagements and the new genetic knowledge that they produce are not pre-determined by geneticists' intentions, conclusions, or framings. The reworkings of indigenous DNA in practice that I look at in this chapter instead offer a way to understand not just the limits of postapartheid genetics, with its racial classificatory hauntings, but also what else genetic knowledge can be and do when it is resituated, recirculated, reimagined, and decentered.

My focus in this chapter on indigenous DNA in practice aligns with what Alondra Nelson calls "the social life of DNA" and what I call "genetic after-lives": the "social life of DNA" emphasizes how DNA takes on new political and legal significance as it circulates as an object of knowledge, and "genetic afterlives" emphasizes how former research subjects in genetic ancestry studies create new genetic knowledge as they themselves circulate, situate, and interpret geneticists' published work derived from their bodies.[3]

In what follows, I consider four cases of anti- and postapartheid grappling with genetic ancestry in South Africa that seek alternative ways to engage indigenous DNA other than simply by reproducing race classification. Three examples consider postapartheid heritage projects, but the first example instead considers the actions taken by geneticists beginning in the 1960s who were compelled to grapple with DNA and state projects of race classifi-cation when they were approached for help by individuals seeking to appeal their legal race designation. I start here because this example demonstrates the apartheid landscape of race classification and genetics that shaped the emergence of twenty-first-century South African indigenous DNA. The three postapartheid examples each demonstrate very different relationships to indigenous DNA: a Johannesburg museum called the Origins Centre that opened in 2006 and was revamped in 2020; a 2007 ceremonial reburial of thirteenth-century remains that had been unearthed in the 1930s at Mapun-gubwe, known as the earliest stratified kingdom in southern Africa and a UNESCO World Heritage site since 2004; and the !Khwa ttu San Heritage Cen-tre, a San-designed and -operated Western Cape land reclamation and cul-tural connection project opened in 2006.

In these examples, the scale of South African indigeneity shifts: it can encompass all humanity when it signals human origins in southern Africa, as in the Origins Centre; it can advance an idea of state indigeneity as an expression of postcolonial power emerging from colonial oppression when it signals all who were present before European colonization, as in the

Mapungubwe reburial; and it can delineate San or Khoisan peoples as distinctly and emblematically indigenous, a more narrow and specific scale of indigeneity that implicates San people not only at !Khwa ttu but also at every site invoking indigenous DNA.

None of this is politically neutral. Indeed, indigeneity is a political category, but its politics—always tied to race, place, history, and power—can become obscured when the marker "indigenous" is used as a way to differentiate some people's and peoples' DNA from others. Science and technology studies (STS) scholars argue that the idea of indigenous DNA builds upon and bolsters old racist ideas about primitive versus modern societies and colonialism as a civilizing project rather than one founded in dispossession, exploitation, and death.[4] Asking about the relationship between indigeneity and DNA is also then necessarily asking about the relationship between race classification, biological evidence, and histories of dispossession. This is why I begin with an account of colonial and apartheid race classification and proceed to consider how these legacies matter in the multiple, multiscalar manifestations of indigeneity in South Africa today. Ultimately, this chapter demonstrates that South African indigenous DNA must be engaged but that the legacies of race classification need not determine what it evidences and for whom.

Subverting Apartheid Race Classification, Making Indigenous DNA

There were two kinds of race classification in effect in apartheid South Africa that were relevant to what would later come to be understood as indigenous African: one, legal race categories established through wide-ranging and often contradictory apartheid legislation, and two, emerging genetic interpretations of South African race typologies that differentiated Bantu and Khoisan in the name of the study of human origins and variation. Both have their roots in the convergence of colonial science and colonial bureaucracy, an entanglement that continues to haunt ideas about indigenous DNA today.[5]

The issue for early South African scientists and administrators alike was how to identify racial types and keep them separate from one another: their common preoccupation was the idea of racial purity.[6] All of the settler colonies that went on to become South Africa through settler independence in

1910 differentiated between Whites and Natives, a distinction that was a central technology of colonial subjugation and rule.[7] Meanwhile, a colonial science of human origins and racial typology emerged among anatomists in South Africa and in Europe that measured and dehumanized Africans; they paid particular attention to southern African hunter-gatherer and herder communities who were together designated "Khoisan" (also expressed as "Khoesan," "Khoi-San," and "Khoe-San") and described as distinct from and more primitive than other Africans, who they otherwise imagined as the lowest ranks in civilizational hierarchies.[8] Apartheid race-based legislation emerged from these by then well-developed South African preoccupations with racial purity and technologies of racial segregation. But apartheid categories reframed indigeneity: the 1950 Population Registration Act, the foundation of apartheid race legislation, replaced the colonially inherited category "Native" with "Bantu," and those who were known to scientists as "Khoisan" became "Coloured," joining a wide range of others of very different backgrounds who were also so categorized.[9] The South African sociologist Zimitri Erasmus points out that erasing the category "Native" effectively undermined Black claims to land and citizenship, and the STS scholars Geoffrey Bowker and Susan Leigh Star note that the choice of "Bantu" rather than "African" to replace "Native" served "to underscore [Afrikaner] Nationalist desires to be recognized as 'really African.'"[10] Though "indigenous" was not yet a salient political identity as such, "Native" and "African" had shifted from settler justifications for dispossession to potential indications of legitimate belonging, setting the stage for indigeneity to emerge as an idiom of postapartheid restitution and contestation.

One of the core goals of apartheid was to codify whiteness and to racially classify and ethnically divide all who fell outside its boundaries. But apartheid racial distinctions were marked by ambiguities of classification in practice. They relied more on what the historian Deborah Posel has called "race as common sense" than on any method of scientific measurement or analysis, and the apartheid bureaucrats responsible for race classification were entirely aware that these distinctions were arbitrary.[11] Furthermore, different legislation created different racial categories, such that an individual might fall into more than one at the same time depending on the law in question, and these categories also changed over time.[12] Race classification boards were set up to interrogate the hard-to-classify and account for the misclassified. They drew not on any kind of biological criteria but rather on social, cultural, and class-based distinctions such as how others generally regarded the person in question and what language they spoke in their home.[13]

Scientists became involved in race classification not because administrators sought their assistance—they did not—but rather at the behest of people seeking to appeal a board's assessment of their race. Some refused these requests because they were wary of legitimizing apartheid race classification. Others who were likewise against apartheid policies agreed to the requests: in these cases, they would present their opinions either by saying there is no scientific evidence to support a person's classification in the racial category they were seeking to leave or by saying that evidence supports that person's classification in their preferred category.[14]

The human geneticists Trefor Jenkins and George Nurse were first enlisted in race-reclassification efforts in the late 1960s by a woman seeking reclassification as white so that she could legally marry a white man. The South African physical anthropologist Alan Morris, based on an interview with Nurse, explains: "Jenkins and Nurse reported to the board that nothing appeared in her genetic spectrum to suggest she was not white, but the board rejected their evidence because the people in her family did not look white. Nurse and Jenkins wrote to the newspapers saying that the minister was disregarding science. In response, the minister wrote to tell them that race was political, not a scientific issue, and . . . neither Nurse nor Jenkins was asked to testify again."[15] When I interviewed him in 2013, Jenkins told me that he nevertheless continued to be consulted by applicants in the decades that followed: every time, he would produce documents for the applicants stating that the genetic data is consistent with their self-proclaimed race, and every time, his expertise was rejected.

Geneticists like Jenkins aligned themselves with nonracialism and against apartheid policies.[16] But they nevertheless were steeped in the inherited logics of race typology, and their interest in Khoisan genetics as especially important for the study of human origins carried forward earlier imaginaries of Khoisan peoples as remnants of the past.[17] So at the very same time that Jenkins was working to subvert apartheid racial categories, he and others, like his student Himla Soodyall, who went on to become one of the most prominent human geneticists in southern Africa, pursued the kind of research on human origins and variation that has become most associated with indigeneity and with bolstering fantasies of purity—the foundation of scientific racism—through the language of admixture.[18]

Zimitri Erasmus has argued that the early obsessions with racial purity are the epistemological bedrock of much contemporary genetic ancestry work in South Africa, particularly where the genetic category Khoi-San is concerned. For her, ongoing and recent genetic ancestry work in this vein is antipolitical, antihistorical, and divisive: "Khoi-San is used to distinguish between Coloureds who can claim First Nation status and those who cannot," she explains.[19] Other STS scholars have likewise critiqued postapartheid South African geneticists' work as depoliticizing and therefore as failing to adequately account for colonial and apartheid legacies of racism that continue to profoundly shape contemporary South Africa, often in direct contrast to how geneticists frame their work.[20] This is especially the case with public-facing genetics projects. For example, in her analysis of a 2007 public engagement project that positioned as antiracist its aims to produce a portrait of South African genetic diversity, Laura Foster argued that what was produced instead was a *nonracial* and therefore antipolitical South African genome.[21] In another example, Katharina Schramm argues that genetic archives reinscribe apartheid racial archives, particularly when white Afrikaaners use personal genetic ancestry testing results that include Khoisan markers to naturalize their claims to authentic Africanness.[22] Both Schramm and Erasmus emphasize that Khoisan genetic ancestry in particular has offered individuals new ways to articulate their South African belonging against, respectively, White or Coloured race categories.[23]

These critiques emphasize the failures of postapartheid South African genetics to adequately break from the legacies of race classification, let alone be transformative toward a better future. As inheritors of histories of scientific racism, South African geneticists must grapple with colonial and apartheid histories of race classification. But their research and especially their public-facing work like the South African national projects of reconciliation described by Foster and the global genetic diversity projects described by Schramm contribute both to understandings of indigeneity as genetic—a racial definition rather than one based on land, relations, and sovereignty—and to the global prominence of genetics as *the* privileged site of personal identity and racial identification.[24]

The example of geneticists' enlistment in race-reclassification appeals under apartheid offers an antiapartheid counterpoint to postapartheid geneticists' variously antiracist and nonracial framings of their twenty-first-century research—ironically one that coincided with these same geneticists' research into human difference that STS scholars argue reinscribes race as genetic.[25] Even as these geneticists were working to subvert apartheid racialization, they were also then, through their research, producing African indigenous DNA—a category that took on much greater significance in postapartheid South Africa as human origins research and a national politics of nonracialism collided.

The central tension of postapartheid genetics is that its practitioners are acutely aware of how science has shaped race classification in the past and also deeply committed to using science to work against those legacies in the present, even when their questions and methods of doing so rely on and so necessarily reinscribe racial categories. Given their antiracist convictions and histories of using genetics to subvert apartheid categories when called on to do so by people protesting their categorization, it is ironic that South African geneticists' work has provided more weight to and renewed investments in the idea of biologically distinct human groups, particularly in the South African context, where under apartheid it was widely understood that race categories were political, social, and often arbitrary rather than scientific, biological, and objective.[26] However, while geneticists' efforts to produce antiracist genetic ancestry may necessarily fail, where they have succeeded is in positioning genetics as essential evidence for understanding human history.

Supplanting Classification: All Humans
Are African at the Origins Centre

Indigenous DNA took center stage at the Origins Centre when the museum on the University of the Witwatersrand campus in Johannesburg opened in 2006. On the surface, this was an anthropology museum with a combined emphasis on southern African rock art and the use of genetics to study human migration histories. But it was also a postapartheid museum: through its exhibits, which featured a blend of artifacts, scientific explanations, and interpretive art installations, it self-consciously embraced emerging national ideals of nonracialism. This was an optimistic moment for postapartheid geneticists. They hoped that by showing the centrality of South Africa to all of human history, they might simultaneously affirm the significance of Africa, avoid the traps of classification and reified race, and move visitors away from their own racialized thinking. The latter goal was supported by the availability for purchase of personal genetic ancestry testing services; this was perhaps a testament to their faith in the transformative power of genetic knowledge to supplant apartheid ideas of race, but it also represents a sharp reversal from the antiapartheid race-reclassification uses of genetics, in which the answer of the genetic truth of someone's race was always to amplify the category they were appealing to join.[27]

The Origins Centre didn't advertise itself as a museum of indigeneity per se, but its emphasis on the role of Africa as the origin of all people, alongside multiple exhibits dedicated to histories and experiences of Khoe and San peoples in particular, pointed toward two of the ways that indigeneity had become meaningful in postapartheid South Africa. The first meaning of African indigeneity was Africa's holding an originary place for humanity in general as the place where *Homo sapiens* emerged and from where they migrated.[28] The second was a singled-out Khoe and San indigeneity as First Peoples not just in the context of South Africa but also the larger world: the second meaning underwrote the first.

The museum, which I visited for the first time in 2013, consolidated and intertwined these two meanings of indigeneity toward a goal of presenting all humans as originally African, a message that dominated the main floor. Framing the entrance to the Origins Centre was an art installation, *World Map*, by Walter Oltmann. The enormous aluminum wire map greeted visitors at the door with Africa front and center stretched floor to ceiling in the middle of the foyer, with the other continents stretching along the ceiling and walls and strands of copper wire woven in to show an origin in Africa and pathways of outmigration across the world. The effect was strikingly powerful, and the message was clear—and clearly directed at visitors from elsewhere: "Welcome Home," the sign beside the installation read, referencing the scientific consensus consolidated through genetic ancestry research that humans emerged in Africa.

This narrative of Africa as all of humanity's home and the genetic methodology that underpinned it grew more elaborate as one moved further into the building. One large sign declared that "the paleontological, archaeological and genetic data all show that modern human beings developed in Africa and then spread to the rest of the world." And at the back of the large space that was the museum's entrance, beyond *World Map*, was the DNA section, designed in consultation with Himla Soodyall, one of the geneticists mentioned earlier in this chapter. A floor-to-ceiling colorful beaded model of a double helix created visual interest, but the real focus of the section was the two backlit panels on the wall that explained what DNA was, how geneticists were using it to reconstruct migration histories, and the implications for how to think about human relatedness. Titled "One World, One People," the main panel explained that "we all share 99.9% of our DNA" and also that "the earliest mtDNA lineage is named L1 and is found in the ancestors of Khoesan and Biaka (West Pygmies)." So on one hand, the text elaborated, "the oldest mitochondrial DNA (mtDNA) lineages are found in Africa," but on the other hand, "it is important to recognize that different mtDNA lineages do not signify substantial differences. . . . Underneath it all, we all share 99.9% of our DNA. This common genetic code and our common ancient history as hunter-gathers make us one people."

Threading the needle between shared genetics among all people, Africa as humanity's origin, and Khoesan and Biaka in particular as inheritors of the oldest genetic ancestries is not easy. Consider this plea from Alan Morris, the South African physical anthropologist discussed earlier, as he describes the importance of southern Africa to molecular genetic research:

> Southern Africa has held a special place in this research because of all the world's populations, it is the Khoesan of southern Africa who harbour the greatest genetic diversity. . . . This places them near the root of modern humans. . . . *Please do not think this makes them some sort of remnant primitive in the manner of Raymond Dart.* . . . What molecular DNA research has provided us is a method to understand the complexity of human variation, without having to use outmoded explanations like racial types. . . . This has shown us the commonality of humanity.[29]

But at the Origins Centre, this hoped-for interpretation that aimed to mitigate the potential racializing effects of singling out Khoesan DNA was complicated by an adjacent hallway display declaring "Southern Africa's San: A Creative & Innovative People." Displays of San material culture were accompanied by panels labeled "Gathering," "A Flourishing Art," and "Hunting," each with old black-and-white photos of San people, rock art, and artifacts, in a visual mode with clear lines of inheritance from colonial and apartheid museum depictions.[30]

Shortly before the emergence of the COVID-19 pandemic in 2020, the Origins Centre secured a grant to do a major renovation and redesign.[31] The massive wire map framing the entrance remained in place, but the sign that greeted visitors with "Welcome Home" had been removed.[32] Furthermore, the DNA section was entirely gone. It was replaced with a hallway-length quotation by the feminist STS scholar Donna Haraway, which read, "The human story is never finished, neither in the direction of the future nor of the past. The origin . . . is ever receding, not only because new fossils are found and reconstituted, but also because the origin is precisely what can never really be found; it must remain a virtual point, ever reanimating the desire for the whole."

This shift away from DNA, with its precision, complexity, and contradictions about unity, diversity, and ancient inheritances, struck me as an enormous change, particularly in relation to its replacement: Haraway's emphasis on indeterminacy, unknowability, and desire seemed an especially self-reflexive choice for a museum devoted to origins. But when I spoke to the curator about it in 2022, she explained that eliminating the genetic framing of the museum wasn't self-consciously intentional. Rather, the panels were removed not only to make way for the Haraway quotation and a new display emphasizing human innovations of the Middle Stone Age but also because they hadn't been updated since 2006, and in the meantime, many additional genetic studies had been published that added more complexity to the story the panels told. In a genetic twist on Haraway's first reason that the origin is ever receding, the curator explained that they didn't want to misrepresent the current state of the field. In particular, the arrows on the map of haplotype movement could be misleading and simplifying.

What was intentional, though, was the revamped adjacent hallway that still displayed some but not all of the same material culture objects from the old San exhibit but that otherwise had a completely different framing: the new main panel was titled "Misrepresenting People and the Past," with a large explanation that read: "The Southern African peoples referred to as the 'San' and 'Khoe' have been the subject of racist myths and misperceptions. They have been persecuted in the past and many remain politically marginalized in the present. Today, 'Khoe' and 'San' groups are fighting against their exploitation and asserting their presence in politics and among southern Africa's diverse societies." Whereas the previous exhibit's photos depicted San people with no context to situate the time or circumstances of their production, the new panel's photos from the Raymond Dart collection clearly showed Dart and others posing with San people as well as an image of a San man with a written specimen label around his neck and the number 4 pinned to the wall behind him.[33] The photos were accompanied by text:

> During the 19th and 20th centuries, linguists and anthropologists in southern Africa often treated the indigenous people as relics from the past. . . . They were studied, displayed at exhibitions and fairs, and their clothing and other items from daily life were placed on show in the natural history sections of museums . . . in contemporary museum practice, curators draw on a range of archeological and documentary evidence and oral histories to build nuanced displays about early life.

When I asked the curator about this new display in 2022, she explained that they had redone that section to respond to critiques from visitors, including those from descendant communities, who felt that Khoe and San people were being objectified and framed as one static group of ancient people—a framing that was out of step not only with how they wished to be represented but also with the focus elsewhere in the museum on a diversity of Khoe and San cultures and art. The effect of the absence of a genetic narrative on the overall framing of the museum, coupled with the Haraway quotation and the Khoe and San update, which seemed to literally be a response to critics, as if quoting their criticisms and writing back directly, felt profound. The retained Oltmann sculpture still invoked genetic narratives for those already familiar with lines and arrows demonstrating the movement of people traced through haplogroups superimposed on continental imagery, but in the absence of the genetic panels to explain its meaning, the effect was more evocative than authoritative.[34]

At the Origins Centre, indigenous DNA in practice initially bolstered the postapartheid antipolitics of nonracialism by reframing the deep past and the present as our common emergence and therefore also as our common inheritance. This emphasis on commonality rather than difference was an attempt to supersede histories of race classification, but invoking Khoesan

DNA beside an exhibit that objectified San people and positioned them in the past points to the challenge of using one scale of indigenous DNA (San people) to build another (all people as indigenous to Africa). Although the curator with whom I spoke in 2022 emphasized that the removal of all mention of DNA was not intentional and that they did plan to find a way to update the panels and the audio guides and reincorporate genetic data somewhere in the space, the momentary absence of indigenous DNA from permanent exhibits and the overall decentering of genetic narratives from the Origins Centre suggested that the path forward from South Africa's intertwined histories of race science and race-based oppression would not be led by genetic ancestry.

The museum in its initial design can be read as the physical manifestation of the paradox of antiapartheid geneticists who simultaneously subverted apartheid categories through offering race-reclassification assistance and solidified indigenous DNA through their human origins and variation research: this legacy was directly inherited by the consulting DNA expert Himla Soodyall, trained by Trefor Jenkins while apartheid laws still limited where she, as a personally racially classified Indian, could live while pursuing her studies with him at the University of the Witwatersrand. The initial centering of DNA evidence was an attempt to supplant ideas of race classification with ideas of human commonality, but it nevertheless relied on ideas of Khoisan difference that located the people so designated in the past. This contradiction was especially apparent because the 2020 redesign not only decentered genetic narratives but also updated how San people were represented in the adjacent section. The redesign—a partial, accidental, and likely temporary decentering of indigenous DNA—can be read as an ambivalence: perhaps a recognition of some of the many critiques of genetic ancestry as reinscribing race, but an uncertainty about how to proceed in a way that, like the San exhibit redesign, might answer those criticisms directly.

Deferring Genetic Inquiry: Mapungubwe, Reburial, and State Indigeneity

In the same year that the Origins Centre opened its doors in Johannesburg and made the case, bolstered by genetic ancestry evidence, for Africa as original home to all people, heated negotiations were taking place between the South African government and multiple groups of people about how to proceed with reburying precolonial remains that had been unearthed in the 1930s from Mapungubwe and stored at the University of Pretoria ever since.[35] Mapungubwe, now a UNESCO World Heritage site, was in the twelfth and thirteenth centuries a stratified, trade-based kingdom. The Mapungubwe reburial was an important postapartheid project: by elevating the site as a national symbol, the postapartheid state was telegraphing its legitimacy backward to Mapungubwe, as if the two were a continuous line of African states.[36]

The reburial itself was not controversial. In fact, reburial of remains was in that moment widely perceived as an act of postcolonial justice. But the national reparative narrative was not the only one circulating in relation to the reburial efforts: each group involved had their own senses of historical injustices that could be at least in part addressed through this reburial. Controversy arose among the groups about whether they should all in fact be part of the reburial process and what exactly that process should be. These groups were designated by the government as "stakeholders," "claimant communities," and "indigenous." Imagined as interchangeable, these labels meant to claim the members of the groups involved as descendants of the people who had lived at Mapungubwe and therefore as having a stake in the proper treatment of the remains of their ancestors.[37]

The government task force that had been formed to determine the reburial process strongly felt that all of these indigenous stakeholder claimant communities must partake of one common claim and agree on one process. But some groups objected. Representatives from the Vhangona Cultural Movement felt that the Lemba Cultural Association, the initial advocates for the reburial and so for a time the only claimants, should be excluded from the process. Lemba people are known as Black Jews and have an oral history of ancient migration from Judea, famously backed up by genetic studies in the 1980s and 1990s—the first of which was authored by Trefor Jenkins along with one of his graduate students.[38] The objectors felt that Lemba history and identity necessarily excluded them from being indigenous. For their part, some leaders of the Lemba Cultural Association declared at the time that they were in fact "the most indigenous" because their oral history clearly accounted for their presence at Mapungubwe and their role in the trade economy that made the kingdom flourish, while others' oral historical links were more ambiguous or absent. The same group that objected to

Lemba inclusion, the Vhangona Cultural Movement, felt that their own claim should be elevated above the others because they were the only group who was still, in the twenty-first century, living and had continuously lived in the areas around what was once the Mapungubwe Kingdom. Other groups felt that the remains should be divided between royals and others but that they alone were the descendants of the royals: in this view, the other groups could only claim descent from the kingdom's nonroyal inhabitants and therefore should not have an equal say in how to proceed.[39]

At stake in the controversies and in the government's efforts to keep the controversies contained so as to proceed with one single, united claim was the definition, scope, and political implications of "indigeneity." Some claimant groups like the San Council had longstanding ties with international indigenous rights activism, but most took on the label "indigenous" for themselves for the first time once the term was applied to them by the government as part of the reburial planning efforts. As they did so, they contributed their own concerns to the emerging definition of indigeneity in South Africa. For the Lemba Cultural Association, for example, taking on the label "indigenous" was a powerful way to articulate to others their own sense of their history, in which they were both Africans *and* Jews, and of Jewish history, in which it was foundationally African.

Despite their differences, for all of the claimant groups, indigeneity was a resonant concept through which to make a claim as a minority on the state. In other words, like it had for others in other African contexts who were part of international indigenous rights efforts, for claimant groups, indigeneity opened a pathway to seek redress against the state. To continue with the Lemba example, this redress was their recognition as a distinct group when under apartheid they had not been recognized as such and instead assigned to citizenship in the Bantustan "homelands" defined through others' ethnic identities. Unlike for the claimant groups for whom indigeneity was a means to advance their minority interests against a postcolonial state in which they were marginalized, for government representatives organizing the joint claim, singular state indigeneity was a means to assert postcolonial power via continuity with precolonial pasts.

Enter DNA. In 2006 as these ideas about indigeneity vis-à-vis Mapungubwe were being worked out through stakeholder planning meetings to prepare for the reburial ceremony, rumors began circulating that DNA tests had been conducted on the remains still held at the University of Pretoria and that they were found to be Lemba ancestors. It wasn't immediately clear that these were rumors: everyone from a Lemba government official to a white archaeologist who was part of the reburial task force personally called the people at the University of Pretoria involved in storing the remains to ask when these tests had been conducted, who had authorized them, and why they were done. They all got the same answers: no DNA had been tested, nor were there any immediate plans to do so.

In this case, while the idea of indigenous DNA threatened to derail the project or at minimum to exclude some of the claimants, it never materialized. The DNA rumors in which genetic ancestry affirmed a specific and direct link between ancient remains and one descendent group aligned with the idea that indigeneity could be a way for a minority group to make a claim against a postcolonial state. But the lack of DNA testing in practice—the realization that genetic ancestry here was a rumor only—aligned with state indigeneity in which all Black South Africans were indigenous, bolstering the indigenous authenticity of the postapartheid state. However, in addition to the state and the claimant groups, with their different investments in whether or not DNA testing should be done, scientists were also considered stakeholders. When deciding the technical logistics of how to rebury the remains, they successfully advocated for the remains to be housed not in coffins but in polyethylene containers that would preserve the bones from further damage in the event that they might be unearthed again in the future for further study.[40]

In the Mapungubwe reburial, indigenous DNA in practice was an absent presence such that specific pasts could be generalized into singular state futures as a method for postcolonial racial repair.[41] Here it was lack of DNA that was most significant: lack of DNA testing cleared the way for the single claim to proceed and with it state indigeneity, while the method of reburial cleared a future for the promise of science to address yet unasked questions that perhaps might aim to link specific remains to specific descendant groups, as was imagined in 2006, or perhaps would aim to address other questions entirely. This deferral of genetic study to unknown research questions in an unknown future opens the possibility of forms of engagement with genetics other than the limited imagination of evidencing or denying connection through the presence or absence of particular genetic markers of difference. It also necessarily and perhaps indefinitely defers questions about whether or not the now reburied human remains should be unearthed again and studied. Presumably all of the designated stakeholders, now collectively known as the indigenous people of Mapungubwe but in practice still very different groups of people with different interests, would need to agree to any future disturbances of the remains of their ancestors.

If the temporary absence of indigenous DNA in the renovated Origins Centre's permanent exhibits speaks to the difficulty of reframing genetic narratives in ways that can both account for San experiences of scientific racism and also articulate an all-encompassing African indigeneity for all people, its absence from the planning for the Mapungubwe reburial speaks to how limited genetic technologies are when they are used to repeatedly ask the same identificatory questions. Nevertheless, the decision to defer possible research to the future speaks to the ongoing power of genetic answers for scientists and state officials alike, even when the questions and the permissions are not yet clear.

Re-Storying Indigenous DNA: !Khwa ttu San Heritage Centre and Building Indigenous Futures

In the postapartheid examples discussed so far, indigenous DNA indicated that all humans or that all Black South Africans are indigenous. Both of these scales of indigeneity folded San people into their indigenous visions, but neither could account for what indigenous DNA might mean for the San people upon whose bodies it rested. The !Khwa ttu San Heritage Centre not only centers San experiences; it also reimagines indigenous DNA by enveloping it with other San stories.

I visited !Khwa ttu in 2022, participating in guided tours outdoors and through the three main buildings, First People, Encounters, and Way of the San. Another visitor on my tour told me and our guide as we began that her professor at the University of Cape Town, when she mentioned that she was coming to !Khwa ttu, insisted that there are no San people in South Africa, a possible reference to and embrace of the colonial myth of vanishing indigenous people.[42] Our guide, a San woman who first joined !Khwa ttu as an intern and had since become a full-time employee, didn't comment directly.

But it was clear both in the tone of the question—critical of the professor—and from our setting, surrounded by San people in South Africa, that the visitor felt the site itself was already the response she was looking for.

!Khwa ttu is a land and history reclamation and cultural renewal project that seeks to reimagine and revitalize San futures.[43] It emerged from the Working Group of Indigenous Minorities (WIMSA), founded in 1996 as a way to coordinate and facilitate newly possible and ultimately successful post-apartheid San legal efforts like the ≠khomani San land claim in the Kalahari and a claim for profit sharing against pharmaceutical companies who had patented the molecular structure of the plant *Hoodia gordonni*.[44] !Khwa ttu traces its roots to a 1998 "San Village" set up by the South African San Institute, a contentious and heavily critiqued model of cultural tourism that was quickly abandoned.[45] In 1999, led by the Swiss anthropologist Irene M. Staehelin in consultation with the San communities with whom she had long worked, the Swiss-based Ubuntu Foundation partnered with WIMSA to purchase what is described at the site as a "run down wheat farm" with the Afrikaans name Grootwater; not explicitly stated but implied was that the 850-hectare farm had been owned and operated by white South Africans, and so the sale was part of a larger postapartheid strategy of indigenous land reclamation, this time through NGO-partnered purchase rather than through land claims courts.

The overall framing of the site is that the ongoing significance of San people lies in strengthening San knowledge to anchor human futures rather than in studying San bodies to understand human pasts. For example, a text display lining the pathway from the parking lot to the reception building reads: "In a time of global environmental crisis, the San provide an exceptional example of how it was once possible to live sustainably. Their knowledge of the use and management of natural resources offers us critical wisdom and inspiration for a better future." This sign and others throughout the site have multiple addressees in mind: !Khwa ttu is open to and partially relies on revenue from tourists, but one of its innovations is that it hosts San people from throughout southern Africa for residencies through which to engage in language, culture, and knowledge exchange with other San people, some of whom are ongoing staff members there. The "us," then, is an inclusive one: tourists and visiting San people alike can access and benefit from the critical wisdom of San histories toward imagining sustainable futures.[46]

In spite of the ecological futures framing, San pasts—including those involving the exploitation of San bodies in the name of science—are engaged throughout the site's three main buildings. This is especially apparent in the first two, First People, which emphasizes San senses of who they are in ways that blend oral histories and passed-down stories with archaeology and genetics, and Encounters, which emphasizes San subjectivities and agency in the face of colonization, genocide, and scientific racism. But throughout, the past is multivalent, and scientific ways of understanding who San people are in relation to others are deeply situated in San self-presentations and commentaries. The effect is a counterhistory to one in which death and extraction dominate: here, those stories are unflinchingly told but enveloped by other San stories so that they are not defining: they have neither the first nor the last word.

Like the Origins Centre, !Khwa ttu is framed by art. But instead of evoking genetic ancestry as Oltmann's *World Map* sculpture does, visitors to the first building, First People, are ushered into a gallery titled Old Time, which displays two sets of stunningly beautiful and colorful paintings, the first commissioned for !Khwa ttu from artists who are part of a Naro San art collective in Botswana and the second produced between 1990 and 2000 by members of !Xun and Khwe San art collectives in South Africa who were resettled there from Angola in 1990; all artists are identified by name. These San framings of multivalent origins—the first "about the earliest times when animals were people" and the second about "memories of growing up before war and urbanization changed their lives forever"—is carried forward into a room about storytelling as knowledge that links the Old Time gallery to the final gallery of the building, Archaeology, History, Genetics.

This first building is the primary place at !Khwa ttu where indigenous DNA emerges: here, it is one story in a dialogue in which San people must participate. At !Khwa ttu, there are no dedicated genetic displays as there were in the original design of the Origins Centre. Instead, genetics is mentioned only a handful of times, and in all cases as situated, small parts of larger displays that emphasize San links to southern African land, even as they also echo aspects of the Origins Centre narrative of all people as originally African. For example, in one display that stretches along an entire wall of the room, a series of connected circles that alternately contain explanatory text, color portraits of smiling (and named) San people, and quotations from the people pictured begins with a similar narrative to that which framed the old DNA exhibit at the Origins Centre: "*Homo sapiens*: The species that all living human beings belong to evolved in Africa. All of us come from Africa. We, the San, are sometimes referred to as 'First People,' because our DNA indicates that our ancient ancestors evolved in southern Africa and stayed here, while other small groups of *Homo sapiens* emigrated out of Africa about 70,000 years ago and eventually populated the rest of the world." But an adjacent circle immediately clarifies: "We, the San, consider ourselves to be First People because our ancestors were the earliest inhabitants of southern Africa. Our stories and knowledge hold us and place us in our landscape in ways that other people cannot claim." The contrast between genetic and land-based meanings of First People points to emergent San strategies for managing others' ongoing interests in them, a theme addressed explicitly in the next circle.[47] Titled "Building Bridges: Archaeology and the San," it points to a possible way forward between San people and scientists: "Ongoing archaeological research and genetic testing of San and Khoekhoen currently places us at the heart of debates about human origins. It is therefore all the more important that we are included in these conversations." The overall effect of this resituating of genetics as part of a San dialogue rather than definitive knowledge in and of itself highlights the inevitability of indigenous DNA in South Africa and how it is managed by indigenous people. San people *must* be part of these conversations, and they *must* respond to existing genetic work, because the alternative is that it goes on without their consultation, consent, and framing.

Genetic knowledge in the First Peoples building—which emphasizes both origins and identity—seems to decouple it from scientific racism. However, indigenous DNA emerges again at the end of the exhibits in the second building, Encounters, which moves from San dehumanization, genocide, and violent exploitation in the name of science in the nineteenth and twentieth centuries to San empowerment through taking control of scientific research with the 2017 San Code of Research Ethics, to which !Khwa ttu is a signatory. Genomics reemerges only in the final display in the Encounters building, where the text of the San Code of Research Ethics is displayed. The first tenet of the code is "Respect," and genomics is invoked as the primary example of past violations of this principle: "We have encountered lack of respect in many instances in the past. In Genomics research, our leaders were avoided, and respect was not shown to them," it reads.

The Encounters building is in some ways a progress narrative (from nine-teenth- and early-twentieth-century grave robbing and specimen making to twenty-first-century limited, collaborative research with sanctions for those who violate San ethics), but it also emphasizes how scientific and polit-ical violence have reinforced each other and how San people have resisted such violence through whatever means were possible in different political moments. For example, one exhibit, "Violence of Representation," empha-sizes how the body and face casts that were made in the 1930s by Hans Lich-tenacker were possible directly through the support of police, colonial administrators, and white farmers and traders: the casts were made at police stations, and phonograph recordings that were made at the same time record not only San songs—which was their goal—but also San people's protesta-tions, fear, and anger.[48] A few photos of cast making in progress appear, but so does a photograph of a boy who, as the caption explains, "refused to have his cast taken." These exhibits underscore the significance of the San Code of Research Ethics: why it is needed and how it provides institutionalized support to carry forward refusals that in the past could only be individual and were not guaranteed.[49]

Indigenous DNA in practice can reframe all of humanity or only the descendants of southern Africa's precolonial inhabitants as indigenous to Africa. The first case directly relies on San DNA. The second includes San people, but it also obscures the specificity of how San bodies have bolstered others' narratives (as in the first case). At the !Khwa ttu San Heritage Cen-tre, San people and their significance for all of humanity are once again cen-tered, but it reads differently, for two reasons. One, like in the Origins Cen-tre redesign, in which DNA narratives are now absent, genetic ancestry is clearly not the focus of why and how San people matter. Although at !Khwa ttu genetic evidence does appear, it is framed as one story among many of how to understand people and place. Two, the San authorship and owner-ship of the site, along with its design, not only condemn the mutually reinforcing practices of colonial land loss, dehumanization, and extractive science; they also move beyond witnessing that devastation to enact a model of reinvestment in San life.

San DNA has been a valuable resource for genetic narratives that enact African indigeneity as something to which all humans have a claim, some-thing to which all Black Africans have a claim, or something to which some but not all people who became Coloured through apartheid laws have a claim. At !Khwa ttu, all of these scales of African indigeneity are embraced, but they each play supporting rather than central roles in the story of San indigenous histories leading the way to better futures for everyone. Furthermore, whereas the making of indigenous DNA is haunted not only by racist typological thinking but also by coercive and dehumanizing practices, at !Khwa ttu it is also re-storied as evidence for San land-based emergence, continuities, and reclamations and as a form of research that can be refused if the San Council finds the ethics questionable or the benefits insufficient or unclear.

Conclusion

This chapter has shown that while race classification legacies necessarily haunt indigenous DNA, and while indigenous DNA has become an inevitable feature of postapartheid heritage, indigenous DNA itself is multivalent, with different meanings depending on how it is framed and by whom.[50] Indigenous DNA reframings are attempted by geneticists when they use genetic ancestry to work against the classificatory logics that shaped the intertwined projects of colonial science and colonial rule; this is, as others have demonstrated, harder than it seems and perhaps even doomed to failure as the same classificatory logics carry forward through new technologies. In the original design of the Origins Centre, the attempt to supplant apartheid categories with an indigenous DNA-centered narrative in which we all have indigenous DNA because we are all originally African—but also that we know what we know about human emergence because of San DNA specifically—was an ambiguous reframing that struggled to fully land among visitors and curators alike. In the current design as of this writing, indigenous DNA is artform, an enveloping entry sculpture without commentary, and is otherwise in a state of temporary absence as museum directors work out what kinds of reframings might better reflect the still intertwined science and politics of the current South African and global moment. At Mapungubwe as well, indigenous DNA is an absent presence, a rumor followed by a deferral, but in both cases a potentiality that cannot be foreclosed. And at !Khwa ttu, indigenous DNA is theirs: their legacy of scientific racism to grapple with, their bodies used to answer others' questions, but also one among many ways of knowing and showing the profundity of their southern African connections and their abilities to change the power balance with researchers going forward.

It is striking that in each of the three examples of indigenous DNA in post-apartheid South Africa, the most significant aspect of it is its absence. At the Origins Centre, this is most likely a temporary suspension that nevertheless provides an opportunity to reconsider what indigenous DNA should be in the future. At Mapungubwe, it may also be a temporary suspension, but what seems clear is that any future production there of indigenous DNA would need to tell a different story than one of dividing those with a legitimate claim from those without. And at !Khwa ttu, it isn't the actual absence but the potential of refusal of the ongoing production of indigenous DNA that punctuates the re-storying carried out throughout the site. Here genetic ancestry is present as it once was at the Origins Centre in Johannesburg, but it is resituated in two crucial ways: one, it is decentered, and two, when DNA is invoked, it is presented as one strand among many for how to understand the past in a way that supports San futures. This is a form of genetic afterlives through which DNA can be a witness to San significance and perseverance, but only if San people frame the narrative. All scales of indigenous DNA discussed in this chapter (all humanity as African, all Africans as indigenous, and San people specifically as indigenous) ultimately build on and therefore implicate San bodies and histories. It remains to be seen how future iterations of indigenous DNA at each of these scales might account both for that foundational violence and for new stories that indigenous DNA could tell.

Notes

Many thanks to Eram Alam, Dorothy Roberts, and Natalie Shibley for their guidance and support through the long process of writing and revising this chapter. Deep gratitude to Dorothy Roberts and Juno for providing comments on earlier drafts and to Kelly Gillespie for bringing !Khwa ttu to my attention. I also thank the late Dr. Trefor Jenkins, Dr. Himla Soodyall, Dr. Tammy Hodgskiss, and the leadership of the Lemba Cultural Association, and especially Pandelani Mutenda, Rudo Mathivha, and the late M. J. Mungulwa, for their time and willingness to share their histories and perspectives in interviews, and I thank Michael Daiber for affirming permission to write about !Khwa ttu.

1. On the global appetite for indigenous DNA, see Amade M'charek, *The Human Genome Diversity Project: An Ethnography of Scientific Practice* (Cambridge: Cambridge

University Press, 2005); Catherine Nash, *Genetic Geographies: The Trouble with Ancestry* (Minneapolis: University of Minnesota Press, 2015); Jenny Reardon, *Race to the Finish: Identity and Governance in an Age of Genomics* (Princeton, NJ: Princeton University Press, 2005); Jenny Reardon and Kim TallBear, "'Your DNA Is Our History': Genomics, Anthropology, and the Construction of Whiteness as Property," *Current Anthropology* 53, no. S5 (2012): S233–S45; and Kim TallBear, *Native American DNA: Tribal Belonging and the False Promise of Genetic Science* (Minneapolis: University of Minnesota Press, 2013). On South African geneticists' framings, see Laura A. Foster, "A Postapartheid Genome: Genetic Ancestry Testing and Belonging in South Africa," *Science, Technology & Human Values* 41, no. 6 (2016): 1015–36; and Katharina Schramm, "Race, Genealogy, and the Genomic Archive in Post-Apartheid South Africa," *Social Analysis* 65, no. 4 (2021): 49–69.

2. See Zimitri Erasmus, *Race Otherwise: Forging a New Humanism for South Africa* (Johannesburg: Wits University Press, 2017).

3. See Alondra Nelson, *The Social Life of DNA: Race, Reparations, and Reconciliation After the Genome* (Boston: Beacon, 2016); and Noah Tamarkin, *Genetic Afterlives: Black Jewish Indigeneity in South Africa* (Durham, NC: Duke University Press, 2020).

4. See Reardon and TallBear, "'Your DNA is Our History'"; and Kim TallBear, "Genomic Articulations of Indigeneity," *Social Studies of Science* 43, no. 4 (2013): 509–33.

5. For more on haunting, science, and indigeneity, see Emma Kowal, *Haunting Biology: Science and Indigeneity in Australia* (Durham, NC: Duke University Press, 2023).

6. Alan G. Morris, *Bones and Bodies: How South African Scientists Studied Race* (Johannesburg: Wits University Press, 2022).

7. Mahmood Mamdani, *Citizen and Subject: Contemporary Africa and the Legacy of Late Colonialism* (Princeton, NJ: Princeton University Press, 1996).

8. Saul Dubow, *Scientific Racism in Modern South Africa* (Cambridge: Cambridge University Press, 1995). In this chapter, my choice of the spelling variations "Khoisan," "Khoesan," "Khoi-San," and "Khoe-San" follows the respective choices made by each site and scholar whom I engage because these variations can evoke subtle political differences that I do not wish to obscure. Generally, "Khoisan" and "Khoesan" emphasizes that Khoe and San groups are better understood in genetic terms as one continuous population, and "Khoi-San" and "Khoe-San" emphasizes that these genetic categories reference actual groups of people who understand themselves as separate from one another.

9. Erasmus, *Race Otherwise*.

10. Erasmus, *Race Otherwise*; Geoffrey C. Bowker and Susan Leigh Star, *Sorting Things Out: Classification and Its Consequences* (Cambridge, MA: MIT Press, 1999). "Black" for Erasmus means all who are not white. This is the political definition of Black most associated with Steve Biko and the Black Consciousness movement: it rejects apartheid distinctions in favor of antiracist solidarity.

11. Bowker and Star, *Sorting Things Out*; Deborah Posel, "Race as Common Sense: Racial Classification in Twentieth-Century South Africa," *African Studies Review* 44, no. 2 (2001): 87–113.

12. Bowker and Star, *Sorting Things Out*; Posel, "Race as Common Sense"; Martin West, "Confusing Categories: Population Groups, National States, and Citizenship," in

South African Keywords: The Uses and Abuses of Political Concepts, ed. Emile Boonzaier and John Sharp (Cape Town: D. Philip, 1988).

13. Bowker and Star, *Sorting*; Posel, "Race as Common Sense."
14. Morris, *Bones and Bodies*.
15. Morris, *Bones and Bodies*, 266.
16. Trefor Jenkins, "Medical Genetics in South Africa," *Journal of Medical Genetics* 27, no. 12 (1990): 760–79; Morris, *Bones and Bodies*; Phillip V. Tobias, *Tobias in Conversation: Genes, Fossils, and Anthropology* (Johannesburg: Wits University Press, 2008).
17. See Dubow, *Scientific Racism*; Christa Kuljian, *Darwin's Hunch: Science, Race, and the Search for Human Origins* (Johannesburg: Jacana, 2016); and Morris, *Bones and Bodies*.
18. On admixture and fantasies of purity, see Duana Fullwiley, "The Biologistical Construction of Race: 'Admixture' Technology and the New Genetic Medicine," *Social Studies of Science* 38, no. 5 (2008): 695–735; María Fernanda Olarte Sierra and Adriana Díaz del Castillo Hernández, "'We Are All the Same, We All Are Mestizos': Imagined Populations and Nations in Genetics Research in Colombia," *Science as Culture* 23, no. 2 (2013): 226–52; and Peter Wade, Carlos López Beltrán, Eduardo Restrepo, and Ricardo Ventura Santos, *Mestizo Genomics: Race Mixture, Nation, and Science in Latin America* (Durham, NC: Duke University Press, 2014).
19. Erasmus, *Race Otherwise*, 114. Though for a reading of the same phenomenon as empowering for those making these claims, see also Katharina Schramm, "Casts, Bones, and DNA: Interrogating the Relationship Between Science and Postcolonial Indigeneity in Contemporary South Africa," *Anthropology Southern Africa* (2016).
20. See Foster, "A Postapartheid Genome"; Schramm "Casts, Bones, and DNA"; and Schramm, "Race, Genealogy, and the Genomic Archive."
21. Foster, "A Postapartheid Genome."
22. Schramm "Race, Genealogy, and the Genomic Archive."
23. Erasmus, *Race Otherwise*; Schramm "Casts, Bones, and DNA"; and Schramm, "Race, Genealogy, and the Genomic Archive."
24. See TallBear, "Genomic Articulations of Indigeneity."
25. See Nadia Abu El-Haj, "The Genetic Reinscription of Race," *Annual Review of Anthropology* 36, no. 1 (2007): 283–300; Troy Duster, "The Molecular Reinscription of Race: Unanticipated Issues in Biotechnology and Forensic Science," *Patterns of Prejudice* 40, no. 4–5 (2006): 427–41; Joan H. Fujimura and Ramya Rajagopalan, "Different Differences: The Use of 'Genetic Ancestry' Versus Race in Biomedical Human Genetic Research," *Social Studies of Science* 41, no. 1 (2011): 5–30; Duana Fullwiley, "The 'Contemporary Synthesis': When Politically Inclusive Genomic Science Relies on Biological Notions of Race," *Isis* 105 (2014): 803–14; Amy Hinterberger, "Publics and Populations: The Politics of Ancestry and Exchange in Genome Science," *Science as Culture* 21, no. 4 (2012): 528–49; Emma Kowal, "Race in a Genome: Long Read Sequencing, Ethnicity-Specific Reference Genomes, and the Shifting Horizon of Race," *Journal of Anthropological Sciences* 97 (2019): 1–16; and M'charek, *The Human Genome Diversity Project*.
26. I don't mean to imply that these concepts (political and scientific, social and biological, and arbitrary and objective) are mutually exclusive. Indeed (for

example), science is always political whether or not its politics are acknowledged or legible.

27. The genetic testing services may have instead or additionally been motivated by the desire to procure more African samples for global genetic diversity projects; it isn't clear to me whether the genetic testing offered at the Origins Centre from 2006 to 2019 and at least initially overseen by Himla Soodyall was part of Soodyall's work on the Genographic Project that Katharina Schramm critiques; see Schramm, "Race, Genealogy, and the Genomic Archive."

28. This model of everyone as African has been rightfully critiqued (see, for example, TallBear, "Genomic Articulations of Indigeneity"), including by the current director of the Origins Centre, Amanda Esterhuysen; see Amanda Esterhuysen, " 'If We Are All African Then I Am Nothing': Hominin Evolution and the Politics of Identity in South Africa," in *Interrogating Human Origins: Decolonisation and the Deep Human Past*, ed. Martin Porr and Jacqueline Matthews (New York, Routledge, 2020), 279–91. However, it does take on a slightly different tone when it is part of an African move to demonstrate the continent's global significance.

29. Morris, *Bones and Bodies*, 282.

30. See Annie E. Coombes, *History After Apartheid: Visual Culture and Public Memory in a Democratic South Africa* (Durham, NC: Duke University Press, 2003); Martin Legassick and Ciraj Rassool, *Skeletons in the Cupboard: South African Museums and the Trade in Human Remains, 1907–1917*, 2nd ed. (Cape Town: Iziko Museum, 2015); Sarah Nuttall and Carli Coetzee, eds., *Negotiating the Past: The Making of Memory in South Africa* (Oxford: Oxford University Press, 1998); and Leslie Witz, Gary Minkley, and Ciraj Rassool, *Unsettled History: Making South African Public Pasts* (Ann Arbor: University of Michigan Press, 2017).

31. According to the Origins Centre curator, the aim of the redesign was to include a section on the Middle Stone Age innovations in South Africa.

32. According to the Origins Centre curator, the removal was part of an agreement with another South African heritage site, Maropeng, which uses "welcome home" as a central part of its branding.

33. This same photograph appears in the "Encounters" building of !Khwa ttu: there the photo appears as one of a series of eight photographs of San people with specimen tags around their necks and numbers behind them: each person is photographed twice, once facing the camera and once in profile, in a style reminiscent of police mug shots.

34. Genetic ancestry still appeared at the museum in 2022 as part of a temporary exhibit, but the overall emphasis of that exhibit was archeological evidence of settlements on the Cape coast from as far back as 120,000 years ago, emphasizing southern African land and water as spaces of human emergence.

35. For a more extensive discussion of the Mapungubwe reburial and the controversies that arose in the planning process, see Tamarkin, *Genetic Afterlives*.

36. Himal Ramji, "Mapungubwe Imagined," in *Archives of Times Past: Conversations About South Africa's Deep History*, ed. Cynthia Kros, John Wright, Mbongiseni Buthelezi, and Helen Ludlow (Johannesburg: Wits University Press, 2022), 240–51.

37. Tamarkin, *Genetic Afterlives*.

38. See Amanda B. Spurdle and Trefor Jenkins, "The Origins of the Lemba 'Black Jews' of Southern Africa: Evidence from P12f2 and Other Y-Chromosome Markers," *American Journal of Human Genetics* 59, no. 5 (1996): 1126–33. Himla Soodyall, also Jenkins's student at the time, was not a coauthor for this particular study but also worked with Lemba DNA as part of her doctoral research and later published a reanalysis of Lemba DNA; see Himla Soodyall, "Lemba Origins Revisited: Tracing the Ancestry of Y Chromosomes in South African and Zimbabwean Lemba," *South African Medical Journal* 103, no. 12, suppl. 1 (2013).

39. Tamarkin, *Genetic Afterlives*.

40. Johan Nel, "'Gods, Graves, and Scholars': The Return of Human Remains to Their Resting Place," in *Mapungubwe Remembered: Contributions to Mapungubwe by the University of Pretoria*, ed. Sian Tiley-Nel (Pretoria: University of Pretoria Press), 230–39.

41. Amade M'charek, Katharina Schramm, and David Skinner, "Technologies of Belonging: The Absent Presence of Race in Europe," *Science, Technology & Human Values* 39, no. 4 (2014): 459–67.

42. See TallBear, "Genomic Articulations of Indigeneity."

43. My description of the site and its history is based on exhibits and explanatory text on display there in 2022.

44. Laura Foster, *Reinventing Hoodia: Peoples, Plants, and Patents in South Africa* (Seattle: University of Washington Press, 2018); Daniel Huizenga, "Documenting 'Communitiy' in the ╪Khomani San Land Claim in South Africa," *PoLAR: Political and Legal Anthropology Review* 37, no. 1 (2014): 145–61; Steven Robins, "Ngos, 'Bushmen,' and Double Vision: The Khomani San Land Claim and the Cultural Politics of 'Community' and 'Development' in the Kalahari," *Journal of Southern African Studies* 27, no. 4 (2001), 833–53; Noah Tamarkin and Rachel Giraudo, "African Indigenous Citizenship," in *The Routledge Handbook of Global Citizenship Studies*, ed. Engin F. Isin and Peter Nyers (New York: Routledge, 2014), 545–56.

45. See, for example, Robins, "Ngos, 'Bushmen,' and Double Vision"; John L. Comaroff and Jean Comaroff, *Ethnicity, Inc.* (Chicago: University of Chicago Press, 2009).

46. For a discussion of indigenous temporality that dovetails nicely with !Khwa ttu's reimagining of past and future, see Lewis R. Gordon, "On the Temporality of Indigenous Identity," in *The Politics of Identity: Emerging Indigeneity*, ed. Michelle Harris, Martin Nakata, and Bronwyn Carlson (Sydney: University of Technology, Sydney, 2013).

47. TallBear, "Genomic Articulations of Indigeneity."

48. Katharina Schramm notes this emphasis on San recorded voices in exhibits at the Museum of South Africa as well and notes discussions of the conditions of cast making and San resistance; see Schramm, "Casts, Bones, and DNA". See also Anette Hoffman, ed., *What We See: Reconsidering an Anthropometrical Collection from Southern Africa: Images, Voices, and Versioning* (Basel: Basler Afrika Bibliographien, 2009).

49. On the significance of refusal, see Audra Simpson, *Mohawk Interruptus: Political Life Across the Borders of Settler States* (Durham, NC: Duke University Press, 2014).

50. See also Marisol de la Cadena and Orin Starn, eds., *Indigenous Experience Today* (New York: Berg, 2007); and Schramm, "Casts, Bones, and DNA," for arguments about indigeneity as multivalent.

TWELVE

The South Asian Heart Disease Paradox

History, Epidemiology, and Contested Narratives of Susceptibility

ALYSSA BOTELHO AND DAVID S. JONES

OVER THE PAST fifty years, a remarkable claim has gained traction in the medical literature: of all the world's populations, South Asians are at the greatest risk of coronary artery disease (CAD). Some advocates warn that the CAD hospitalization rate of South Asians is "four times higher than any other ethnic population."[1] Others assert that South Asians account for 60 percent of all cardiovascular patients worldwide.[2] Neither claim is true, but that did not stop the *Wall Street Journal* from describing how "Heart Disease Snares South Asians."[3] As the narrative goes, South Asians succumb to heart disease in higher numbers and at younger ages than other ethnic groups—even in the absence of conventional risk factors such as high cholesterol, high blood pressure, or smoking. San Jose resident Mahendra Agrawal, who received an unexpected CAD diagnosis at age sixty-three, fit this profile. "Despite his good habits," the *New York Times* explained, "there was one important risk factor Mr. Agrawal could not control: his South Asian ancestry."[4] Researchers call this phenomenon the "South Asian paradox": a high prevalence of CAD without the expected risk factors.[5]

Epidemiologists have looked far and wide to uncover the causes of this South Asian predisposition. One investigation, the MASALA study, has tracked the cardiovascular health of more than nine hundred South Asians in Chicago and San Francisco since 2010.[6] The *New York Times* hailed MASALA as a badly needed "Framingham-type study" of an at-risk group that had not been included in the landmark study of cardiovascular risk in white New

Englanders.[7] Indeed, the MASALA website notes that there are "only a handful of long-term studies of South Asians," and even fewer that investigate "behavioral, social, cultural, and clinical risk factors" for heart disease.[8]

As MASALA progressed, physicians eager to address this health inequity established specialty clinics to serve South Asian patients. In 2014, the cardiologist Rajesh Dash helped found the Stanford South Asian Translational Health Initiative (SSATHI), a play on the Hindi word (sathi), or "partner." Its doctors invite South Asian patients in Silicon Valley to work with them toward a "heart healthy life."[9] The South Asian Cardiovascular Center near Chicago "serves South Asian people through a special combination of community outreach, advanced clinical services and research."[10] Even though these clinicians—many of whom identify as South Asian—have different opinions on the drivers of heart disease in this population, none doubt that this ethnic susceptibility exists. "We all have someone in our first-degree circle that has either died suddenly or had premature cardiovascular disease," said the MASALA investigator Alka Kanaya.[11] She launched MASALA after seeing many of her friends and family die young from heart disease.

Any claim of racial disease susceptibility warrants attention from social scientists.[12] This case is no different. Since the 1960s, clinicians and researchers have sought to characterize the unique features of South Asian CAD in hopes of offering better treatments for it. However, despite decades of research and a dizzying number of genetic, developmental, and environmental hypotheses, little consensus has emerged. It might actually be the case that some South Asians have a true susceptibility to heart disease, but recent estimates of the magnitude of the increased risk are much lower than they used to be: not four times higher but possibly just 1.43 times for South Asian men and 1.12 times for South Asian women.[13] Many researchers now attribute the elevated CAD risk to an underlying elevated risk of insulin resistance and diabetes.[14] But that just displaces the question: why is *that* risk elevated?

The literature on South Asian CAD poses important questions. Why do claims of risk that are greatly exaggerated circulate so widely? Why have the claims often moved far beyond the evidence behind them? What work do the claims do for those who make them? Why has elucidation of the causes proven so difficult? What other modes of thinking has the rhetoric allowed researchers to elide? The enduring popularity of the South Asian paradox provides a valuable case study to explore the powerful draw of race-based thinking in health care.

At its core, this susceptibility discourse relies on a specific population category: "South Asian." But South Asians, who make up a quarter of the world's population, have long transfixed researchers for their *diversity*, not their sameness. Social scientists have studied the entrenched religion and caste divisions in South Asian communities. Geneticists have investigated the Indian subcontinent's complex demographic history, which includes many migrations over thousands of years. These diverse populations, both within South Asia and throughout their diaspora, have a wide range of diets, lifestyles, and environmental exposures. How, then, can "South Asians" be held together as a distinct group that shares a unique susceptibility to CAD? And why? Physicians describe the South Asian paradox as the disproportionate burden of CAD in the absence of conventional risk factors. We suggest that the true paradox at the heart of this story is that researchers continue to search for the cause of "South Asian" susceptibility despite historical, epidemiological, and genetic evidence that no such coherent group exists. Only recently have CAD researchers begun to grapple seriously with the heterogeneity of South Asian populations, both within South Asian countries and in the diaspora.[15]

Clinicians and epidemiologists who search for the answer to the "South Asian paradox" run the risk of recapitulating colonial assumptions about South Asians as a distinct race and portraying South Asian bodies as inherently pathological. In their haste to treat what they perceive as racial susceptibility, they may overlook other ways of understanding how our biology is shaped by the world we live in. This is not the first time that cardiology researchers have fallen into the trap of racialized thinking. In fact, theories on ethnic predispositions to metabolic disorders such as diabetes, obesity, and heart disease have a long and infamous history. Prominent discussions of Hispanic, African American, and Indigenous populations' high risk for these conditions have prompted scrutiny from social scientists.[16] In a famous 2005 case, the U.S. Food and Drug Administration approved the drug BiDil for the treatment of heart failure in people who self-identify as Black. The NAACP and the National Medical Association lauded BiDil as a victory for the African American community. The *Times of India* even wondered whether something similar could be done for Indians: "With the entry of a heart drug in the market meant exclusively for the Blacks, shouldn't there be attempts to try out a formulation targeting Indians and their 'higher' risk for contracting cardiac diseases?"[17] History illuminates the futility of such quests.

Despite BiDil's "much-trumpeted release," the *New York Times* wrote years later, "patients did not request the medication and practicing doctors did not prescribe it." Embroiled in accusations of shoddy science and exploitative marketing, NitroMed—the company that developed BiDil—folded in 2009. Today, BiDil is best known "as one more misstep in medicine's long history of race-related disasters."[18] It remains to be seen whether initiatives such as MASALA and the SSATHI clinic will be able to move past a simple notion of "South Asian" risk and develop—and implement—sophisticated understandings of the nature of disease susceptibility and identity.

South Asian CAD initiatives do redirect resources toward minority populations in a historically white-centered biomedical enterprise—and that is a worthy endeavor. Many South Asian health advocates point out that conventional practices in the United States, which focus on basic race and ethnic categories (e.g., Black, white, Hispanic), ignore the experiences of immigrants from South Asia, the Middle East, and North Africa, all of whom are lumped into the category of "white." However, such efforts also introduce the risk of overtesting, overtreating, and otherwise exoticizing South Asian bodies when all that might be needed is business-as-usual heart disease prevention and care.

We trace the winding, contested discourse of South Asian susceptibility to CAD over the past century, paying attention to how western and South Asian researchers grappled with ideas about race and modernity on the Indian subcontinent. We then reflect on how these ambiguous but powerful imaginations of South Asian CAD are intertwined with promises of cutting-edge, personalized medicine for South Asians—at least for those who can afford it. We call on researchers and clinicians to be more conscious of the risks inherent in pursuing a simplistic discourse of racial susceptibility and to submit their own work to strict scrutiny to ensure that the claims they make are justified.

The Rise of Heart Disease, and Cardiologists, in India

CAD, or at least a clinical syndrome that would earn that diagnosis from a modern doctor, has existed in India for centuries. Allan Webb, an anatomy professor in Calcutta in the 1840s, described "an old Bengallee" whose "coronary arteries are quite ossified, like quills."[19] In 1891, an army surgeon in Hyderabad examined an Indian soldier who collapsed with severe chest pain, shortness of breath, a feeble pulse, and clammy skin. Autopsy revealed a fatty heart with narrowed coronary arteries.[20] But the disease came slowly to public attention. The *Times of India* first reported an Indian dying of a heart attack in 1926.[21] When physicians discussed CAD, they emphasized its rarity in South Asia. In 1941, for instance, a Bombay pathologist reported the results of 4,335 autopsies performed at a local hospital between 1926 and 1939. While coronary atherosclerosis caused 8.4 percent of all deaths in the United States at that time, it was only responsible for 1.3 percent of deaths in Bombay.[22] A Delhi cardiologist reasoned that India "has the lowest incidence of coronary artery disease in the world" because Indians died too young to develop hypertension and atherosclerosis. Indian life expectancy in 1951 was just thirty-two years.[23]

The low prevalence of CAD, coupled with the higher prevalence of syphilitic and other infectious forms of heart disease, fueled discourses on race and civilization in India.[24] The Bombay physician Rustom Jal Vakil wrote in 1949 that while rates were low in all Indians, Hindus "appear relatively more immune." "Whether this is due to their being . . . 'strict vegetarians' in their dietetic habits or to their more 'placid' dispositions with 'less sensitive' nervous systems," Vakil noted, "one cannot say."[25] The Calcutta physician Gerard Kelly added that many Indian patients could not articulate their symptoms: their "educational level" and "discriminating and descriptive powers" of pain were "totally inadequate" for a doctor to diagnose CAD.[26] Such claims, which can shock modern readers as racist, seemingly raised no concerns at the time.

Dismissals of Indian CAD mirrored the discourse on race and heart disease in the United States in the 1930s and 1940s. American cardiologists wrote extensively about the rarity of CAD among African Americans and the much higher prevalence of syphilitic and rheumatic heart disease. This fueled a fraught discourse on race, civilization, and primitiveness in American cardiology.[27] Indian physicians knew this literature well. Vakil compiled data from American studies to illustrate huge variations in CAD incidence, from 6.3 percent in "negroes" in Texas to 35.7 percent in predominantly white New England. These reports, he stressed, "revealed the important role of the 'racial factors' in the incidence and behavior of heart disease."[28] Coronary disease was a "disease of civilization."[29] Neither African Americans nor South Asians were civilized enough—in the eyes of Indian and American physicians—to suffer it.

Increasing concern with the problem of rheumatic heart disease in India in the 1930s motivated the emergence of a small cadre of Indian heart specialists. Some relied on their own experience in India. Other physicians sought cardiac expertise by going abroad for training. Rustom Jal Vakil received his medical degree from St. Thomas's Hospital Medical School in London in 1934 and then worked at the Heart Hospital in Liverpool before returning to Bombay in 1938 to establish his private practice; he limited himself to the care of patients with heart disease.[30] By 1941, Calcutta's U. P. Basu reported that "there are now a band of capable men who have taken up cardiology as special branch for their life work." He hoped that this would facilitate the professionalization of Indian cardiology: "It is incumbent upon us to combine together and form an association on the lines of the Cardiac Society of Great Britain and Ireland, or the American Heart Association and establish a national heart hospital in this first City of India."[31]

World War II put these plans on hold. After the war, however, professionalization moved quickly. In January 1946, members of the medical community in Calcutta voted to establish the Cardiological Society of Bengal.[32] When India won its independence in August 1947, this group moved to become a national organization. On April 4, 1948, it formally established the Cardiological Society of India, and the "long march to a glorious future was flagged off!"[33] The society published the first issue of its *Indian Heart Journal* in January 1949. Its editor, J. C. Banerjea, launched the journal with a statement of cardiological nationalism: "Free India, today, in its size, its teeming millions, exceeding 300 odd, its wide variation in climate and altitude, its variety of races and customs—visitors and residents of the old world and the New—is a miniature world. Climate and altitude alone would afford ample scope for comparative study of all classes of people, indigenous and foreign." India's climate fostered both temperate and tropical diseases, thereby allowing the study of both. Its cardiologists were certain to make many valuable contributions: "It is clear, therefore, that no apology is needed for the burden of adding one more medical journal on Cardiology to our colleagues far away." They sought their space on the international stage: "Before the medical and scientific forum of the world, we make our respectful bow."[34]

These South Asian physicians increasingly began to see CAD as a threat. In 1949, Vakil acknowledged "the unanimous opinion of most of our senior colleagues in this country that coronary disease is on the increase."[35] The next year, an *Indian Medical Gazette* editorial noted that the rise of CAD "has caused much surprise and alarm among the well-to-do in our country recently."[36] The author identified many disease drivers, all of which followed from an improved standard of living. By 1955, Vakil described CAD as a "growing menace."[37]

Increased attention to CAD during this period reflected concerns that a modernizing India would encounter a growing burden of chronic disease. Theorists of the epidemiological transition assumed that developing countries would follow Western trends, with infectious disease epidemics giving way to a new burden of chronic conditions—first among elites, and then the general population.[38] Indian cardiologists worried that the broader population would soon be at risk of CAD. In 1960, the physician Krishna Mathur noted that "with the advance of modernization, industrialization, and urbanization . . . people are exposed to greater stress and strain. These may be important factors in the increasing incidence of coronary heart disease."[39] Ancel Keys, famous for his studies of diet and heart disease, predicted at the Fifth World Congress of Cardiology in Delhi in 1966 that "as India becomes more prosperous, here too coronary heart disease will become your No. 1 health problem unless effective prevention is developed."[40]

India's leaders worried particularly about the impact of CAD on the young nation's elites. In 1955, the Indian government wrote to the World Health Organization: "Coronary disease has a predilection for intellectuals, business executives and people occupying key positions in various walks of life . . . it causes a considerable amount of invalidism and consequent economic loss to the State."[41] Paul Dudley White, a prominent American cardiologist, warned the Indian public in 1964 that CAD "is common among the more prosperous business and professional men and so a threat itself to the future of India."[42] White's fears were soon realized. In May 1964, Jawaharlal Nehru, India's prime minister, collapsed and died. Even though his doctors believed he had suffered a ruptured aortic aneurysm, the press reported it as a heart attack.[43] In either case, cardiovascular disease had become a force that shaped India's history.

The South Asian Paradox

The rise of CAD in India prompted a flurry of epidemiological studies in the second half of the twentieth century. Many were inspired by the

Framingham Heart Study, which identified the classic risk factors for CAD, including diet, exercise, tobacco use, and family history.[44] Some Indian studies echoed Framingham findings.[45] But many did not. S. L. Malhotra compiled data from 1.15 million workers in the Western Railway service and found the highest rates of CAD in southern India. This was unexpected, since it was in northern Punjab where Indian habits most resembled an American lifestyle, with high consumption of fat, meat, and cigarettes. South Indians, in contrast, were more likely to be nonsmoking vegetarians. In another "unexpected and extraordinary finding," mortality was lower in sedentary clerks than in the physically active mechanics. The conclusion seemed inescapable: "The current hypothesis of diet, the decline of physical activity, and smoking and stress does not find support in this study." Malhotra searched widely to explain his findings, examining socioeconomic factors and subtleties of Indian diets. In the end, he could not "disentangle from the thicket of environmental factors those which relate with certainty to the causation of ischaemic heart disease."[46]

Researchers also found substantial disparities in CAD rates between rural and urban populations (e.g., 3 percent versus 9 percent in the 1960s) and observed that rates in both regions were changing over time.[47] This suggested that environmental factors played a key role in Indian CAD and that the disease could be addressed if "appropriate preventive strategies" were identified.[48]

However, confusion deepened as epidemiologists studied the South Asian diaspora. Studies of Japanese immigrants to Hawaii and California had shown that CAD risk increased as they adopted the diet and customs of their new homes.[49] These convinced researchers that "between-population differences" in CAD rates "are environmentally rather than genetically determined."[50] South Asian immigrants, however, responded differently. Studies in Singapore, Fiji, South Africa, England, Uganda, and Trinidad found that South Asian immigrants had the highest rates of CAD of any group in those places.[51] Perplexingly, the risk existed regardless of the degree to which the immigrants adopted local customs. Michael Marmot and Paul McKeigue studied South Asians in London and found that they had higher rates of CAD than the native population whether they were rich or poor, smokers or nonsmokers, or vegetarians or meat eaters. This posed a serious puzzle. "Any general explanation of the high rates of coronary heart disease in south Asians overseas," they wrote, "must invoke some factor that is common to the diverse communities that make up the Asian population in Britain."[52] As to what this factor might be, they were not sure. "Indian populations overseas originate from several different parts of the subcontinent and are likely to be genetically dissimilar," they wrote. "It is not obvious how the consistent findings of high rates of diabetes and [CAD] can be reconciled with either environmental or genetic explanations."[53]

The Proliferation of Hypotheses

These epidemiological findings presented researchers with many conundrums—and they responded with a bewildering range of possible explanations.

Many researchers turned to genetics. As the authors of a 2016 review wrote, the failure of traditional risk factors to explain the high CAD prevalence in South Asians "suggests a strong genetic causation that needs active characterization."[54] Researchers identified many possible genetic suspects. One of the first to emerge was the gene for lipoprotein(a), a blood protein involved in cholesterol metabolism. A 1995 study screened 247 Punjabi migrants in West London and 117 of their siblings in India and compared them to white Londoners. They found that Punjabis had higher serum concentrations of lipoprotein(a) wherever they lived. When Punjabis encountered the abundant calories of London, they developed high rates of CAD.[55] Something similar happened with Asian Indian physicians in the United States and their families.[56] The conclusion seemed obvious to the study's authors: "A genetic predisposition to CAD, mediated by high levels of lipoprotein(a), markedly magnifies the adverse effects of traditional risk factors related to lifestyle and best explains the South Asian Paradox."[57]

A second candidate soon emerged: cardiac myosin binding protein C, or MYBPC3. One polymorphism in the MYBPC3 gene found only in South Asians was found to be associated with cardiomyopathies. The researcher who characterized the mutation hoped that it "will help in identifying and counseling individuals predisposed to cardiac diseases in this region."[58] A 2014 review claimed that the mutation, "very common in the Indian subcontinent," was "responsible for 45% of sudden heart attack deaths."[59] This narrative quickly ran into trouble. The mutation, found in just 4 percent of people with Indian ancestry, hardly seems common.[60] And no published reports had documented a link between *MYBPC3* and CAD or heart attacks.[61] As with the claim that South Asians account for 60 percent of all cardiac patients worldwide, something about the constellation of race, South Asians, and heart disease fosters the circulation of implausible claims.

Frustrated by the limited predictive power of single genetic variants, researchers refocused their attention onto polygenic risk scores based on millions of sites in the human genome. These can predict risk for diseases like CAD. However, their efficacy appears to be limited to the population in which they were developed. Most, so far, have been developed in people with European ancestry. Because of this problem, a multinational team of researchers turned to the UK Biobank, with its 8,024 South Asian participants (out of over 500,000 overall), to develop new risk scores "tuned" to people of South Asian Ancestry.[62] While this approach no longer implicates specific genes, it relies on the assumption that there is something significantly different, and actionable, in South Asian genomes.

Researchers also found an evolutionary rationale for the supposed genetic differences: the thrifty gene hypothesis. In 1962, James Neel speculated that ancient human populations had evolved specific genes to survive periods of food scarcity. When food became abundant with modern economic development, these genes became a liability.[63] Neel had written about diabetes, but the analogy to CAD was clear. When researchers identified lipoprotein(a) as a possible genetic cause of South Asian CAD, they invoked thrifty genes immediately.[64] As a *Lancet* commentary explained in 2004, "genetic variants that were protective and enhanced survival at a time when calories were less abundant are now harmful. . . . Such alleles could be more common in south Asians and predispose them to central obesity, insulin resistance, and diabetes mellitus, which are important risk factors for CHD in this population."[65] But it was not clear whether such alleles actually were more common in South Asians or why that might be. Recurrent famine had been the bane of many human populations for millennia, not just those in India.

Others honed in on diabetes—which can drive inflammation in the circulatory system—as the underlying driver of CAD in South Asians. When Marmot and McKeigue studied South Asian immigrants in London, they settled on high rates of insulin resistance and diabetes as the underlying risk.[66] Researchers continue to pursue this question, wondering about the "unexplored pathways" of South Asian diabetes risk.[67] But why would Indian immigrants everywhere, regardless of acculturation, have a high risk of diabetes?[68] Researchers again invoked the thrifty gene hypothesis, as they did to explain high rates of diabetes in American Indians and Mexicans. But in each case, the hypothesis simply fueled debate about the relative role of genetic and social factors in fostering diabetes.[69] Neel himself recanted the thrifty gene explanation for diabetes in 1989, but by that time the theory had spread far and wide.[70]

Competing explanations soon appeared. In the 1980s, the epidemiologist David Barker conducted studies showing that malnutrition *in utero* was associated with an increased risk of hypertension and CAD in adulthood.[71] In one study he conducted in Mysore, where pregnant women weighed on average just forty-seven kilograms, "mothers with the lowest weights had the smallest babies who as adults had the highest rates of coronary heart disease." Such findings suggested that "adaptations made by the fetus in response to undernutrition lead to persisting changes in metabolism."[72] Adult South Asians who enjoyed better nutrition than their malnourished mothers succumbed to new metabolic risks. The argument here is that fetal malnutrition, not ancestral malnutrition, drove the high prevalence of CAD in South Asians. The modern epidemic, therefore, was not the result of ancient famine but of rapid, recent economic development. Initial versions of the Barker hypothesis were vague about how fetal exposures influenced metabolic setpoints. Epigenetic research now offers a possible explanation: the children of malnourished South Asian mothers could have been exposed to epigenetic changes *in utero* that predisposed them to metabolic abnormalities later in life.[73]

These two types of explanation—genetic (thrifty gene hypothesis) and developmental (Barker hypothesis)—rely on different conceptions of time. Genetic explanations suggested that selective pressure over evolutionary time frames (many thousands of years) created the disparity in CAD (or diabetes) rates that we see today. Developmental theories of South Asian CAD susceptibility, in contrast, offered a more proximal temporality. While the thrifty gene explanation is genetic and the Barker hypothesis developmental, both rely on an epidemiological transition: a mismatch between ancestral or fetal and contemporary environments. In contrast to genetic theories, however, the Barker hypothesis offered actionable advice. "Prevention of the rising epidemic of the disease in India," Barker advised in 1996, "may require improvements in the nutrition and health of young women."[74] This would be a different kind of Indian heart paradox: to prevent CAD, seen as a disease of overweight older men, India needed to improve the nutrition of young women. The Barker hypothesis also makes a testable prediction: as economic conditions in South Asia or the diaspora improve and South Asian mothers are no longer malnourished, then the elevated risk should dissipate.

Meanwhile, many cardiac surgeons, both in India and the United States, believed that the problem is anatomic. They claimed that Indians had peculiarly narrow coronary arteries, which increased the consequences of atherosclerotic plaque buildup. The Chennai surgeon M. R. Girinath remembers that in the 1970s and 1980s, before India had substantial capacity in cardiac surgery, wealthy Indians flew to Houston, Cleveland, or London for bypass surgery. Surgeons at those hospitals called him to complain about the difficulty of operating on Indians' small arteries. Girinath and his Indian colleagues redoubled their own efforts and perfected their skills. Now, Girinath said, "Indian patients get better care in India."[75]

Do Indians actually have narrow coronaries? Attempts to confirm whether surgeons' impressions are true have yielded inconsistent results. Some studies reported that while South Asians do have slightly narrower coronary arteries than whites, the difference disappears after adjusting for body size.[76] A 2005 study in New York, however, found that South Asian coronaries were smaller even after correcting for body size (the difference was subtle: 2.2 versus 2.3 mm).[77] A 2011 study from Philadelphia reported inconsistent results—some South Asians had narrower coronary arteries than whites, but others had wider.[78] More recent findings, especially those using the new technique of CT coronary angiography, have generally found that South Asians do seem to have smaller coronary artery diameters and volumes than Europeans and East Asians.[79] But even when clinicians have agreed that the difference in artery size exists, they disagree on its causes. Anatomic variation is influenced by genetic, developmental, and environmental factors.

Most researchers today put forward some mix of the thrifty gene and Barker hypotheses to explain the South Asian paradox, bringing distinct lines of genetic, evolutionary, and developmental thought together. These investigators believe that South Asians have a genetic predisposition to heart disease (or underlying diabetes) that is unmasked by the environmental stresses of a Western diet and lifestyle and especially by the rapid shift in ways of life over the past generation, as seen most dramatically in immigrant diasporas. The genetic burden, they hypothesize, is likely rooted in a con-stellation of genes that regulate insulin secretion, fat deposition, and inflam-mation.[80] Harking back to Neel, Marmot, and McKeigue, they argue that these genetic pathways predispose South Asians to diabetes and obesity, and these, in turn, predispose South Asians to CAD.

While such mechanisms are plausible, researchers must tread carefully. Genetic explanations for high rates of diabetes in other minority groups have sparked fierce debates about the harms of biologically reductionist think-ing, especially when assessing disease risk in historically disenfranchised groups. A 2022 report recommended screening Black, Hispanic, and Asian Americans for diabetes at a lower body mass index than white Americans but acknowledged that the cause of this increased risk remained unclear and that there was substantial heterogeneity within each of those groups.[81] Given researchers' acknowledgments of South Asian diversity, the basic factual claim that South Asians have an increased susceptibility to CAD might not be true for all of them—or at least may be more complicated than it seems.

What *Is* South Asian?

The SSATHI clinic in Silicon Valley welcomes anyone with roots in India, Pak-istan, Bangladesh, Nepal, Bhutan, Maldives, or Sri Lanka—places that rep-resent a quarter of the world's population. They are also places famous in the public imagination for entrenched distinctions of religion, caste, and ethnicity. Does it even make sense to conceptualize South Asians as a natu-ral kind that shares a unique susceptibility to CAD? The question of South Asian identity, so troublesome in this discourse on disease susceptibility, has deep roots in Indian colonial history.

The bewildering complexity of South Asian populations has long stymied rulers, ethnographers, and medical researchers. Colonial ethnologists in eighteenth-century Calcutta struggled to fit Indians into their taxonomies of human populations: Indians were both same (i.e., Caucasian) and different (e.g., darker complexions).[82] By the nineteenth century, most British writ-ers described Indians as racially distinct.[83] Yet Indians still defied simple efforts at classification, with identities that were grounded in lineage, locale, religion, occupation, and military affiliation. As the historian Nicholas Dirks explains, the British encountered a highly stratified society: "In precolonial

India, the units of social identity had been multiple, and their respective relations and trajectories were part of a complex, conjunctural, constantly changing, political world." Colonial scholars debated whether ancestry, culture, diet, or environment mattered most for demarcating social difference. As the British imposed stricter control on its Indian subjects after the rebellion of 1857, they focused on caste as the key "to know and to rule India." But caste itself was a British construct. As Dirks argues, "It was under the British that 'caste' became a single term capable of expressing, organizing, and above all 'systematizing' India's diverse forms of social identity, community, and organization."[84] Under British rule, Indians were a natural kind, to be organized and controlled with the colonial tool of caste.

As Indians fought for independence from the British, the question of whether Indians were one race became fraught. Britain's exit strategy—partition—triggered one of the century's great human tragedies. What was once British India is now many countries: India, Pakistan, Bangladesh, Nepal, Bhutan, Myanmar, the Maldives, and Sri Lanka. Despite a constitutional ban on caste, India remains stratified by caste, class, ethnicity, language, and other markers of difference. Are South Asians one people? Few in South Asia today would likely agree.

Cardiac researchers have long recognized and valued South Asian diversity. As Sivaramakrishnan Padmavati wrote in 1962: "India is fertile ground for the cardiovascular epidemiologist." It offered "great variety in ethnic and food habits." Northern Indians descended from "a mixture of the original Indo-Aryans with successive waves of invaders from the Northwest," while southern Indians came from "a large Dravidian element." Across the subcontinent, there were "many so-called aboriginals with primitive ways of life."[85] These differences, she stressed, contributed to wide variations in CAD incidence across India.

Some Western researchers who studied Indian diaspora communities also attended to this multiplicity. Marmot and McKeigue's team observed that "smoking rates range from very low in Gujarati women in Brent and Harrow to high in Bangladeshi men in Tower Hamlets" and that "most Asians in Brent and Harrow are vegetarian whereas the Moslem communities of Tower Hamlets and Waltham Forest are generally not."[86]

Other epidemiologists rejected the utility of the category "South Asian" altogether.[87] A 1994 review reminded its readers: "Words such as South Asian, Asian, Indo-origin are 'umbrella' terms. They do not accurately describe a group of people who are diverse not only in geographical origin, but also in religion, culture and dietary habits."[88] In 2004, Khunti and Samani warned in the *Lancet* that "South Asians are not a uniform group but include ethnic subgroups with different cultures and practices."[89]

Geneticists, like cardiologists, saw this diversity as opportunity. A 2014 overview described India as "a treasure for geneticists and evolutionary biologists due to its vast human diversity, consisting of more than 4,500 anthropologically well-defined populations (castes, tribes and religious groups). Each population differs in terms of endogamy, language, culture, physical features, geographic and climatic position and genetic architecture." But this overview, titled "Unity in Diversity," sent mixed messages. It noted that previous comparisons of Indian populations to other groups showed Indians to be a distinct cluster, "indicating genomic unity." But it also acknowledged that many studies found differentiation within India "large enough to warrant care in selecting patients and the controls for genomic/epigenomic investigations."[90]

Even when researchers acknowledged South Asians' heterogeneity, they often lumped them together. Khunti and Samani warned against grouping South Asians but then argued that "South Asian" ethnicity should be considered in CAD risk calculations.[91] A 2006 *Circulation* study noted that "it is important to recognize that the term 'South Asian' refers to a heterogeneous population, with important differences in diet, culture, and lifestyle." Later, however, it elided these differences: "Multiple studies of migrant South Asian populations have, however, confirmed a 3- to 5-fold increase in the risk for myocardial infarction and cardiovascular death as compared with other ethnic groups."[92] Unity in diversity indeed.

Ongoing research on South Asian genomics has not resolved this problem. A 2019 study described genetic mixture between two ancient populations, the Ancestral North Indians and Ancestral South Indians, as "the main gradient of genetic variation in South Asia today."[93] But another recent analysis by Indian researchers sampled 367 individuals from twenty locations and found evidence not of two ancestral populations (the Aryans and Dravidians) but four ancestral populations on the subcontinent and a fifth on the Andaman and Nicobar Islands.[94] Mainland populations mingled for most of their history, though mixing decreased among some upper-caste Indo-European speakers over the past seventy generations. Yet another report examining the incidence of rare genetic diseases in South Asians emphasized that "the more than 1.5 billion people who live in South Asia are correctly viewed not as a single large population but as many small endogamous groups," some of which were the result "of founder events more extreme than those in Ashkenazi Jews and Finns."[95] This array of findings could support narratives of unity or disunity. Research groups continue to decipher the genetic substructures of the Indian subcontinent, but for now it is clear that the genetics of South Asian populations are complex, with multiple founder events.[96]

Researchers in England have been more attentive to this heterogeneity than those in the United States (empowered by British demographic data that often distinguish South Asians, and sometimes different South Asian communities, from other groups). A 2021 study of CAD risk in individuals in the UK Biobank found an increased risk of 2.03 times for South Asians compared to Europeans. But there were wide variations within that group. The risk was 3.66 times higher for people with Bangladeshi ancestry, 2.45 times higher for Pakistani ancestry, and 1.83 times higher for Indian ancestry.[97] Reducing these people to a single category, South Asian, obscures these disparities and their potential causes. Given that the region's evolutionary and genetic histories remain fiercely contested, claims that South Asians are a "natural kind" with broadly shared genetic risks must be carefully examined.

Contested Pasts and Promissory Claims

Despite acknowledgments of South Asia's vast diversity, South Asians are still presented—and marketed to—as a monolithic other, even by the well intentioned. Though SSATHI and other specialty clinics utilize racialized medical theory and practice, they are undoubtedly benevolent endeavors that seek to provide customized care to a minority group. South Asian health activists welcome the attention that their group, which has generally been neglected in health statistics in the United States, has finally received. Doing so, however, is scientifically and rhetorically complex. In the face of anxious patients, physicians must distill incomplete datasets into actionable narratives and treatments. We empathize with the decisions they face. But even if race appears to correlate with medical outcomes, it is another matter to justify its use in medical risk prediction and diagnosis of individual people.

"Race corrections" abound in risk calculators and other clinical algorithms, but researchers have shown that race is often factored into these tools without robust evidence or clear justification that such modifications would ameliorate racial health inequity.[98]

In this case, the consequences of race-based risk assessment and cardiac care for South Asians are not well understood. In 2018, the American Heart Association and other medical groups revised cholesterol guidelines that, for the first time, recommended that physicians factor in a patient's race and ethnicity while assessing CAD risk. Citing the MASALA study and others, the guidelines highlighted South Asian as a "high-risk ethnicity" for developing cardiovascular disease and argued that South Asian patients in the United States are "stronger candidates for statin therapy."[99] In another scientific statement issued that same year, however, the American Heart Association took a more ambivalent stance.[100] Though the organization stood by the claim that South Asians in the United States have a higher CAD risk compared to other U.S. populations, it maintained that the biology of the disease "is no different in South Asians than in any other racial/ethnic group." The majority of the increased CAD risk in South Asians, the statement went on, could be explained by the increased prevalence of known risk factors, especially those related to insulin resistance. "No unique risk factors in this population have been found," the authors concluded, though they added that the existence of such factors "cannot be ruled out." The MASALA researchers, meanwhile, have become interested in various forms of heterogeneity in their research subjects. In addition to attending to nuances of ancestry (e.g., Indian, Pakistani, or Bangladeshi), they have also noted significant differences in acculturation, with some immigrants fully adopting practices of their new culture, others maintaining their heritage, and most doing some combination of both.[101]

Any claim of race or ethnic susceptibility also carries the risk of stigmatization. The team that developed the new polygenic risk score for South Asian CAD celebrated its achievement as "an important public health opportunity, particularly given the increased rates of a sedentary lifestyle and reluctance to take medicines frequently encountered in South Asian individuals."[102] It is striking how casually the authors introduce stigmatizing assertions about South Asian health behaviors.

Perhaps reflecting the equivocation about the drivers of South Asian CAD susceptibility, the SSATHI clinic's services do not differ greatly from the familiar standard of care. Patients are interviewed and examined before they receive a treatment plan, including (as needed): weight loss and exercise goals, prescriptions to treat hypertension and high cholesterol, and referrals to Stanford's interventional cardiologists and cardiac surgeons.[103] Where SSATHI services diverge is in their coaching on "unique" South Asian cardiovascular risks (including more vigilant monitoring of weight, cholesterol, blood pressure, and blood sugar) and nutritional guidance from a dietician familiar with South Asian cuisine. They also receive an ethnic-screening test for lipoprotein(a)—though its clinical significance remains unclear, and there are not yet any drugs on the market that target it. [104]

Could this special coaching make a difference? There is evidence that it might not be necessary and that all cardiac patients, regardless of ethnic origin, should follow the same treatment guidelines. A prominent 2004 study that tracked heart attacks in 170,000 people in fifty-two countries found that nine traditional risk factors, none of which involved ethnicity, accounted for 90 percent of all people's attributable risk of heart attack.[105] A 2006 commentary on this study, written by three Canadian physicians of South Asian ancestry, argued that "there is no evidence to suggest that treatment targets should differ between ethnic groups." A 2022 study similarly found that seven risk factors accounted for 85 percent of the risk: diabetes, smoking, depression, hypertension, household income, family history, and hypercholesterolemia.[106] For now, it seems that South Asian patients, like everyone else, should eat in moderation, exercise, abstain from tobacco, and control their blood pressure, cholesterol, and blood sugar.[107] Of course, if SSATHI's special attention to these issues in South Asian patients leads more of them to achieve these goals, then these people will benefit.[108]

Susceptibility and Policy

SSATHI physicians are using a racialized narrative of risk to carve a niche in a competitive cardiac marketplace. Operationalizing race in this way *can* be useful when race is an accurate proxy for social and environmental experiences in a world that is structured (unjustly) around race and racism. But Silicon Valley's tech entrepreneurs and Bihar's subsistence farmers have little in common aside from their ancestry. Could it possibly make sense to lump them all into a unified risk category?

This conundrum has stymied India's public health responses to CAD. Srinath Reddy, then president of the Public Health Foundation of India, blamed India's neglect of CAD on the perception of CAD as a disease of elites. "The lack of a public health response so far," he said in 2007, "was mostly due to perceptions among both policymakers and the public that [CAD] is largely a problem of the urban rich." He also worried that migrant studies were "misinterpreted to mean that the conventional risk factors did not matter in the Indians." Reddy rejected both assumptions. He felt that conventional risk factors were still relevant for Indians, especially if superimposed on "as yet undefined factors that contribute to ethnic susceptibility." More significantly, Reddy believed that the poor, rather than elites, were now "the dominant victims" of the disease.[109] India had transitioned epidemiologically, he argued, and the government had to respond.

Others disagreed. The Harvard epidemiologist S. V. Subramanian and colleagues in England and India insisted that CAD deaths continued to "occur disproportionately among the more economically advantaged groups."[110] They warned that formulating Indian health policy on the misconception that CAD was a disease of the poor would divert resources from the more urgent problems of maternal and child health, malnutrition, and infectious diseases. This is again a debate about unity and disunity—but featuring class, rather than race.

What should be done? The *Wall Street Journal* encourages Silicon Valley's South Asian entrepreneurs to manage their risk at boutique South Asian heart clinics. Indian elites can access similarly state-of-the-art cardiac care in any major Indian city. The poor are caught in a temporal disjuncture. Are they still mired in diseases of the past and therefore safe from CAD, or do genetic and developmental factors put them at risk when they encounter modernity at home or abroad? Do their shared histories from the Indian subcontinent create a kinship of cardiac risk that ties all South Asians together, across discrepancies of nationality, ethnicity, religion, caste, class, and diet? Or did migration somehow solidify the category of South Asian such that a new group of patients became legible in a white-centric American biomedical enterprise?

In this setting of uncertainty, science and technology studies scholars, like physicians, cannot provide all of the answers. But we can help clinicians and researchers grapple with the complexity of race-based risk assessment in medicine. There is recurring evidence that South Asians appear to be at high risk of CAD. Yet there is overwhelming evidence that "South Asian" is not a coherent category for medical intervention. The South Asian paradox endures, as researchers continue to finesse the complexities of genetics and history and make sense of this claim of racial predisposition.

Notes

1. "The Benefits of South Asian-Focused Care," a subheading of "Stanford South Asian Translational Heart Initiative," https://stanfordhealthcare.org /medical-clinics/stanford-south-asian-translational-heart-initiative.html. This figure is likely based on a study that reported that South Asians in California had a risk 3.7 times higher of being hospitalized for CAD compared to whites—on the basis of *nine* South Asian admissions. See Arthur Klatsky et al., "The Risk of Hospitalization for Ischemic Heart Disease Among Asian Americans in Northern California," *American Journal of Public Health* 84, no. 10 (1994): 1672–75.
2. Jyoti Madhusoodanan, "MASALA Study Examines South Asian Heart Disease Risks," University of California–San Francisco, April 6, 2016, https://www.ucsf .edu/news/2016/04/402316/masala-study-examines-south-asian-heart -disease-risks. According to the Institute of Health Metrics and Evaluation, of 12.0 million deaths from heart disease in 2019, just 2.4 million (20 percent) were in the eight South Asian countries; https://www.healthdata.org/results /gbd_summaries/2019. It is not plausible that the South Asian diaspora accounts for another 40 percent of global heart disease deaths. If you try to follow the 60 percent figure back to its source, there is none to be found. Some authors cite a 2006 WHO report, but its authors provide no source for this claim: Thomas A. Gaziano et al., "Cardiovascular Disease," in *Disease Control Priorities in Developing Countries*, ed. Dean T. Jamison et al., 2nd ed. (Washington, DC: World Bank, 2006), 650. Others cite Abdul Ghaffar et al., "Burden of Noncommunicable Diseases in South Asia," *BMJ* 328 (2004): 807–10. However, this source makes a more plausible claim: not that South Asians make up 60 percent of all CAD deaths but that noncommunicable diseases, including CAD, cause 60 percent of all deaths worldwide.
3. Vauhini Vara, "Heart Disease Snares South Asians," *Wall Street Journal*, March 3, 2011, https://www.wsj.com/articles/SB10001424052748703409304576167003904805530.
4. Anahad O'Connor, "Why Do South Asians Have Such High Rates of Heart Disease?," *New York Times*, February 12, 2019, https://www.nytimes.com/2019/02/12 /well/live/why-do-south-asians-have-such-high-rates-of-heart-disease.html.
5. The first time the phrase "South Asian Paradox" appeared in the medical literature may have been in R. B. Singh et al., "Coronary Artery Disease and Coronary Risk Factors: The South Asian Paradox," *Journal of Nutritional and Environmental Medicine* 11 (2001): 43–51.

6. MASALA stands for "Mediators of Atherosclerosis in South Asians Living in America." https://www.masalastudy.org/.

7. Sandeep Jauhar, "The Heart Disease Conundrum," *New York Times*, November 28, 2015, https://www.nytimes.com/2015/11/29/opinion/sunday/the-heart-disease -conundrum.html.

8. "About the Study," MASALA, https://www.masalastudy.org/about.

9. "Stanford South Asian Translational Heart Initiative," *Stanford Health Care*, https://stanfordhealthcare.org/medical-clinics/stanford-south-asian-trans lational-heart-initiative.html.

10. "South Asian Cardiovascular Center," *Advocate Health Care*, https://www .advocatehealth.com/health-services/advocate-heart-institute/preventive -care/south-asian-cardiovascular-center.

11. O'Connor, "Why Do South Asians Have Such High Rates of Heart Disease?"

12. This has always been true, but the debates were reinvigorated after the murder of George Floyd in 2020 and the rise of the global Black Lives Matter movement. See Ruha Benjamin, "Assessing Risk, Automating Racism," *Science* 366 (2019): 421–22; Dorothy Roberts, "Abolish Race Correction," *Lancet* 397 (2021): 17–18; Bram Wispelwey and Michelle Morse, "An Antiracist Agenda for Medicine," *Boston Review*, March 17, 2021; Jessica P. Cerdeña, Jennifer Tsai, and Vanessa Grubbs, "APOL1, Black Race, and Kidney Disease: Turning Attention to Structural Racism," *American Journal of Kidney Disease* 77 (2021): 857–60.

13. P. O. Jose et al., "Cardiovascular Disease Mortality in Asian Americans," *Journal of the American College of Cardiologists* 64, no. 23 (2014): 2486–94.

14. Sandeep Krishnan, "South Asians and Cardiovascular Disease: The Hidden Threat," *Cardiology*, May 17, 2019, https://www.acc.org/latest-in-cardiology/articles/2019 /05/07/12/42/cover-story-south-asians-and-cardiovascular-disease-the-hidden -threat.

15. The MASALA study is a follow-up study that breaks the category down by country of origin (e.g., India, Bangladesh, Pakistan), sociocultural factors, and degree of acculturation, to see if differences in risk factor profiles (e.g., diet, exercise) or protective factors can be discerned across the South Asian subgroups. See Namratha Kandula, "Understanding Health Disparities in Pakistani, Bangladeshi and Asian Indian Immigrants: The Masala Expansion Study," October 22, 2021, https://www.masalastudy.org/blog/2021/10/25/presentation-by -dr-kandula.

16. David S. Jones, "The Persistence of American Indian Health Disparities," *American Journal of Public Health* 96, no. 12 (2006): 2122–34; Jonathan Kahn, *Race in a Bottle: The Story of BiDil and Racialized Medicine in a Post-Genomic Age* (New York: Columbia University Press, 2014); Michael Montoya, *Making the Mexican Diabetic: Race, Science, and the Genetics of Inequality* (Berkeley: University of California Press, 2011); Anne Pollock, *Medicating Race: Heart Disease and Durable Preoccupations with Difference* (Durham, NC: Duke University Press, 2012).

17. "Heart Drug for Indians," *Times of India*, July 10, 2005.

18. Abigail Zuger, "From Bang to Whimper: A Heart Drug's Story," *New York Times*, December 24, 2012, https://www.nytimes.com/2012/12/25/science/from -bang-to-whimper-a-heart-drugs-story.html. In 2019, however, a different

pharmaceutical company launched a campaign to resurrect BiDil: "Arbor Pharmaceuticals Teams Up with Shaquille O'Neal to Help African Americans 'Get Real' About Heart Failure," *BioSpace*, March 14, 2019, https://www.biospace.com /article/releases/arbor-pharmaceuticals-teams-up-with-shaquille-o-neal-to -help-african-americans-get-real-about-heart-failure/.

19. Allan Webb, *Pathologia India* (London: Thacker, 1848), liv.

20. Patrick Hehir, "Angina Pectoris, with *Post-Mortem* Examination: Fatty Degeneration of Heart," *Indian Medical Gazette* 26 (1891): 268.

21. "Obituary: Retired Hyderabad Judge," *Times of India*, February 8, 1926.

22. P. Raghavan, "Aetiological Incidence of Heart Disease in Bombay—Analysis of 4335 Autopsies," *Journal of the Indian Medical Association* 10 (1941): 365–70.

23. S. Padmavati, "The Cardiac Patient in Underdeveloped Countries," *American Heart Journal* 58 (1959): 423.

24. Mark Harrison, *Climates and Constitutions: Health, Race, Environment, and British Imperialism in India* (Oxford: Oxford University Press, 1999); David Arnold, "Diabetes in the Tropics: Race, Place, and Class in India, 1880–1965," *Social History of Medicine* 22 (2009): 245–61; Suman Seth, *Difference and Disease: Medicine, Race, and Locality in the Eighteenth-Century British Empire* (Cambridge: Cambridge University Press, 2018).

25. R. J. Vakil, "A Study of Coronary Heart Disease in India," *Indian Heart Journal* 1 (1949): 208.

26. Gerard Kelly, "More Notes on Clinical Heart Disease," *Indian Medical Gazette* 75 (1940): 718.

27. Pollock, *Medicating Race*.

28. Vakil, "A Study of Coronary Heart Disease," 208.

29. Charles Rosenberg, "Pathologies of Progress: The Idea of Civilization as Risk," *BHM* 72 (1998): 714–30.

30. Sumit Isharwal and Shubham Gupta, "Rustom Jal Vakil: His Contributions to Cardiology," *Texas Heart Institute Journal* 33 (2006): 161–70. See also "Dr. R. J. Vakil Dead," *Times of India*, November 21, 1974.

31. U. P. Basu, "Rheumatic Heart Disease," *Indian Medical Gazette* 76 (1941): 15.

32. "Notes and News: History and Development of the Cardiological Society of India," *Indian Heart Journal* 1 (1949): 95–96.

33. "About CSI," Cardiological Society of India, https://csi.org.in/history/.

34. "Ourselves," *Indian Heart Journal* 1 (1949): preface.

35. Vakil, "A Study of Coronary Heart Disease," 202.

36. "Sudden Heart Failure," *Indian Medical Gazette* 85 (1950): 557.

37. R. J. Vakil, "Modern Approach to Cardiovascular Disease," *Journal of the Indian Medical Association* 25 (1955): 521.

38. George Weisz and Jesse Olszynjo-Gryn, "The Theory of Epidemiologic Transition: The Origins of a Citation Classic," *Journal of the History of Medicine and the Allied Sciences* 65 (2010): 287–326.

39. Krishna S. Mathur, "Environmental Factors in Coronary Heart Disease: An Epidemiologic Study at Agra (India)," *Circulation* 21 (1960): 688.

40. Ancel Keys, quoted in "Heart Diseases 'Related to Prosperity,'" *Hindustan Times* (New Delhi), November 2, 1966, in the Paul Dudley White Papers (PDW Papers),

Box 19, Folder 69, HMS c36, Harvard Medical Library, Francis A. Countway Library of Medicine, Boston.

41. Government of India, "Cardiovascular Diseases and Hypertension," included as attachment to A. T. Seshadri to Director-General, December 6, 1955, in "9th. Session of the World Health Assembly—Agenda," WHO Archives (Geneva), W 3/87/4 (9).

42. Paul Dudley White, "A Message About Heart Disease to the Laymen and Women of India," July 14, 1964, PDW Papers, Box 19, Folder 66.

43. "Jawaharlal Nehru Is Dead: Sudden End Follows a Heart Attack," *Times of India*, May 28, 1964; Dinesh C. Sharma, "Panditiji Bled to Death in a Medical Mess," *Mail Today*, January 2, 2008.

44. Gerald M. Oppenheimer, "Profiling Risk: The Emergence of Coronary Heart Disease Epidemiology in the United States (1947–70)," *International Journal of Epidemiology* 35 (2006): 720–30.

45. S. G. Sarvotham and J. N. Berry, "Prevalence of Coronary Heart Disease in an Urban Population in Northern India," *Circulation* 37 (1968): 939–53.

46. S. L. Malhotra, "Epidemiology of Ischaemic Heart Disease in India with Special Reference to Causation," *British Heart Journal* 29 (1967): 900, 904, 903.

47. Enas A. Enas, "Dyslipidemia in South Asian Patients," *Current Atherosclerosis Reports* 9 (2007): 367.

48. P. M. McKeigue et al., "Coronary Heart Disease in South Asians Overseas: A Review," *Journal of Clinical Epidemiology* 42 (1989): 605.

49. M. G. Marmot et al., "Epidemiologic Studies of Coronary Heart Disease and Stroke in Japanese Men Living in Japan, Hawaii and California: Prevalence of Coronary and Hypertensive Heart Disease and Associated Risk Factors," *American Journal of Epidemiology* 102 (1975): 514–25.

50. McKeigue et al., "Coronary Heart Disease in South Asians Overseas," 605.

51. T. J. Danaraj et al., "Ethnic Group Differences in Coronary Heart Disease in Singapore: An Analysis of Necropsy Records," *American Heart Journal* 58 (1959): 516–26; M. G. Marmot et al., "Lessons from the Study of Immigrant Mortality," *Lancet* (1984): 1455–57; McKeigue et al., "Coronary Heart Disease in South Asians Overseas"; Milan Gupta and Stephanie Brister, "Is South Asian Ethnicity an Independent Cardiovascular Risk Factor?" *Canadian Journal of Cardiology* 22 (2006): 193–97.

52. P. M. McKeigue and M. G. Marmot, "Mortality from Coronary Heart Disease in Asian Communities in London," *BMJ* 297 (1988): 903.

53. McKeigue et al., "Coronary Heart Disease," 605.

54. Jessica Kraker et al., "Recent Advances in the Molecular Genetics of Familial Hypertrophic Cardiomyopathy in South Asian Descendants," *Frontiers in Physiology* 7 (2016): 10. This, of course, ignores the fact that CAD had been rare in India just decades before.

55. Deepak Bhatnagar et al., "Coronary Risk Factors in People from the Indian Subcontinent Living in West London and Their Siblings in India," *Lancet* 345 (1995): 405–9.

56. Enas A. Enas et al., "Coronary Heart Disease and Its Risk Factors in First-Generation Immigrant Asian Indians to the United States of America," *Indian Heart Journal* 48 (1996): 343–53.

57. Enas A. Enas et al., "Dyslipidemia in South Asian Patients," *Current Atherosclerosis Reports* 9 (2007): 367.
58. Perundurai S. Dhandapany et al., "A Common *MYBPC3* (cardiac myosin binding protein C) Variant Associated with Cardiomyopathies in South Asia," *Nature Genetics* 41 (2009): 187.
59. Sarabjit S. Mastana, "Unity in Diversity: An Overview of the Genomic Anthropology of India," *Annals Human Biology* 41 (2014): 297.
60. Dhandapany et al., "A Common *MYBPC3* Variant."
61. Researchers have only linked the mutation to cardiomyopathies and Noonan syndrome, a rare pediatric disorder.
62. Minxian Wang et al., "Validation of a Genome-Wide Polygenic Score for Coronary Artery Disease in South Asians," *Journal of the American College of Cardiology* 76, no. 6 (August 11, 2020): 703–14.
63. James V. Neel, "Diabetes Mellitus: A 'Thrifty' Genotype Rendered Detrimental by 'Progress,'" *American Journal of Human Genetics* 14 (1962): 353–62.
64. McKeigue et al., "Coronary Heart Disease in South Asians Overseas," 605.
65. Kamlesh Khunti and Nilesh J. Samani, "Coronary Heart Disease in People of South-Asian Origin," *Lancet* 364 (2004): 2077.
66. McKeigue et al., "Coronary Heart Disease in South Asians Overseas."
67. K. M. Venkat Narayan and Alka M. Kanaya, "Why Are South Asians Prone to Type 2 Diabetes? A Hypothesis Based on Underexplored Pathways," *Diabetologia* 63 (2020): 1103–9.
68. See Arnold, "Diabetes in the Tropics," for a historical analysis of the widely circulated claimbn that South Asians are predisposed to diabetes.
69. Montoya, *Making the Mexican Diabetic*; Jones, "The Persistence of American Indian Health Disparities."
70. James V. Neel, "Update to 'The Study of Natural Selection in Primitive and Civilized Human Populations,'" *Human Biology* 61 (1989): 811–23.
71. D. J. P. Barker et al., "Fetal and Placental Size and Risk of Hypertension in Adult Life," *BMJ* 301 (1990): 259–62; D. J. P. Barker et al., "The Relation of Small Head Circumference and Thinness at Birth to Death from Cardiovascular Disease in Adult Life," *BMJ* 306 (1993): 422–26. For a discussion of Barker's hypothesis, see Sarah S. Richardson, *The Maternal Imprint: The Contested Science of Maternal-Fetal Effects* (Chicago: University of Chicago Press, 2021).
72. C. E. Stein et al., "Fetal Growth and Coronary Heart Disease in South India," *Lancet* 348 (1996): 1272, 1269.
73. Therese Tillin et al., "The Relationship Between Metabolic Risk Factors and Incident Cardiovascular Disease in Europeans, South Asians, and African Caribbeans," *Journal of the American College of Cardiology* 61, no. 17 (2013): 1785.
74. Stein et al., "Fetal Growth and Coronary Heart Disease in South India," 1269.
75. M. R. Girinath, Interview with David Jones and Kavita Sivaramakrishnan, March 19, 2013, Chennai, India. Other Indian surgeons have told us similar stories.
76. J. Dhawan and C. L. Bray, "Are Asian Coronary Arteries Smaller than Caucasian? A Study on Angiographic Coronary Artery Size Estimation During Life," *International Journal of Cardiology* 49 (1995): 267–69.

77. Amgad N. Makaryus et al., "Coronary Artery Diameter as a Risk Factor for Acute Coronary Syndromes in Asian-Indians," *American Journal of Cardiology* 96 (2005): 778–80.

78. Rani K. Hasan et al., "Quantitative Angiography in South Asians Reveals Differences in Vessel Size and Coronary Artery Disease Severity Compared to Caucasians," *American Journal of Cardiovascular Disease* 1 (2011): 36.

79. For findings of difference, see Amgad N. Makaryus et al., "Comparison of the Diameters of the Major Epicardial Coronary Arteries by Angiogram in Asian-Indians Versus European Americans <40 Years of Age Undergoing Percutaneous Coronary Artery Intervention," *American Journal of Cardiology* 120 (2017): 924–26; Abdul Rahman Ihdayhid, "Ethnic Differences in Coronary Anatomy, Left Ventricular Mass and CT-Derived Fractional Flow Reserve," *Journal of Cardiovascular Computed Tomography* 15 (2021): 249–57; Tanya Welch, "Comparison of Cardiac CT Angiography Coronary Artery Dimensions and Ethnicity in Trinidad: The CADET Pilot Study," *Open Heart* 9 (2022): e001922. However, these studies potentially confuse the issue, since they rely on studies of people with CAD. They might be detecting more severe atherosclerotic disease and not smaller arteries. A decisive study on people with normal coronary arteries has not yet been published.

80. For examples (from the MASALA researchers), see K. M. Venkat Narayan, "Type 2 Diabetes: Why We Are Winning the Battle but Losing the War?," *Diabetes Care* 39 (2016): 653–63; Alka M. Kanaya, "India's Call to Action—Prioritize Chronic Cardiovascular Disease," *Journal of the American Medical Association* 178 (2018): 373; Narayan and Kanaya, "Why Are South Asians Prone."

81. Rahul Aggarwal et al., "Diabetes Screening by Race and Ethnicity in the United States: Equivalent Body Mass Index and Age Thresholds," *Annals of Internal Medicine* (2022).

82. Shruti Kapila, "Race Matters: Orientalism and Religion, India and Beyond c. 1770–1880," *Modern Asian Studies* 41 (2007): 472.

83. Harrison, *Climates and Constitutions*.

84. Nicholas B. Dirks, *Castes of Mind: Colonialism and the Making of Modern India* (Princeton, NJ: Princeton University Press, 2011), 13, 123, 5.

85. S. Padmavati, "Epidemiology of Cardiovascular Disease in India: II. Ischemic Heart Disease," *Circulation* 25 (1962): 711.

86. McKeigue and Marmot, "Mortality from Coronary Heart Disease," 903.

87. Raj Bhopal, "Is Research Into Ethnicity and Health Racist, Unsound, or Important Science," *BMJ* 314 (1997): 1751–56.

88. Naeem Shaukat and David P. de Bono, "Are Indo-Origin People Especially Susceptible to Coronary Artery Disease?" *Postgraduate Medical Journal* 70 (1994): 315.

89. Khunti and Samani, "Coronary Heart Disease in People of South-Asian Origin," 2077.

90. Mastana, "Unity in Diversity," 287, 297.

91. Khunti and Samani, "Coronary Heart Disease in People of South-Asian Origin," 2077.

92. Milan Gupta et al., "South Asians and Cardiovascular Risk: What Clinicians Should Know," *Circulation* 113 (2006): e924.

93. Vagheesh M. Narasimhan et al., "The Formation of Human Populations in South and Central Asia," *Science* 365 (2019).
94. Analabha Basu et al., "Genomic Reconstruction of the History of Extant Populations of India Reveals Five Distinct Ancestral Components and a Complex Structure," *Proceedings of the National Academy of Sciences* 113 (2016): 1598.
95. Nathan Nakatsuka et al., "The Promise of Discovering Population-Specific Disease-Associated Genes in South Asia," *Nature Genetics* 49, no. 9 (2017): 1403.
96. David Reich, *Who We Are and How We Got Here: Ancient DNA and the New Science of the Human Past* (New York: Pantheon, 2018), 123–54.
97. Aniruddh P. Patel et al., "Quantifying and Understanding the Higher Risk of Atherosclerotic Cardiovascular Disease Among South Asian Individuals," *Circulation* 144 (2021): 410–22.
98. Darshali A. Vyas, Leo G. Eisenstein, and David S. Jones, "Hidden in Plain Sight—Reconsidering the Use of Race Correction in Clinical Algorithms," *New England Journal of Medicine* 383 (August 27, 2020): 874–82.
99. Scott M. Grundy et al., 2018, "AHA/ACC/AACVPR/AAPA/ABC/ACPM/ADA/AGS/APhA/ASPC/NLA/PCNA Guideline on the Management of Blood Cholesterol: A Report of the American College of Cardiology/American Heart Association Task Force on Clinical Practice Guidelines," *Circulation* 139 (2018): e1082–e1143; American Heart Association News, "Ethnicity a 'Risk-Enhancing' Factor Under New Cholesterol Guidelines," American Heart Association, January 11, 2019, https://www.heart.org/en/news/2019/01/11/ethnicity-a-risk-enhancing -factor-under-new-cholesterol-guidelines.
100. Annabelle Santos Volgman et al., "Atherosclerotic Cardiovascular Disease in South Asians in the United States: Epidemiology, Risk Factors, and Treatments: A Scientific Statement from the American Heart Association," *Circulation* 138 (2018): e1–e34.
101. Volgman et al., "Atherosclerotic Cardiovascular Disease in South Asians," e1, e22, e13.
102. Wang et al., "Validation," 713.
103. "South Asians and Heart Disease Q&A," *Stanford Health Care*, May 18, 2015, https://stanfordhealthcare.org/medical-clinics/stanford-south-asian -translational-heart-initiative.html/presentation-mode/stanford-health-now /2015/south-asians-heart-disease-qa.
104. A new drug has recently been shown to reduce lipoprotein(a), but studies of clinical outcomes have not yet been reported. See Sotirios Tsimikas et al., "Lipoprotein(a) Reduction in Persons with Cardiovascular Disease," *New England Journal of Medicine* 382 (January 16, 2020): 244–55.
105. Salim Yusuf et al., "Effect of Potentially Modifiable Risk Factors Associated with Myocardial Infarction in 52 Countries (the INTERHEART Study): Case-Control Study," *Lancet* 364 (2004): 937–52.
106. Yuan Lu et al., "Sex-Specific Risk Factors Associated with First Acute Myocardial Infarction in Young Adults," *JAMA Open* 5 (May 3, 2022): e229953.
107. Milan Gupta et al., "South Asians and Cardiovascular Risk."

108. For testimonials that exemplify this hope, see Krishnan, "South Asians and Cardiovascular Disease."

109. K. Srinath Reddy, "India Wakes Up to the Threat of Cardiovascular Diseases," *Journal of the American College of Cardiology* 50 (2007): 1370, 1371.

110. S. V. Subramanian et al., "Jumping the Gun: The Problematic Discourse on Socioeconomic Status and Cardiovascular Health in India," *International Journal of Epidemiology* 42 (2013): 1414.

THIRTEEN

Roots of Coincidence
The Racial Politics of COVID-19

BANU SUBRAMANIAM

All our phrasing—race relations, racial chasm, racial justice, racial profiling, white privilege, even white supremacy—serves to obscure that racism
is a visceral experience, that it dislodges brains, blocks airways, rips muscle, extracts organs, cracks bones, breaks teeth. You must never look away
from this. You must always remember that the sociology, the history, the
economics, the graphs, the charts, the regressions all land, with great violence, upon the body.

—TA-NEHISI COATES, *BETWEEN THE WORLD AND ME*

We are our viruses.

—LYNN MARGULIS, *SYMBIOTIC PLANET*

IN DECEMBER 2019, A cluster of pneumonia cases was reported at a
wholesale seafood market in Wuhan. It was soon linked to a coronavirus—a
single-stranded RNA, twenty-six to thirty-two kilobases in length, encased
in a protein coat.[1] The novel virus was named SARS-CoV-2 and identified as
the cause of COVID-19, a severe acute respiratory syndrome in humans.[2] Like
many, I was transfixed by the virus. I scanned COVID-19 dashboards obses-
sively and began to track the epidemiology of the virus and its disparate,
unequal, and varied demographic toll. I traced the wake of the virus as it
swept the world—media representations, the rhetoric deployed, visual representations, epidemiological models, health protocols, government actions
and inactions, R_0 numbers, mortality rates, and the language of pandemic
narratives. This chapter was written in November 2020. Much has transpired

since, but the fundamental patterns outlined here have not, as COVID remains an ongoing crisis. In particular, I closely followed two countries, one in which I was born (India) and the other where I now reside (United States).

To understand the pandemic—be it through science, medicine, public policy, health care, language, political economy, cultural politics, news, or social media—one must understand the politics of race in all its complex manifestations. As noted in the introduction to this volume, race is best understood as a "formation" that comes into being through processes of racialization, that is, how categories, identities, and objects acquire racial typologies. Racialization tracks how groups are stratified and mediated by other hierarchies, including class, caste, gender, sexuality, nationality, religious identity, and geopolitics. A focus on racialization makes visible how white and Hindu nationalisms have racialized new and potent landscapes during the pandemic.

Far from being a leveler, COVID-19 has revealed the "interwoven threads" of inequality and health produced by longer racialized histories of nature, science, and medicine.[3] Racism, as Ruth Gilmore argues, is specifically the "state-sanctioned or extralegal production and exploitation of group-differentiated vulnerability to premature death."[4] The coronavirus illuminates our racial landscapes of premature death. To understand the virus, to grasp its epidemiology and demographic impact, to attend to the rhetoric of pandemics, we need to explore the conceptual links between the natural, biological, and social, all steeped in a politics of racial exclusions and inclusions. In particular, I trace the rhetorical power of language—expert medical knowledge, health policy, and social scripts—because it emerges as a critical mediator between human values and our descriptions of nature. As Evelyn Fox Keller argues, to understand how science functions, we have to "look at the language of science and see how that works, how the traffic between ordinary and technical language works as a carrier . . . of ideology into science."[5]

The pandemic was not inevitable—the title "roots of coincidence" is a methodological and interdisciplinary intervention to signal that the global patterns of COVID-19 are not accidental. Rather, the hauntingly familiar script across species and nations is part of the same unfolding horror story. Ideas of "democracy" hide the vast racialized infrastructures that shape pandemic statistics and inequities. As in health crises of the past, I fear some will refuse to grasp these racial histories while others will forget them yet again. What will remain are claims of the natural—scientific languages of populations, comorbidities, and vulnerabilities—that erase their roots in the violent wounds of colonization and conquest that have been literally written and carried on the body.

The Naturecultural Politics of COVID-19

Early in the pandemic, news headlines and opinion columns put "nature" squarely at the center of the pandemic. Headlines such as "Is Nature Taking Revenge?," "Nature Strikes Back," and "Is the Coronavirus Pandemic Mother Nature's Revenge?" implied that humans had been irresponsible and disrespectful inhabitants of the planet and were due for a reckoning.[6] As someone who works in the interdisciplinary field of feminist science and technology studies (STS), the familiar framing of nature and culture as mutually exclusive, oppositional, and binary categories always raises concerns. Humans are part of nature, not outside of it; nature and culture are not oppositional but co-constituted and coproduced conceptual categories. There is no nature and culture, only *naturecultures*. But the binary and oppositional framing is long enduring, with profound consequences. Two things in particular are worth noting in this popular narrative. First, even in times of devastation where the natural world is the focus, the "human" remains squarely at the center of the narrative. Second, the generic universal "human" in this crisis elides the unequal global burdens of history. It ignores the complexities of human inequities—how only *some* humans, those who rule(d) colonial nations and enslaved and exploited others are the ones who enjoyed the fruits of colonial spoils and riches, while *other* humans, the conquered and colonized, bore the brunt of colonialism, enslavement, and nature's ostensible revenge. Indeed, across the world, an anodyne "human" obscures the figure I will refer to as "Human," the figure of entitlement, past and present.[7]

Scientific hierarchies of race that ranked human societies from primitive to higher orders provided the racial and civilizational logics for colonization, conquest, and slavery.[8] Although the biological sciences grew watchful about deploying "race" in the immediate aftermath of the World War II era and the horrors of the Holocaust, the term "race" has reemerged. With the Human Genome Project, race and racial logics are prominent again. As

Lisa Gannett reminds us, ideas about biological "race" have been recoded in the terminology of "population" thinking.[9] The ideological commitments remain, albeit in new vocabularies. This history and foundational logic of race remain deeply embedded in the language and methods of biology. Racism and, as we will see, nationalism both undergird all things biological; they are constitutive of biopolitics. There is no abstract nature and culture, only racialized naturecultures. It is within this history that I want to narrate the unfolding pandemic.

The pandemic reveals layered sedimentations of racialization. Racialized bodies—both virus and human—emerged as visible and singular targets of blame for the pandemic and have been met with threats, disdain, and violence. At the root of violent politics in the United States and India are the virulent politics of white nationalism. In the United States, we have seen four years of white nationalists, "Proud Boys," marching down the streets.[10] We have seen tacit support, via dog whistles and code words, for white nationalism through to the presidency, who supported them and called them "fine people." India presents a comparable situation with Hindu nationalists targeting Muslim and Dalits. While beyond the scope of this essay, there is an extensive literature that connects Hindu nationalism with white nationalism, going back to the nineteenth-century idea of an Aryan race. The term "Arya" means "aristocratic" in Sanskrit and was appropriated as "Aryan" to refer to "bioracial" connotations in European discourse, which during colonial India shaped ideologies of caste, religious affiliation, and Hindu supremacy. Caste hierarchies also divided Indians into superior Aryans, who migrated into India, and inferior Dravidians, the original inhabitants.[11] Since colonial times, the long-enduring links between race and caste endure.[12] The links between whiteness and upper-caste Indians is seen as so strong that Dalit activists have redefined themselves as "the Black untouchables of India" and built solidarity with Black resistance movements across the world.[13] There exist the Dalit Panthers, modeled after the Black Panthers of the United States.[14] There is ample evidence that whiteness and upper-caste ideologies are linked in their historical entanglements, genealogies of thought, colonial circulations, and literally in the visible figures of Prime Minster Modi and President Trump embracing, each publicly using the other to bolster their stature.[15] Both leaders are supported by the majority of their communities (white people in the United States and upper castes in India). In discussing the large rallies the two have attended, a recent article summarized:

"The events offered Trump and the Indian Prime Minster Narendra Modi a chance to win political points domestically while cementing their bond as right-leaning nationalist leaders."[16] To understand this convergence, race and white nationalism are key, always simmering in the cauldron of politics and health care.

In pathologizing minorities, the two right-wing nationalist leaders project an aggressive masculinity. Many have read this as a return to an old-fashioned masculinity of machismo and alpha male–dom. In an insightful piece, Susan Faludi argues that, far from this, what President Trump displays is an "ornamental" masculinity characterized by display, "a pantomime of aggrieved aggression."[17] Indeed, this performative mode that hollows out the substance of masculinity perfectly captures the theater of COVID leadership in both countries. Both leaders have taken masterful control of news media and social media and have led their majoritarian communities to seek redress from an imagined injury and grievance. As we shall see, from Twitter tweets to televised performances, both have led their countries through a pandemic landscape of racialized violence. A gendered and racialized masculinity underlies both charismatic leaders.

In what follows, I trace how racial scripts in science and medicine pervade transspecies and transnational "racial" logics and discourses of COVID-19. In particular, what is striking is how racialized discourse translates into a profound deficit in empathy, making racialized bodies expendable and reducing rich lives into individualized rational statistics. I begin with showing how racial logics in virus and human alike transcend species boundaries. Next, I trace transnational contours of the pandemic by examining the United States and India, the world's oldest and largest democracies, and show how histories of racialization have shaped the contours of COVID-19 in both sites.

The Racial "Other": Viral Politics of SARS CoV-2

In a naturecultural view of the world, the human is but one species in a teeming planet of living creatures, albeit a species that is dominant and immensely influential. In a human-centered view of nature, however, anything that harms the Human is rendered the "other," the enemy that needs to be eradicated. As such, while a virus may be natural, SARS CoV-2 is a naturecultural object that has been racialized and rendered a racial "other."

Because of the purported origins of the virus in China, sinophobic and orientalized discourses have pervaded the rhetoric of the virus. Articles have called it the "new yellow peril" and cautioned American to be on "yellow alert."[18] In keeping with orientalist rhetoric, the virus has been rendered sneaky, cunning, an assailant, an unexpected mugger, shifty like a chameleon—an invisible enemy that is pure evil. Not only evil, the virus has transformed into a foreign agent and been met with immense xenophobia. In the United States and India, the virus has been called the Chinese virus, the Wuhan virus, the Kungflu, the anti-Muslim Virus,[19] and the spread of the virus as Corona jihad and Talibani crime.[20] Studies that have quantified the xenophobia circulating in social media argue that coronavirus spread has been driven by racial animus[21] and that those infected have been predominantly racial minorities.

The other central narrative of the pandemic that acknowledges our naturecultural world are theories about the virus and its zoonotic origins.[22] Zoonosis is a term for an infectious disease caused by a pathogen that has jumped from a nonhuman animal to a human. Currently, experts suggest that the virus entered human population through wet markets in Wuhan.[23] In wet markets, meat and produce are sold alongside wildlife. Repeatedly, media accounts of wet markets represent Chinese people as uncivilized barbaric "others" who bring with them dangerous diseases because they consume the "exotic" meat of dogs, cats, and other animals outside the norm of the occidental diets.[24] Fake videos and images have circulated through social media depicting Chinese eating habits as dirty and "weird." Public figures like the Republican senator John Cornyn of Texas blamed China for the coronavirus because it was a "culture where people eat bats and snakes and dogs, and things like that."[25] Sustained focus on the exotic, dirty, weird eating habits of Chinese people speaks more to effective and potent racist tropes than to any credible explanation of the course of the pandemic.

I am also struck by how zoonosis itself has been represented—as an unusual, dangerous event caused by "primitive" people living too close to nature, thus facilitating zoonotic events. Contrary to this mainstream representation, zoonoses are ubiquitous.[26] In fact, three-quarters of infectious diseases are zoonotic spillovers.[27] As zoonotic vectors such as viruses and bacteria move between organisms, they carry genes from one individual to another, and often one species to another, allowing for lateral or horizontal gene transfer. When they infect a host, viruses insert their genetic material into the cellular genome of the host, thus at times transmitting important genes across species. Evolutionary biologists have shown that such events have had a profound impact on evolution, providing novel adaptations for the recipient organism.[28] Indeed, in the course of evolution, such cross-species transmission has been critical for the evolutionary patterns that underlie all living things.[29] For example, it is sequences from viruses that have given humans the amylase gene cluster, enabling us to produce amylase in our saliva, which allows us to eat starchy food.[30] The mammalian placenta is also one such innovation.[31] As Lynn Margulis, a key biologist whose work has inspired such understanding argues, "We are our viruses."[32] To not understand viruses and other microorganisms is to misunderstand evolution on Earth. We need to retheorize zoonosis as an integral part of our naturecultural world. We cannot blame the virus for the pandemic; humanity needs to craft a way of life that prevents future pandemics.

Viral Battles: Fighting a War with SARS CoV-2

What does one do with one's enemy? In a Human naturecultural world, one wages war. Cultural frameworks of war and warfare get deployed into the natural world into naturecultural immunolandscapes of biology. COVID-19 responses emerge as thoroughly naturecultural landscapes with an omnipresent rhetoric of "war"—fight, battle, combat, attack, tackle, defeat, and defend. In COVID-19, both the virus and Human response have been thoroughly militarized. The virus has been declared a military agent—with images of maces and robots bearing spikes ready to crack open our cells. President Trump, a self-proclaimed "wartime president," described the coronavirus as our great enemy and has declared an "all-out war," saying, "The virus will not have a chance against us" and a few weeks later, "We will win this war!"[33] PM Modi in rousing the nation declared: "[The] virus may be an invisible enemy, but our COVID-19 warriors are invincible."[34] More recently, military veterans have urged people enduring COVID-19 to think of themselves as prisoners of war.[35] Metaphors of war frame our conceptual landscape of infection and immunity—the virus, our medicine's response, the body and its immune system.[36]

In both the United States and India, the military model has framed the pandemic response. "Armies" of scientists battle the virus, overtaxed hospitals are "war zones," and protective gear for health care workers is "armor."[37] And yet, feminists have long critiqued the nationalist and militaristic model of the immune system. [38] There are other rhetorical turns, other explanatory schemes, that can represent the unfolding of the pandemic. For example, terminology of physical distancing, safe contact, and cocooning has been offered as alternate vocabulary for pandemic safety measures. The language of "frontline" workers poorly translates the material labor of workers, rendering them into nameless soldiers in a war. It is an impoverished imagination of leadership that touted physical distancing *as* social distancing.

More significantly, militaristic models miss evidence that do not conform to an antagonistic immune model. For example, throughout the pandemic, we have seen cases where patients do not die because of the virus but from the excessively aggressive immune response of their own bodies.[39] Here, in military terms, we see the nation-state overreacting with nuclear weapons to a minor incursion—our worst nightmare! Metaphors enable certain explanatory frameworks but disable others. An excellent case is that of DNA, once the "holy grail" or "blueprint" of life. Now we know that it is not a "master molecule" but only one actor in a vast and complex regulatory system of the cell.[40] Likewise, the immune system is not a military service for an armed fortress. Human health is far more complex, emerging from complex naturecultural biopolitical phenomena.

If the medical and scientific discourse about the virus was racialized, the government actions of the United States and India were utterly politicized. Even early in the pandemic certain aspects of the virus were clear. We knew from previous epidemics that its R_0 (number indicating how contagious it is) is not a natural attribute of the virus but a naturecultural attribute of how the virus functions in a social system. Ignore the "Three C's"—closed spaces with poor ventilation, crowded places, and close contact—and allow the free mingling of people, and R_0 sharply increases. Even minor health precautions such as physical distancing and mask wearing could have saved innumerable lives. Also, understanding the intimate links between health, jobs, and the economy emerges as critical. Nevertheless, neither the United States nor India followed the global protocols or policies that many other "successful" countries did. As of September 2022, they are first and second in the world for COVID-19 cases and mortality rates. In what follows, I examine COVID-19 in the U.S. and Indian contexts to uncover how racialization shapes the landscape of those affected and the racialized discourses of scientists. Ultimately, it is not the virus itself but a human-induced "perfect storm" of politics, health, and economic priorities that created the conditions for a horrendous pandemic that killed the most vulnerable in each nation.

Pox Americana: Racial Politics of COVID-19 in the United States

The election of President Trump, his official support for white nationalism, and an increasingly conservative judiciary remain stark reminders that we live in the midst of a profound backlash against the little racial progress the country has seen. Indeed, the president, Republican members of Congress, and their supporters openly flaunt racist rhetoric, now a mainstay of U.S. politics. No case exemplifies this more than the nation's lack of response to COVID-19. It is not that the country developed the wrong national policy to deal with the pandemic. It had none, and any attempt at one was challenged, protested, refused, and fought at every turn. The Trump administration in fact gutted the pandemic preparedness system they inherited from the Obama administration.[41] Established practices from other countries—policies on compulsory mask wearing, physical distancing, limits on crowds—became rallying cries about freedom. In fact, following safe public health protocols emerged as a sign of compliance, weakness, and a challenge to one's masculinity.[42] Testing was sporadic at best, and the president repeatedly blamed the pandemic on excessive testing.[43] Pax Americana, the term that characterized American power that brought peace to the Western Hemisphere, has been completely dismantled into the new reality of a racialized Pox Americana—a fertile cultural landscape for viral harms.

The virus, itself thoroughly racialized, found an even more racialized American landscape. Black, Indigenous, and People of Color (BIPOC) communities were harder hit in active cases and fatality rates. While local and regional patterns vary, overall Latino and African American residents were three times more likely to be infected than their white neighbors and twice as likely to die from the virus as white people.[44] The top hotspots in the country were Indian lands, devastating community after community of Indigenous people.[45]

Epidemiologists proposed several reasons, all legacies of longstanding dispossession and marginalization. First, some pointed out that people of color, especially women of color, were more vulnerable because of the materiality of their lives at work and home that necessitated interpersonal contact.[46] Many labored as nurses, medical assistants, nursing assistants, home health-aid workers, personal care workers, cashiers, maids, housekeeping cleaners, retail salespeople, teachers, manicurists and pedicurists, and restaurant wait staff: employment requiring direct engagement with publics. This reality was a reminder that in a nation built on the bodies and labor of people of color, the pandemic revealed how the labor infrastructure continues to depend and exploit the labor of the same groups.

The second reason often cited was the existing health profiles and social realities in communities of color.[47] Experts rationalized these high death rates by arguing that marginalized communities had "long standing systemic health and social inequalities." People of color, they argued, were more likely to be in poor health and have higher rates of hypertension, obesity, diabetes, lung disease—conditions that have been linked with increasing mortality rates with the coronavirus.[48] In particular, this argument was prominent when explaining why across the country, African Americans in particular have had an unusually high rate of contracting the virus and of mortality.[49] Repeatedly, comorbidities became a proxy to rationalize the high rates of mortality among marginalized groups and were used to blame victims for their poor health outcomes. Unsurprisingly, this ubiquitous line of argument failed to ask why systemic inequalities occurred in the first place.[50] Further, studies argued that social determinants such as "poverty and health care access, housing, educational income and wealth gaps healthcare access" affect health outcomes. So, again, it should be expected that the most marginalized would experience the highest mortality rates.

Finally, and inevitably, experts invoked the biological because they argued that social factors could not explain the staggering mortality rates in communities of color. They argued that minority communities had higher genetic predispositions, more sensitive receptors that enabled the virus to infect the body, and genetic polymorphisms in the expression levels of ACE2 that increased infectivity and pathogenesis. Experts cited studies on the genetic basis associated with underlying diseases such as hypertension and diabetes and genetic susceptibility for sudden cardiac death as an explanation for the horrific death toll in African American populations.[51]

Whether experts invoke political economic patterns of job segregation, or geographic patterns of housing, or biological explanations of genetics, each explanation rationalizes the high death rates in BIPOC communities as inevitable, albeit regrettable. At each turn, we see familiar arguments that a cursory understanding of the history of public health can easily debunk. It is not a lack of knowledge that perpetrated this health disaster but willful ignorance, signaling that some lives are dispensable. In the Trump administration, this was deliberate. These patterns are a stark reminder of the power of white nationalism, which has cultivated and nurtured a racist naturecultural landscape where medical language naturalizes and biologizes the death of marginalized populations in abstract scientific language of comorbidities, social vulnerability, biological propensity, and genetic predispositions. The sad lesson of history is an endless repetition of this script during each health crisis, even while there is no redress between crises.

COVInDian Dreams: Riding the Corona Waves

Although the United States ignored public health recommendations, India embraced it perhaps too completely. And while race and class significantly affected outcome in the United States, caste and class proved important in the Indian case. Any analysis of contemporary India must reckon with the rise of Hindu nationalism over the last many decades. With deep roots in white nationalism, Hindu nationalism joins the global rise of the far right. India is now governed by a Hindu nationalist party, and its prime minister, Narendra Modi, was nurtured through the ranks of the movement since childhood. Hindu nationalists invoke an ancient Vedic science as a prelude to India's modernity. They have selectively and strategically used rhetoric from both science and Hinduism, modernity and orthodoxy, Western and Eastern thought to build a dangerous vision of India *as* a Hindu nation, what I have called an *archaic modernity*.[52] Hindu nationalists are on a quest to transform the country into a "Hindu" nation rather than the multireligious, secular society imagined by its founders.[53] As such, Indian Muslims have emerged as the ultimate Other of Hindu nationalism.

[305]

Unlike the United States, the unfolding of COVID-19 in India is *not* a story of failure brought on by a hostility toward science or the state. Quite the contrary, India's Hindu right has embraced scientific epidemiology and followed global protocols, often more vigorously than other nations. However—and this is crucial—modern Western science is not all that constitutes "science" in contemporary India; the Indian government and Indian social media have enthusiastically promoted ancient Vedic science *as* modern science.[54] It is precisely the constant blurring of the boundaries between ancient Vedic sciences, modern Western sciences, and pseudosciences that have obfuscated health policy and preventative health information among India's populace.

The initial actions by the government were top down.[55] The most consequential was the March 2020 imposition of a three-week national curfew by PM Modi, which took effect with a four-hour notice. This fateful action was particularly significant for millions of India's migrant workers left stranded in cities and India's informal labor market (80 percent of India's workforce). In addition to the brutal enforcement of the lockdown, for many, physical distancing meant food insecurity. This abrupt measure was followed by innumerable rumors of special buses and trains heading back to villages, resulting in throngs of panicked migrants utterly disregarding lockdown rules.[56] These sudden actions were coupled with widespread rumors, misinformation and disinformation campaigns against Muslims, a targeting of Muslim businesses, violence against allegedly meat eaters, and even violence against health care workers because they were seen as carriers of the virus.[57] Citizens from the northeast of India have long been targeted with racial slurs such as "Chinese," "Momos," "Chowmein"—all suggesting that they do not belong. During the pandemic, "corona" and "coronavirus" were added to this long list.[58] And in several cases, citizens accused of being carriers were forced to leave their homes.[59]

The worst sufferers in the pandemic have been India's internal migrants, especially Dalit migrants—those considered lower caste. Dalit migrants faced ostracism by members of higher castes and were accused of bringing diseases, and while they were en route to their homes, they were refused food, water, and shelter.[60] There was also a surge in atrocities against Dalits, including violent attacks, assault, murder, and rape.[61] Like the rise in xenophobia in the United States, the pandemic certainly exacerbated caste-based violence in India.

However, despite the hurried decision and the many cases of rule break-ing, COVID-19 numbers in India remained relatively low during the early months of the pandemic.[62] If COVID brought out explanations of vulnera-bility, the Indian case was presented as the opposite. Epidemiologists pro-posed many theories—India's population is unusually young compared to other countries, the heat is a deterrent, the mandatory BCG vaccination against tuberculosis in India may provide immunity against the corona-virus, and finally the theory of innate immunity that made Indians excep-tionally resistant to the virus.[63] Nevertheless, as experts predicted, once the curfew regulations were lifted, numbers soared. With the virus housed in their bodies, when migrants reached their home villages, the virus spread rapidly throughout the country.

This was not inevitable. A naturecultural analysis reminds us that the virus needs bodies for transmission, but human bodies also need sustenance. Health and the economy are not mutually exclusive choices. For these rea-sons, some epidemiologists in India advised against the Western model of a national and complete lockdown. They argued that in a country with wide-spread poverty, multigenerational families in a household, and inadequate resources, a stringent lockdown was not practical.[64] Yet, alternatives were possible. The migrants could have been housed and fed in cities or provided safe transportation to their homes early in the pandemic, before its spread, thus preventing viral spread to remote villages. However, the Modi govern-ment chose otherwise. Fueling viral hate instead of actively planning, Modi used the pandemic to reiterate and reinforce networks of Hindu nationalism.[65]

The blurring boundaries of science and pseudoscience are all around India's COVID-19 response—from symbolic offerings and drinking of the sacred cow urine *gaumutra* by the All India Hindu Mahasabha to the wor-ship of new religious deities such as Corona Mata. The government itself issued various advisories during the pandemic that included dubious pre-vention measures and prophylactics to the virus, such as cow urine, ginger, and turmeric. In these claims, what is most striking is the mixing of well-established science and medical knowledge, alternative medical prac-tices, and blatant disinformation. Among other actions, the government introduced a phone app to track COVID. This algorithmic surveillance sys-tem labeled poorer neighborhoods and slums as "high risk" and tracked people if they moved between various risk zones. Unsurprisingly, many of the high-risk "red" zones were in Muslim and Dalit neighborhoods and resulted in creating and enforcing ghettos of the sick. In this way, medical geographies also operated as caste geographies.[66]

While the number of infections in India is high, it is important to note that the mortality rate remains low. Why this is the case remains an open question. One possibility casts widespread doubt on the official mortality rates emerging from India.[67] Indeed, India is not alone here. Suspicion of data has proven to be a hallmark of authoritarian regimes, and COVID-19 data in the United States and India are engulfed in a world of questionable data— intentional and unintentional. However, despite the statistical anomalies, there is a distinct pattern; the poor and marginalized have been the most affected by the virus and its management. As we have seen the world over, COVID-19 is not a "natural" leveler, where all individuals are equally suscep- tible, but a naturecultural actor whose actions are shaped by inequality and hierarchies of race, caste, and class.[68]

Conclusion

The racialized hierarchies of the natural that I began this chapter with are replicated in the hierarchies of the cultural. In each, the racialized others are less worthy and, therefore, more exploitable and expendable. COVID-19 has unfolded in a hauntingly familiar naturecultural script—a viral pan- demic accompanied by xenophobia, fear, distrust, violence, callous govern- ments, misinformation, disinformation, and stratified health care and health outcomes. The virus has revealed how the visible threads of inequality and health form a firm scaffold for racial inequality in science, health, and med- icine. This inequality is vividly in view—in the language of the virus, medi- cal descriptions of the pandemic, national rhetoric, government actions and inactions, and divisive politics, all of which have resulted in a profound racial Othering in the United States and India. The pandemic is not a natural event but fundamentally a naturecultural event in a hierarchical, racialized, and unequal Human world.

There are global political, economic, and naturecultural conditions that produced the pandemic. Chuang persuasively argues that SARS CoV-2 emerged as a global pandemic in China, especially in Wuhan, one of the "four furnaces" of growth in China. It represents an "evolutionary pressure cooker of capitalist agriculture and urbanization. This provides the ideal medium through which ever-more-devastating plagues are born, transformed, induced to zoonotic leaps, and then aggressively vectored through the human population."[69] The global trade infrastructure links and binds the world intimately together.

From the vantage point of feminist STS, the virus is not evil, Chinese, or foreign—it is a single strand of RNA. The virus cannot replicate on its own. The biology of the virus necessitates that it finds a cell to reproduce itself. Planet Earth in the year 2020 proved fertile ground. The virus is successful only because of the world some humans have created—increased colonization of the wild opening new pathways for viruses into human worlds, globalization hubs that transmit goods and people everywhere, and an impoverished health system that renders the virus lethal. Our focus can't be the virus but the oppressive systems that have and will forever enable pandemics of all kinds.

It is a transnational analysis that allows us to see the elisions and erasure of the foundational racial logics that govern our worlds. For example, in examining global patterns, the low initial numbers coming out of India elicited headlines such as "Are South Asians Exceptions?," "30% South Asians Have Neanderthal Gene That Increases Risk of Severe Covid-19," "BCG-Induced Trained Immunity," and "Do South Asians Have Innate Immunity?" Yet in the United Kingdom, South Asians were most likely to die in hospital of COVID-19.[70] In the United States, a staggering one in every 1,020 Black Americans have died because of COVID-19.[71] Yet for all the talk about racial susceptibility to the virus, cases in Africa have remained remarkably small.[72] These comparative data should put any argument about race to rest. Yet it lives on. In tracking trans-species and transnational naturecultural discourses of COVID-19, we see that race is foundational. Race is structural. Race is scaffolding. Race is text. Race is subtext. Racial logics and discourse ground and shape everything, including nature, bodies, and biologies. This feels like an old, familiar—and tiring—story in the history of medicine. It cannot be "racial propensity" if members of any purported group only exhibit those patterns within one nation and not others. One should blame the nation, not the race! The best predictive analyses remain sustained naturecultural and interdisciplinary analyses of the political economic conditions of the local. In short, COVID-19 teaches us that by tracking biological claims of "population" susceptibility, we see the enduring and structural foundational logics of racial othering that continue to undergird our narratives and our theories of health, illness, and Human biology.

But one thing is clear, we need alternative models and visions. As Lynn Margulis reminds us—we are our viruses. Rather than biological models of war with the virus, we need naturecultural models of living on a planet teeming with life. At the heart of the pandemic is an unequal world with concentrated wealth and power. If companies were not after the cheapest bottom line, China would not have emerged as an industrial global hub—and as the epicenter of the pandemic. More diffuse global power would create more diverse and vibrant global networks of trade and travel. Similarly, racialized landscapes of inequality fueled the pandemic. Viruses are ubiquitous, and many nations in the world understood how to realign human social organization to mitigate the spread of the virus. Evolutionary biologists remind us that pathogens usually adapt and evolve toward less virulence, thus saving pathogen and host.[73] Refusing to think with the complex naturecultural histories of humans and pathogens lead to ineffective responses, such as those of the United States and India. The past is indeed prologue. Unless we contend with the histories of colonialism, conquest, and slavery and refuse to hide behind the deceptive screens of democracy and nationalism, we are the prologue for the next pandemic.

THIS ESSAY HAS BENEFITED FROM COMMENTS AND DISCUSSIONS WITH JENNIFER HAMILTON, ANGELA WILLEY AND MEMBERS OF THE RACE AND REPRESENTATION SEMINAR AT THE UNIVERSITY OF MASSACHUSETTS, AMHERST. ALSO THANKS TO THE EDITORS OF THIS VOLUME FOR THEIR FEEDBACK.

Notes

1. Huihui Wang et al, "The Genetic Sequence, Origin, and Diagnosis of SARS-CoV-2," *European Journal of Clinical Microbiology*, April 24, 2020, 1–7.
2. Wuhan Municipal Health Commission, "Report of Clustering Pneumonia of Unknown Etiology in Wuhan City," December 31, 2019, http://wjw.wuhan.gov.cn /front/web/showDetail/2019123108989.
3. Isaac Chotiner, "The Interwoven Threads of Inequality and Health," *New Yorker*, April 14, 2020; Evelynn Hammonds, "A Moment or a Movement? The Pandemic, Political Upheaval, and Racial Reckoning," *Signs*, October 2020, http://signsjournal .org/covid/hammonds/.

4. Ruth Wilson Gilmore, *Golden Gulag: Prisons, Surplus, Crisis, and Opposition in Globalizing California* (Berkeley: University of California Press, 2007).

5. Bill Moyers, "Evelyn Fox Keller: The Gendered Language of Science," May 6, 1990, https://billmoyers.com/content/evelyn-fox-keller/.

6. Alan Weisman, "Is the Coronavirus Pandemic Mother Nature's Revenge," *Boston Globe*, April 22, 2020; Herbert Girardet, "Is Nature Taking Revenge?," *The Ecologist*, April 15, 2020; Jaydev Jana, "Nature Strikes Back," *The Statesman*, April 30, 2020.

7. Sylvia Wynter, "Unsettling the Coloniality of Being/Power/Truth/Freedom: Towards the Human, After Man, Its Overrepresentation—an Argument," *CR: The New Centennial Review* 3, no. 3 (2003): 257–337.

8. Joseph Graves, *The Emperor's New Clothes: Biological Theories of Race at the Millennium* (New Brunswick, NJ: Rutgers University Press, 2003); Londa Schiebinger, *Nature's Body: Gender in the Making of Modern Science* (New Brunswick, NJ: Rutgers University Press, 2004); Kavita Philip, *Civilizing Natures: Race, Resources, and Modernity in Colonial South India* (New Brunswick, NJ: Rutgers University Press, 2003); Harriet Washington, *Medical Apartheid: The Dark History of Medical Experimentation on Black Americans from Colonial Times to the Present* (New York: Anchor, 2008).

9. Lisa Gannett, "Racism and Human Genome Diversity Research: The Ethical Limits of Population Thinking," *Philosophy of Science* 68, no. 3 (2001): S479–S492.

10. Hawes Spencer and Sheryl Gay Stolberg, "White Nationalists March on University of Virginia," *New York Times*, August 11, 2017.

11. Ania Loomba, "Race and the Possibilities of Comparative Critique," *New Literary History* 40, no. 3 (2009): 501–22.

12. Yulia Egorova, "Castes of Genes? Representing Human Genetic Diversity in India," *Genomics, Society, and Policy* 6, no. 3 (2010): 32–49; Kamala Viswesaran, *Un/Common Cultures: Racism and the Rearticulation of Cultural Difference* (Durham, NC: Duke University Press, 2010).

13. V. T. Rajshekar, *Dalit: The Black Untouchables of India* (Atlanta: Clarity, 2009).

14. Gail Omvedt, *Dalit Visions: The Anti-Caste Movement and the Construction of an Indian Identity* (New Delhi: Orient Longman, 1995).

15. Thomas Crowley, "When Hindu Nationalism and White Nationalism Meet," *Jacobin*, September 25, 2019.

16. Joanna Slater, "What the U.S Election Means for India," *Washington Post*, October 29, 2020.

17. Susan Faludi, "Trump's Thoroughly Modern Masculinity," *New York Times*, October 29, 2020.

18. Ibrahim Al Marashi, "Orientalism and the Geopolitics of the Coronavirus Outbreak," *TRT World*, February 14, 2020.

19. Harsh Mander, "The Coronavirus Has Morphed Into an Anti-Muslim Virus," *The Wire*, April 13, 2020.

20. Mehdi Hasan, "The Coronavirus Is Empowering Islamophobes—but Exposing the Idiocy of Islamophobia," *The Intercept*, April 14, 2020.

21. Runjing Lu and Yanying Sheng, "From Fear to Hate: How the Covid-19 Pandemic Sparks Racial Animus in the United States," July 2, 2020, https://arxiv.org/ftp/arxiv/papers/2007/2007.01448.pdf.

22. For a discussion on zoonosis, see Banu Subramaniam, "Zoonosis," *Theory & Event*, special issue: "COVID Keywords," ed. Jennifer Nash and Samantha Pinto (forthcoming).
23. Alonso Aguirre et al., "Illicit Wildlife Trade, Wet Markets, and COVID-19: Preventing Future Pandemics," *World Medical and Health Policy* 12, no. 3 (September 2020): 256–65.
24. Jenny Zhang, "Pinning Coronavirus on How Chinese People Eat Plays Into Racist Assumptions," *The Eater*, January 31, 2020.
25. Calum Marsh, "How the Myths Surrounding Bat Soup Came to Represent Our Collective Fear and Confusion Over COVID-19," *County Weekly News*, April 17, 2020.
26. Slingenbergh et al., "Ecological Sources of Zoonotic Diseases," *Scientific and Technical Review of the Office International des Epizooties* 23, no. 2 (2004): 467–84.
27. Matthew Wills, "The Law and the Coronavirus," *The Daily*, January 30, 2020.
28. Dong-Sheng Chen et al., "Horizontal Gene Transfer Events Reshape the Global Landscape of Arms Race Between Viruses and *Homo sapiens*," *Scientific Reports* 6 (2016).
29. Salvador E. Luria, *Virus Growth and Variation*, ed. A. Isaacs and B. W Lacey (Cambridge: Cambridge University Press, 1959): 1–10.
30. John M. Coffin et al., *Retroviruses* (Cold Spring Harbor Laboratory Press, 1997).
31. Edward B. Chuong, "Retroviruses Facilitate the Rapid Evolution of the Mammalian Placenta," *BioEssays* 35, no. 10 (2013): 853–61.
32. Lynn Margulis, *Symbiotic Planet: A New Look at Evolution* (New York: Basic Books, 1998), 64.
33. Stephen T. Asma, "Does the Pandemic Have a Purpose?," *New York Times*, April 16, 2020.
34. India.com News Desk, "Virus May Be an Invisible Enemy, but Our COVID-19 Warriors Are Invincible: PM Modi," *India.com*, June 1, 2020.
35. Virginia Heffernan, "Metaphors Matter in a Time of Pandemic," *Wired*, May 19, 2020.
36. Emily Martin, "Toward an Anthropology of Immunology: The Body as Nation State," *Medical Anthropology Quarterly* 3, no. 3 (1990): 410–26.
37. Iona Francesca Walker, "Comparing Antimicrobial Resistance and the COVID-19 Pandemic in the United Kingdom," *MAT: Medicine Anthropology Theory* 7, no. 2 (2020).
38. Walker, "Comparing Antimicrobial Resistance and the COVID-19 Pandemic in the United Kingdom."
39. Apoorva Mandavilli, "The Coronavirus Patients Betrayed by Their Own Immune Systems," *New York Times*, April 1, 2020.
40. Evelyn Fox Keller, *The Century of the Gene* (Cambridge, MA: Harvard University Press, 2002).
41. Abigail Tracy, "How Trump Gutted Obama's Pandemic-Preparedness Systems," *Vanity Fair*, May 1, 2020.
42. Anna North, "What Trump's Refusal to Wear a Mask Says About Masculinity in America," *Vox*, May 12, 2020.
43. Will Feuer, "Trump Blames Rise in Coronavirus Cases on Increased Testing, Despite Evidence of More Spread," *CNBC*, June 23, 2020.

44. Richard Oppel et al., "The Fullest Look Yet at the Racial Inequality of Coronavirus," *New York Times*, July 5, 2020.

45. Nicholas Kristof, "The Top U.S. Coronavirus Hot Spots Are All Indian Lands," *New York Times*, May 30, 2020; Mark Walker, "'A Devastating Blow': Virus Kills 81 Members of Native American Tribe," *New York Times*, October 8, 2020.

46. Jocelyn Fruey, "On the Frontlines at Work and at Home," *American Progress*, April 2, 2020, https://www.americanprogress.org/issues/women/reports/2020/04/23/483846/frontlines-work-home/.

47. Ladan Golestaneh et al. "The Association of Race and COVID-19 Mortality," *EClinical Medicine* 25 (August 1, 2020): 100455.

48. Tiffany Ford, Sarah Reber, and Richard Reeves, "Race Gaps in COVID-19 Deaths Are Even Bigger Than They Appear," Brookings Institute, June 16, 2020, https://www.brookings.edu/blog/up-front/2020/06/16/race-gaps-in-covid-19-deaths-are-even-bigger-than-they-appear/.

49. Rashawn Ray, "Why Are Blacks Dying at Higher Rates from COVID-19?," Brookings Institute, April 9, 2020, https://www.brookings.edu/blog/fixgov/2020/04/09/why-are-blacks-dying-at-higher-rates-from-covid-19/.

50. "Health Equity Considerations and Racial and Ethnic Minority groups," CDC, July 24, 2020, https://www.cdc.gov/coronavirus/2019-ncov/community/health-equity/race-ethnicity.html.

51. See Nicole Phillips et al., "The Perfect Storm: COVID-19 Health Disparities in US Blacks," *Journal of Racial Ethnic Health Disparities*, September 23, 2020, 1–8; J. R. Giudicessi et al., "Genetic Susceptibility for COVID-19 Associated Sudden Cardiac Death in African Americans," *Heart Rhythm* 17, no. 9 (2020): 1487–92; Alireza Hamidian Jahromi and Anahid Hamidianjahromi, "Why African Americans Are a Potential Target for COVID-19 Infection in the United States," *Journal of Medical Internet Research*, June 12, 2020.

52. Banu Subramaniam, "Archaic Modernities: Science, Secularism, and Religion in Modern India," *Social Text* 18, no. 3 (Fall 2000).

53. Banu Subramaniam, *Holy Science: The Biopolitics of Hindu Nationalism* (Seattle: University of Washington Press, 2019).

54. Banu Subramaniam and Debjani Bhattacharyya, "A Viral Education: Scientific Lessons from India's WhatsApp University," *Somatosphere*, May 31, 2020.

55. For a more detailed discussion of COVID and Hindu nationalism, see Banu Subramaniam, "Viral Nationalisms: Riding the Corona Waves in India," *Religion Compass* (forthcoming).

56. I. Aravind, "Covid-19: How Healthcare Workers Are Paying a Heavy Price in the Battle," *Economic Times*, April 12, 2020.

57. M. Abi-Habiband Samir Yasir, "For India's Laborers, Coronavirus Lockdown Is an Order to Starve," *New York Times*, March 30, 2020; H. Ellis-Petersen and Shaikh Azizur Rahman, "Coronavirus Conspiracy Theories Targeting Muslims Spread in India," *Guardian*, April 13 2020; S. Pandey, "Saffron Leaders Blame Non-Veg for Coronavirus Outbreak in India, Seek Ban on Sale of Meat," *Deccan Herald*, March 16, 2020; Aravind, "Covid-19."

58. Akshita Jain, "Coronavirus Is Now a Racial Slur in India," *HuffPost India*, April 2, 2020.

59. Naorem Pushparani Chanu and Gorky Chakraborty, "A Novel Virus, a New Racial Slur," *India Forum*, July 10, 2020; Gayathri Mani, "Asked to Vacate Hostels, Racial Slurs: Northeast Indians in Delhi Deal with Daily COVID-19 Hatred," *New Indian Express*, March 30, 2020.

60. Sumit Ganguly, "India's Coronavirus Pandemic Shines a Light on the Curse of Caste," *The Conversation*, June 2, 2020.

61. "Surge in Atrocities Against Dalits and Adivasis Under COVID-19 Lockdown in India Reported," International Dalit Solidarity Network, June 10, 2020, https://idsn.org/surge-in-atrocities-against-dalits-and-adivasis-under-covid-19-lockdown-in-india-reported/; Jeya Rani, "Tamil Nadu Blanket of Silence Over Caste-Based Atrocities During COVID-19 Pandemic," *The Wire*, July 30, 2020.

62. "Why Are COVID-19 Deaths Lowest in Poor, Densely Populated South Asia?," *The Week*, April 26, 2020.

63. A. Rammohan and Mohamed Rela, "COVID-19: Could India Still Escape?," *Journal of Global Health* 10, no. 1 (June 17, 2020).

64. Priyanka Pulla, " 'The Epidemic Is Growing Very Rapidly': Indian Government Adviser Fears Coronavirus Will Worsen," *Nature* 583, no. 180 (June 26, 2020).

65. Subramaniam and Bhattacharyya, "A Viral Education."

66. D. Bhattacharyya and Banu Subramaniam, "Technofascism in India," *n+1*, May 13, 2020.

67. See Pulla, " 'The Epidemic Is Growing Very Rapidly.' "

68. A. Chauhan, "Class, Caste, and COVID-19 Casualties!," *Diplomatist*, June 8, 2020.

69. "Social Contagion: Microbiological Class War in China," *Chuang*, http://chuangcn.org/2020/02/social-contagion/.

70. Haroon Siddique, "South Asians in Britain Most Likely to Die in Hospital of Covid-19, Study Finds," *Guardian*, June 19, 2020.

71. Dylan Scott, "It's True: 1 in 1,000 Black Americans Have Died in the Covid-19 Pandemic," *Vox*, September 29, 2020.

72. Karen Attiah, "Africa Has Defied the Covid-19 Nightmare Scenarios," *Washington Post*, September 22, 2020.

73. Bob Holmes, "How Viruses Evolve," *Smithsonian*, July 17, 2020.

Contributors

JOÃO LUIZ BASTOS currently works as an associate professor in racism and health at Simon Fraser University, Canada. His main research interests include the health effects of intersecting systems of oppression, as well as the development of psychometric scales to assess perceived discrimination.

ALYSSA BOTELHO is a graduate of the Harvard-MIT MD/PhD Program and the Department of the History of Science at Harvard University. She is currently a physician in the Internal Medicine Residency Program at the University of Washington in Seattle. Her research focuses on questions of equity in health and medicine, and her dissertation examines the history of health disparities in the rural United States, with a focus on how efforts to sustain rural hospitals over the twentieth century figured into movements for racial equality, economic justice, and Indigenous sovereignty in rural communities.

ELISE K. BURTON is an assistant professor at the Institute for the History & Philosophy of Science & Technology at the University of Toronto. Her research focuses on the life sciences in the modern Middle East, and she is the author of *Genetic Crossroads: The Middle East and the Science of Human Heredity* (2021).

DENISE FERREIRA DA SILVA is the Samuel Rudin Professor in the Humanities at the Department of Spanish and Portuguese, New York University; adjunct professor at Monash University's Art, Design and Architecture;

and serves as faculty at the European Graduate School. She is the author of *Unpayable Debt* (2022), *Dívida Impagável* (2019), and *Toward a Global Idea of Race* (2007) and coeditor (with Paula Chakravartty) of *Race, Empire, and the Crisis of the Subprime* (2013) and (with Mark Harris) *Indigenous Peoples and the Law* (2019).

SEBASTIÁN GIL-RIAÑO is an assistant professor in the History and Sociology of Science Department and Center for Latin American and Latinx Studies at the University of Pennsylvania. His first book, titled *The Remnants of Race Science: UNESCO and Economic Development in the Global South*, was published by Columbia University Press in 2023.

JAEHWAN HYUN is assistant professor of history of science and technology at Pusan National University, Busan, South Korea. The main area of his research is the history of human genetics, biological anthropology, and race in South Korea, with a focus on transnational connections among Japan, South Korea, and the United States.

DAVID S. JONES teaches history of medicine, medical ethics, and social medicine at Harvard University. Trained in psychiatry and history of science, his research has ranged from the history of epidemics to heart disease, cardiac therapeutics, air pollution, and the health effects of global warming.

AMADE AOUATEF M'CHAREK is professor of anthropology of science at the Department of Anthropology, University of Amsterdam, where she has received prestigious research grants such as the ERC consolidator grant and the ERC advanced grant, as well as membership in the Dutch Royal Academy of Science. She is the PI of the RaceFaceID project, guiding a team in a project on forensic identification and the making of face and race, and PI of the Vital Elements project, a team endeavour in which forensic methods are reappropriated to study migrant death in relation to (post) colonial circulations and extractions.

PAUL WOLFF MITCHELLis a postdoctoral scholar in the Department of Anthropology at the University of Amsterdam in the Netherlands and a member of the project "Pressing Matter: Ownership, Value, and the Question of Colonial Heritage in Museums." Alongside curation of museum exhibitions on scientific racism in the United States and the Netherlands, his current book project concerns the global history and afterlives of human remains collections and ideas of racial origins in the eighteenth and nineteenth centuries.

PROJIT BIHARI MUKHARJI is professor and head of the Department of History at Ashoka University, Sonepat, India. He is also a Guggenheim Fellow. His most recent monograph is *Brown Skins, White Coats: Race Science in India, c.1920-66* (2022).

ERIC REINHART is a political anthropologist, psychoanalyst, and physician based in Chicago.

JULIA E. RODRIGUEZ is professor of history at the University of New Hampshire. She is the author of *Civilizing Argentina: Science, Medicine, and the Modern State* (2006) and editor of the open-source teaching website HOSLAC: History of Science in Latin America and the Caribbean (www.hoslac.org).

RICARDO VENTURA SANTOS is an anthropologist and senior researcher at the National School of Public Health (Escola Nacional de Saúde Pública) of the Oswaldo Cruz Foundation (Fundação Oswaldo Cruz), Brazil; he is also professor in the Department of Anthropology of the National Museum, Federal University of Rio de Janeiro, Brazil.

BANU SUBRAMANIAM is professor of women, gender, and sexuality Studies at the University of Massachusetts, Amherst. Trained as a plant evolutionary biologist, Banu engages the feminist studies of science in the practices of experimental biology.

NOAH TAMARKIN is an associate professor of anthropology and science and technology studies at Cornell University and a research associate at the Wits Institute for Social and Economic Research (WISER) at University of the Witwatersrand in Johannesburg, South Africa. His book *Genetic Afterlives: Black Jewish Indigeneity in South Africa* (2020) received the Jordan Schnitzer Prize from the Jewish Studies Association and an Honorable Mention for the Diana Forsythe Prize from the American Anthropological Association.

ISAAC WARBRICK of Ngāti Te Ata, Te Arawa, and Ngā Puhi tribes in New Zealand is the director of the Taupua Waiora Māori Research Centre at the Auckland University of Technology. His research explores the connection between the environment and Indigenous health and the role of *kōrero tuku iho* (traditional narratives, observations, and teachings) in guiding and promoting health and well-being.

Index

Abel, Wolfgang, 143
Account of the Regular Gradation in Man, An (White), 37
Adam: multiple, 32–36; skin color of, 24
aesthetic judgments: facial reconstructions and, 203; Kant on, 202–203; neoclassical, 26; physiognomy hierarchy and, 25–26
Afropessimism, 53
Age of Exploration, 6
Agote, Luis, 84
Agote Law, 84
Agrawal, Mahendra, 268
Ahmed, Sarah, 104
AIMS. *See* ancestry informative markers
Akbari, Mohammad, 105
Akiyoshi, Suda, 147
albinism, 23
Albinus, Bernhard, 24
Alfaro, Gregorio Aráoz: on children, 75, 82–83; on eugenics, 83
alienation: in language, 54; particular forms of, 56; subjective, 53; universal nature of, 56
All India Hindu Mahasabha, 307
Althusser, Louis, 55

American, as "biological" category, 9
American Psychiatric Association, 59
American Revolutionary War, 28
ancestry informative markers (AIMs), 173
Annandale, Nelson, 152
Anthropological Survey of India, 149, 150
anthropology: Kant on, 57; Ramos on, 90; systemization, 9–10. *See also* seroanthropological studies
apartheid, 10; founding of, 245; goals of, 245; Indigenous DNA and, 243–248; racial classifications and, 244–248; scientific involvement in, 246
appearance classifications, 17n28
archaic modernity, 305
Argentina: child care in, 79–85; economic momentum of, 80–81; exploitative labor in, 82; factories in, 82; fertility rates, 94n13; maternalism in, 79–85; politics of care in, 79–85; as young country, 80
Arish, 218, 220; DNA research on, 221–222; sculpture of, *219*
Aryan race, 204, 298
Asch, Georg Thomas von, 31

Printed and bound by CPI Group (UK) Ltd, Croydon, CR0 4YY

07/05/2024

14498573-0002